From Our Footsteps, Giant Forests Grow.
A Photographic Introspection of British Columbia's Coastal Tree Planting Industry.

Published by DJ Bolivia Inc.
P.O. Box 6074, Sackville, New Brunswick, E4L 1G6.
jonathan.scooter.clark@gmail.com

ISBN 978-1999016814

Hardcover edition, printed by IngramSpark.

Inquiries: www.jonathanclark.ca

Disclaimer

This book is presented solely for educational purposes, and to allow the public to gain a greater understanding of the tree planting industry in Canada. The author and publisher are not offering it as professional, legal, financial, physiological, medical, human resources, or any other type of advice. While best efforts have been used in preparing this book, the author and publisher make no representations or warranties of any kind, and assume no liabilities of any kind with respect to the accuracy or completeness of the contents, and specifically disclaim any implied warranties of appropriateness or fitness of use for any particular purpose.

Neither the author nor the publisher shall be held liable or responsible to any person or entity with respect to any loss or incidental or consequential damages of any kind caused, or alleged to have been caused, directly or indirectly, by the information contained herein. No warranty may be created or extended by website links, photos, videos, or written materials related to or referenced by the author and publisher. The content of any websites referenced in this book may change at any time.

The contents of this book should not be construed as a suggestion or recommendation, explicit or implicit, for any person or persons to seek employment within the forestry sector in general, nor specifically within the tree planting or reforestation sectors. Tree planting is extremely difficult work, and most people are not well suited for employment positions within the industry.

Acknowledgements

Many thanks to all of the people who appear in the various photographs within this book. Without the photographs of planters at work, which are able to demonstrate the human-interest aspects of the coastal tree planting industry, this book would have been pretty boring.

The vast majority of the photos in this book were compiled from two specific years of coastal planting, while I was working on various reforestation projects on Vancouver Island, British Columbia. I've also included a handful of slightly older coastal photos. As a whole, this collection encompassed photos from both spring and fall coastal seasons. Of course, it seems to rain about 95% of the time during any typical spring coastal season, and I don't like getting my cameras wet, so perhaps the photos in this book aren't a true representation of what a spring coastal season is really like. More of these photos were taken during the fall than during the spring. Spring coastal weather is usually much more miserable than you might assume from these photos. Don't be fooled.

The majority of these photos were taken with a Canon 5D Mark IV, using one of several different lenses. Predominantly, I used the EF 24-105, the EF 17-40, or the EF 200 fixed focal length lens. A handful of photos were taken with a Canon T3i or T5i body.

Thanks also to Andrew Ulmer, who occasionally picked up my camera and took photos for me while I was planting. You can find his own professional photography on Instagram at @andrewulmerphoto

The information in this book presents an overview of British Columbia's coastal tree planting industry. For detailed information about the more accessible segment of the planting industry which is practiced throughout the rest of Canada, look for another of my books called "Step By Step: A Tree Planter's Handbook." That book, which is available on Amazon, focuses on finding a summer job and learning specific planting techniques. I also have another tree planting photo book called "Rite Of Passage." That book focuses on traditional summer planting operations in remote tent camps, rather than on coastal projects. Visit www.jonathanclark.ca for more information about these books.

Preface

This book is a project that I've wanted to tackle for quite a while. For years, the most popular part of the Replant website was the collection of photo galleries. I had always thought it would be nice to have a coffee-table style photo book that would feature a solid collection of high-quality coastal planting photos. Such a book would be a good way to show experienced planters what the coastal planting industry is all about, at least as much as it's possible to do that with photography.

So far, I've spent ten years planting on various coastal projects, in addition to my other reforestation work elsewhere in western Canada. For a few of those years, I was only available for coastal work during a short window, but the majority of those years included the full spring and fall seasons. During all those years, I somehow managed to survive numerous cuts, bruises, and "weather events" without taking a single sick day. Perhaps this means that I hate missing an opportunity to immerse myself in misery. I'll be clear though; I have no regrets and would do it all again if I had the choice.

I try not to get my camera too wet, so you'll find a disproportionate number of photos that were taken on sunny days during the fall, when the weather is slightly more civil than during the spring plant. This "weather bias" will unfortunately give a slight misrepresentation of what coastal planting can really be like. On top of that, a two-dimensional photograph cannot truly capture the steepness of the slopes, nor can it truly make you understand what it feels like to spend a day on the side of a cliff, soaking wet in near-freezing temperatures, while surrounded by slash up to your waist. To understand it, you have to do it.

A lot of experienced Canadian planters eventually aspire to seek coastal planting work. My initial tendency would be to scream a warning: Don't do it! It's by far the most challenging planting that you could ever experience in Canada. However, it's an adventure, and I love it.

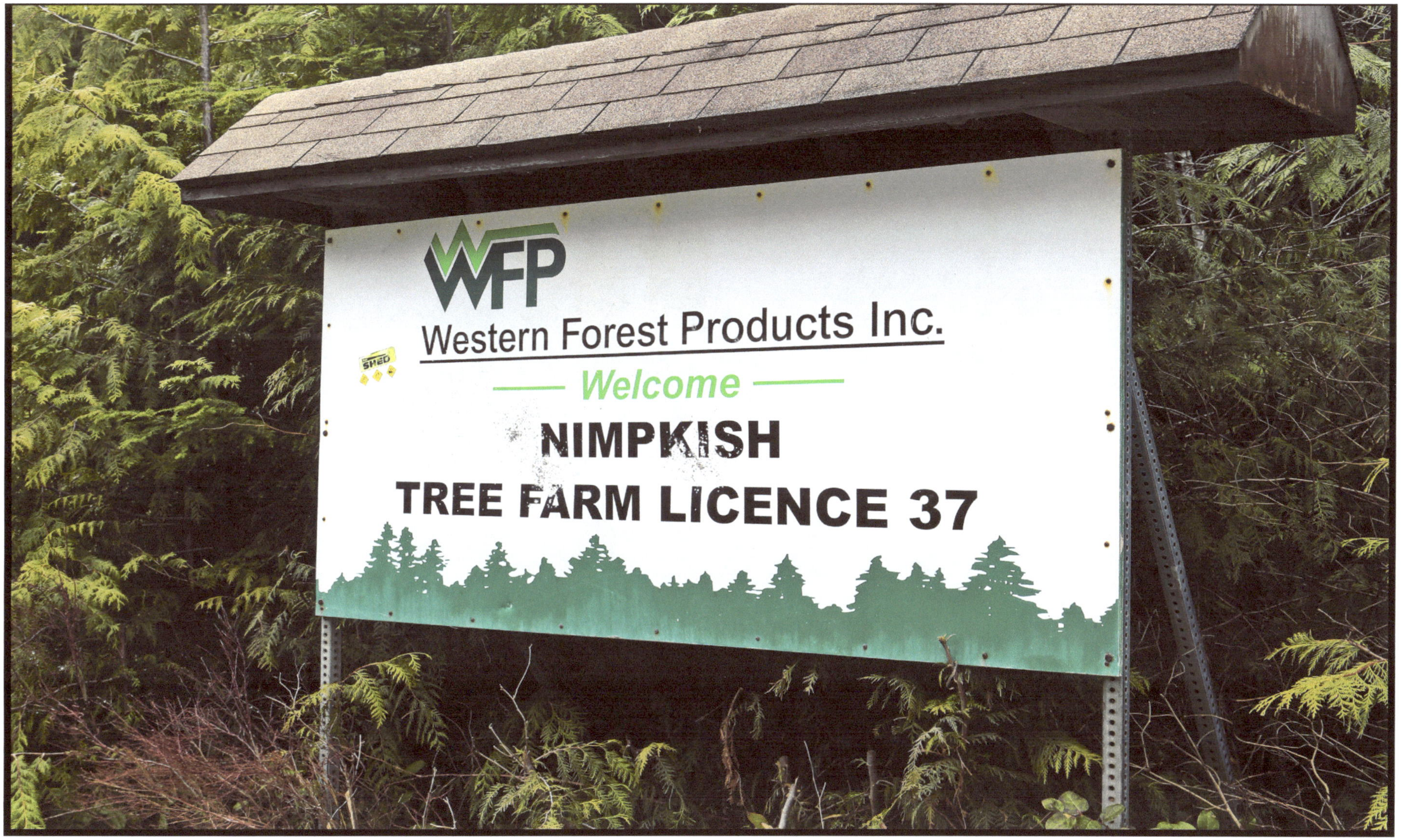

When planters refer to "coastal" planting in British Columbia, they're referring to work on Vancouver Island, the Sunshine Coast, Haida Gwaii, and surrounding areas. Some coastal planting is quite "civilized," with planters working out of motels in towns and cities. Other projects are more inaccessible, with people staying in remote lodges or on barges, and accessing blocks by helicopter, float plane, or water taxi. Western Forest Products is one of the logging companies with a relatively large presence on Vancouver Island, especially on the north part of the Island.

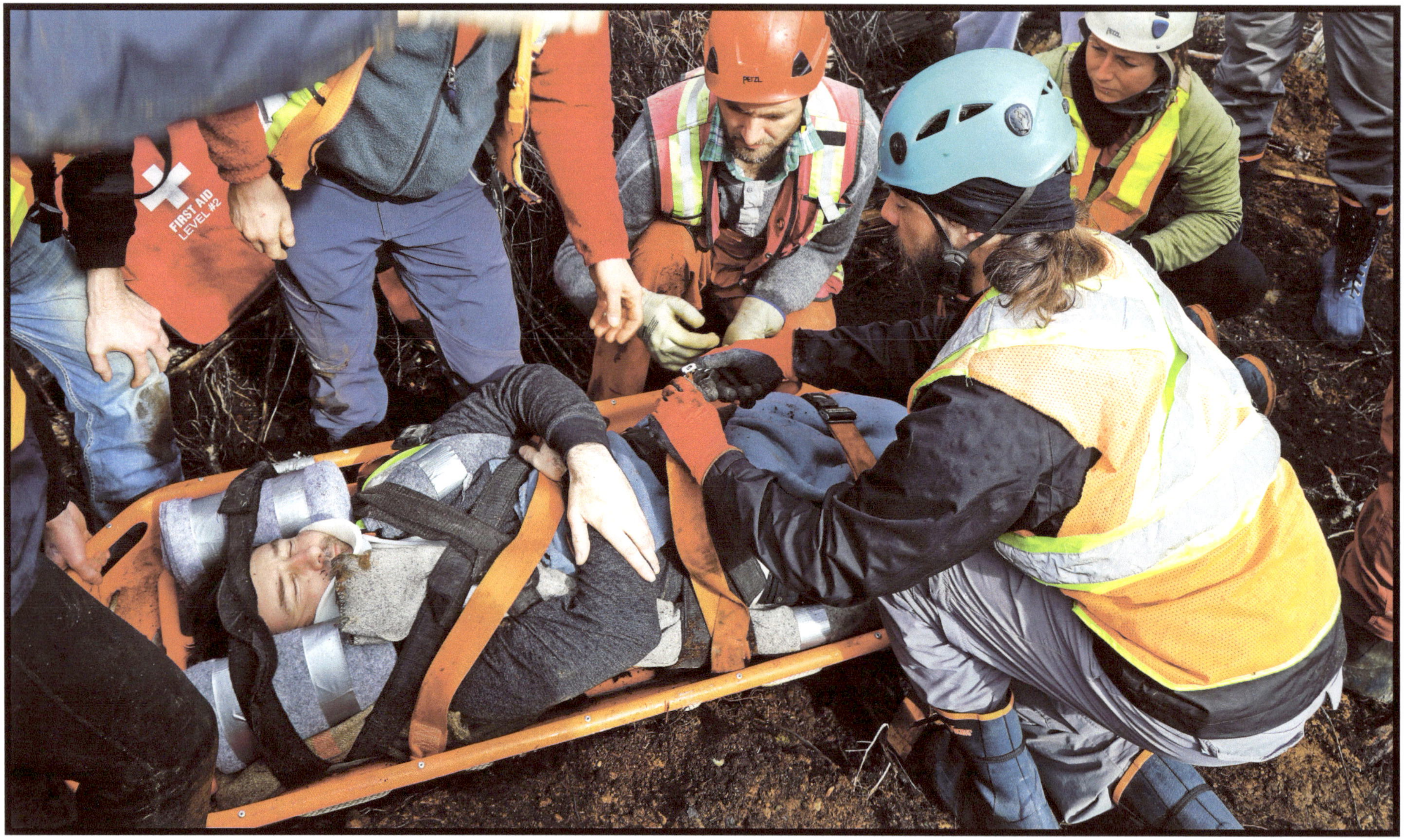

Before starting work, it's very instructive to do a first-aid simulation with a mock emergency evacuation drill. Coastal planters are often working in very remote locations, far from advanced medical aid. It's important to treat safety with the highest priority.

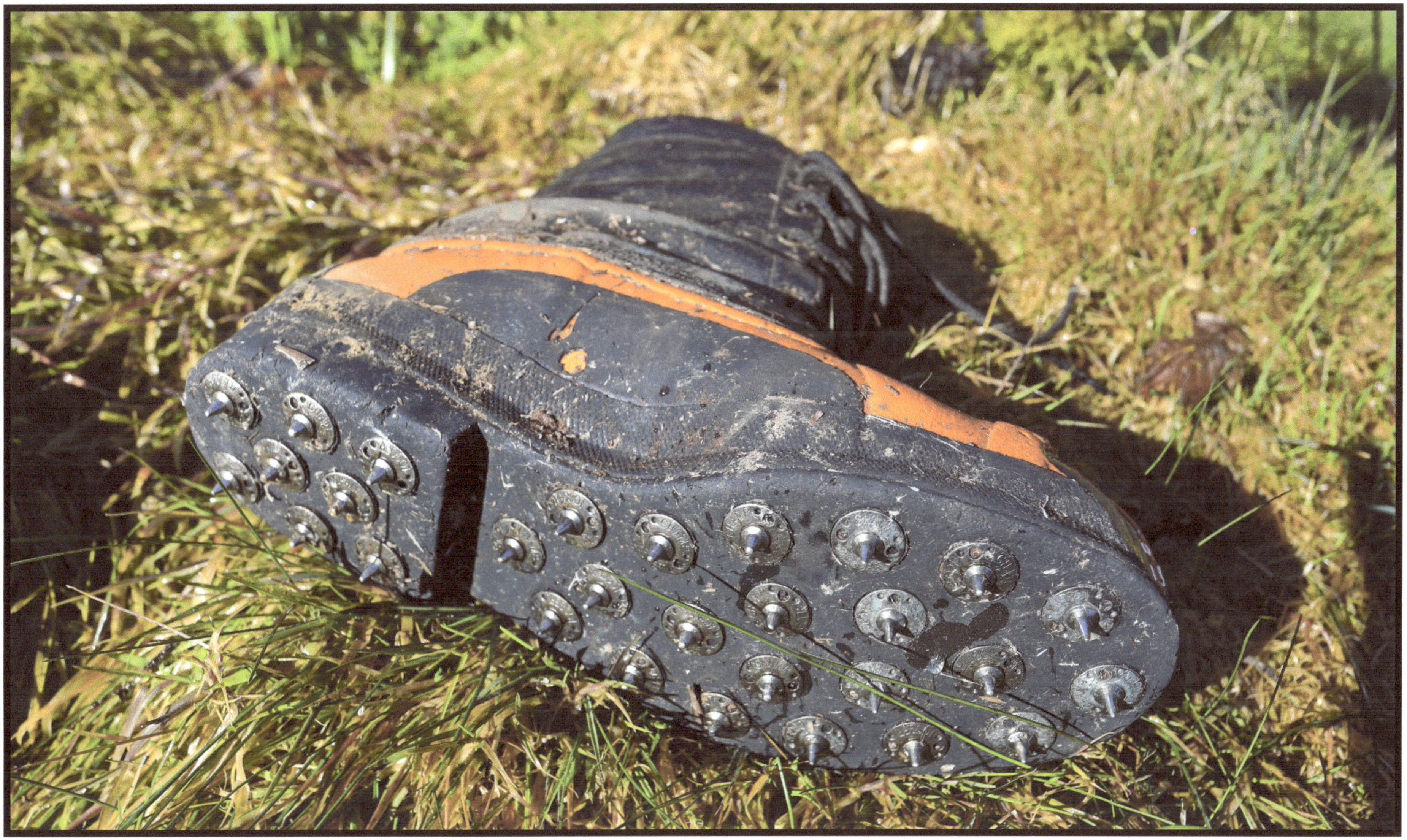

One of the greatest hazards of coastal planting involves walking around on steep coastal blocks which are full of slash. It would be impossible for a planter to keep their balance using normal boots, and since the consequences of slips/trips/falls on coastal blocks can be fairly severe, caulked boots are mandatory. These boots are named caulks (pronounced corks) after the name for the metal spikes on the bottom of the boots. Even with caulks, it's very difficult to keep one's footing on wet cedar slash. I replace half of my caulk spikes after every six to eight days of planting, to ensure that they're always sharp.

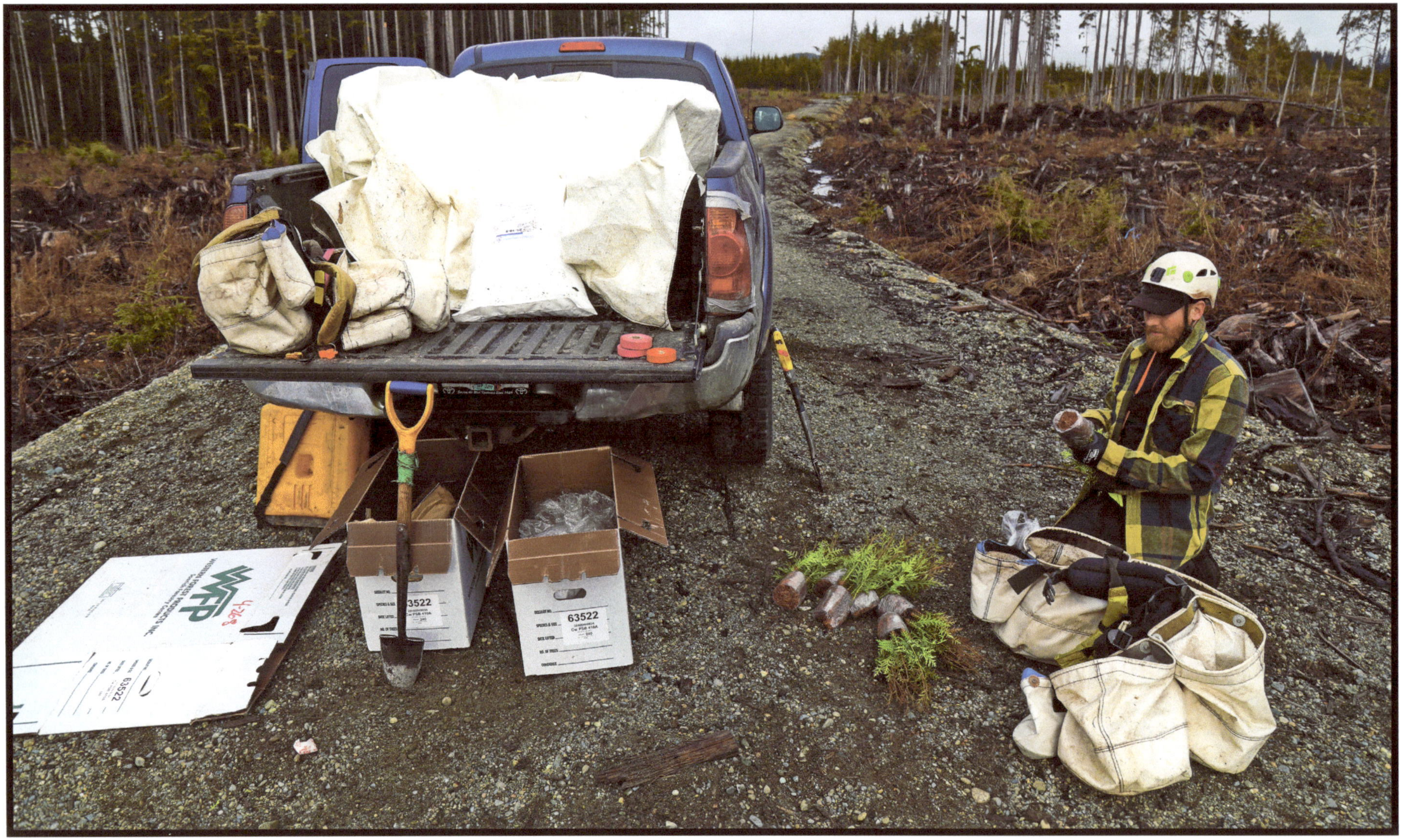

We usually try to work on relatively flat blocks on the first day or two of the season, while planters are starting to get back into shape. Unfortunately, flat blocks are usually a rare commodity on coastal contracts.

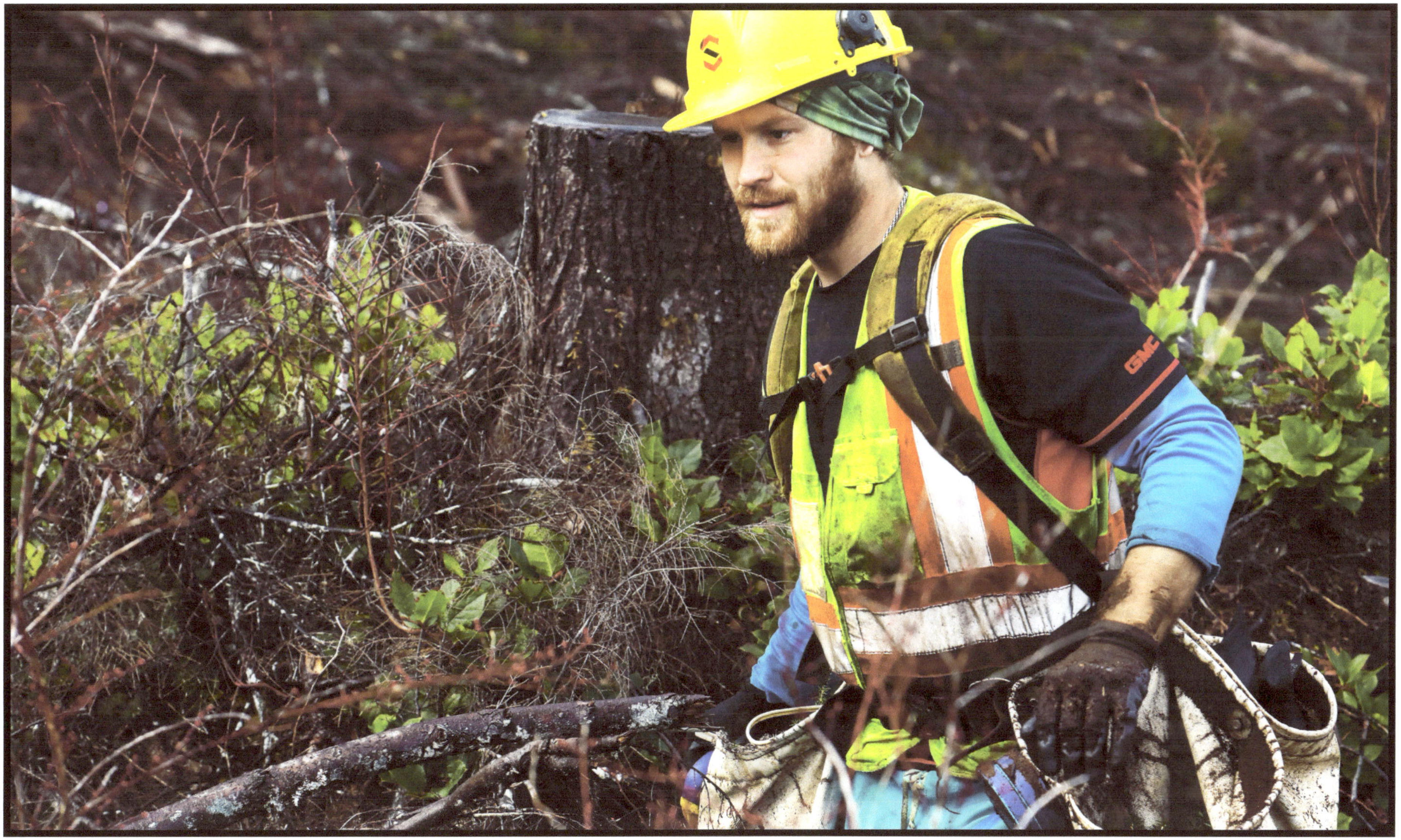

Unlike planting in BC's Interior, most coastal planting contracts require planters to wear full high vis clothing, plus hard hats or CSA-approved climbing helmets. That's due to the significantly higher chance of injury on the blocks, and the difficulty of spotting a planter who has fallen to the ground.

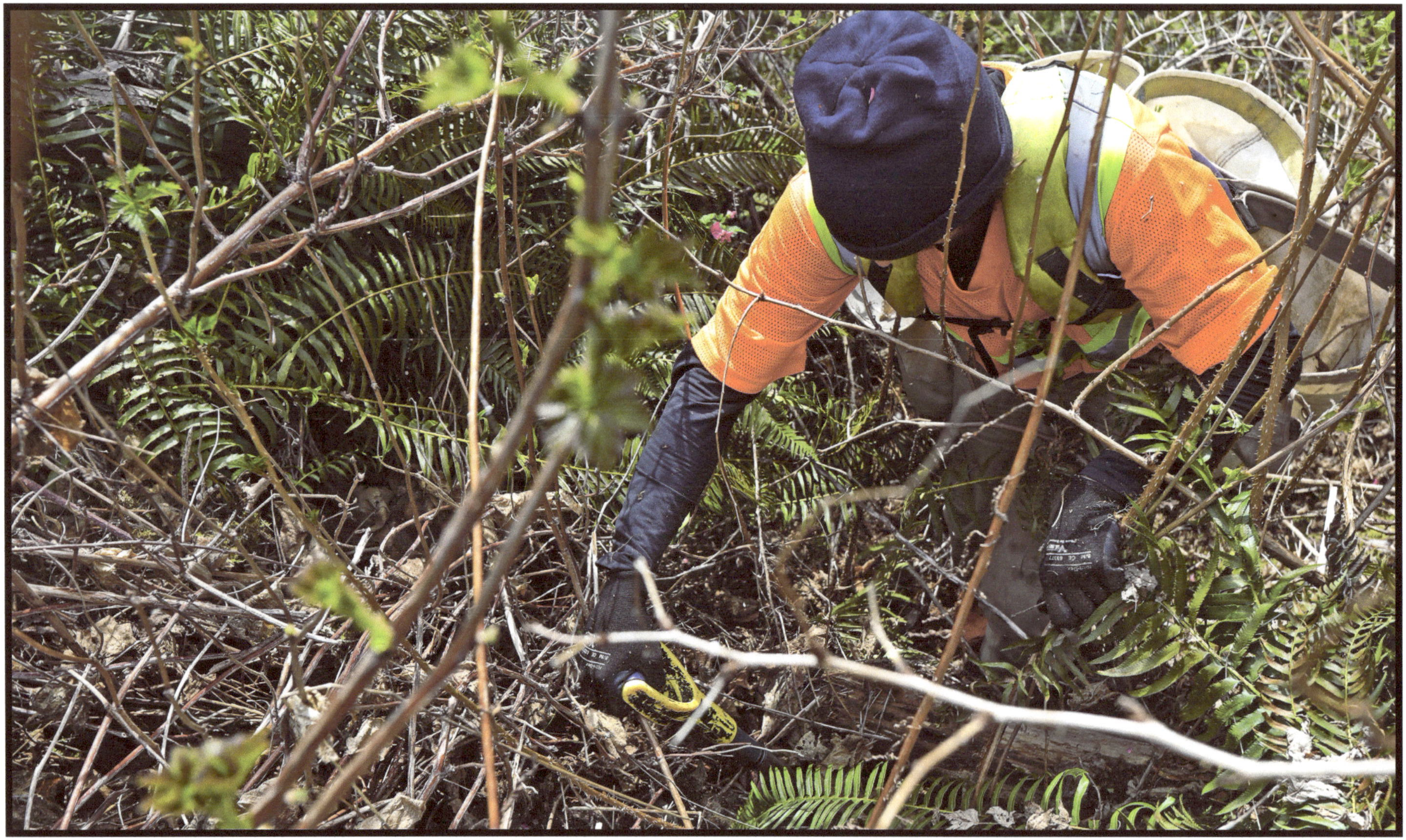

What isn't rare on coastal contracts is slash and brush. Most planters need to accrue several years of experience in the Interior of BC before a company will hire them for coastal planting work. These planters are told to expect some fairly tough ground when they start working on the coast, but many are still surprised at just how challenging the ground really is.

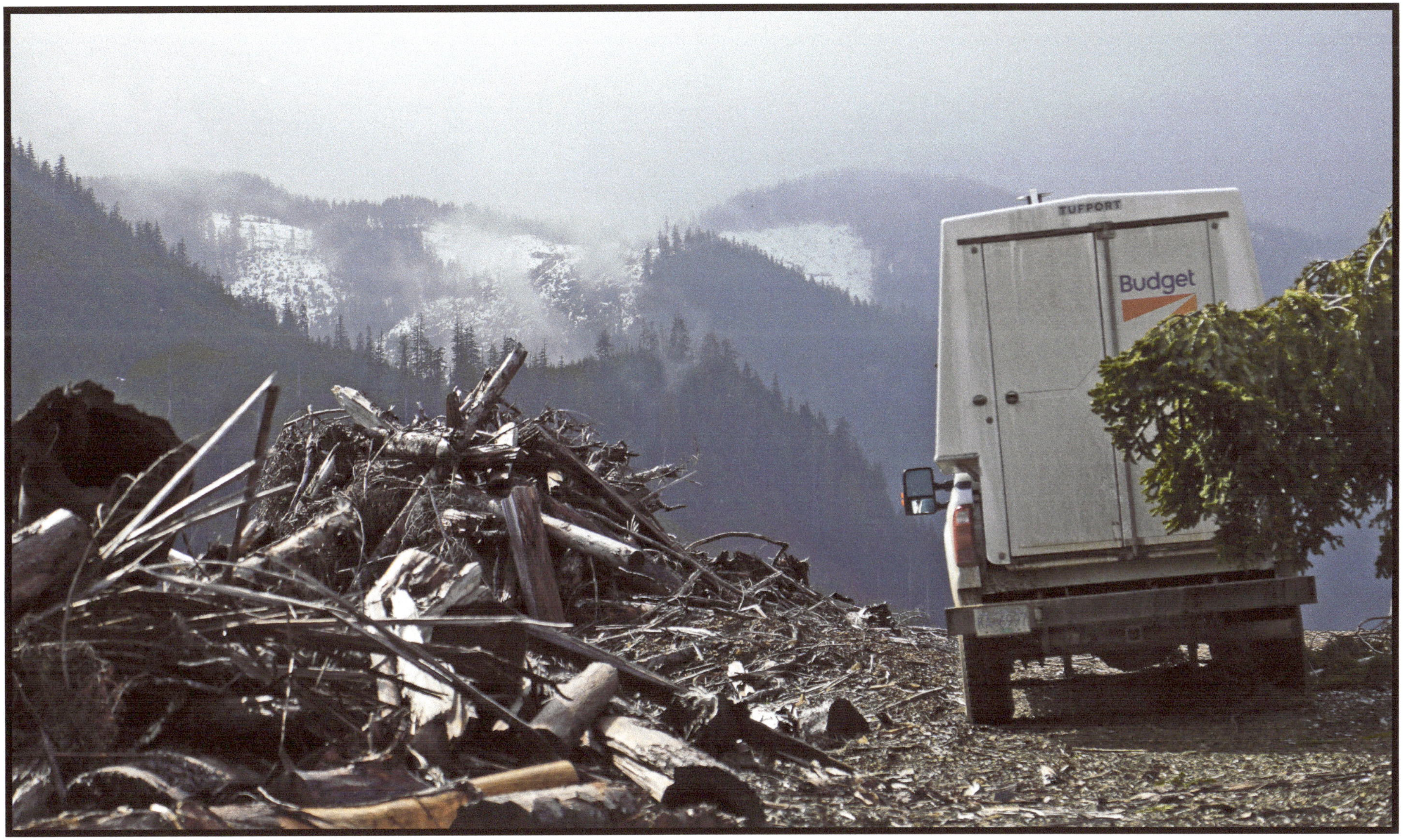

The spring coastal season usually starts in late February to early March, although in some cases, it can start as early as the end of January. Most spring coastal contracts run until mid-April, with a few running into May. However, most companies prefer to be done their coastal work by mid-April so their workforce can move to contracts in the Interior of BC. Snow is a constant problem in the spring, although blocks at very low elevations (within a few hundred meters of sea level) usually sit below the snow line. Planters frequently pay close attention to the daily elevation of the snowline. Days off due to snow are common.

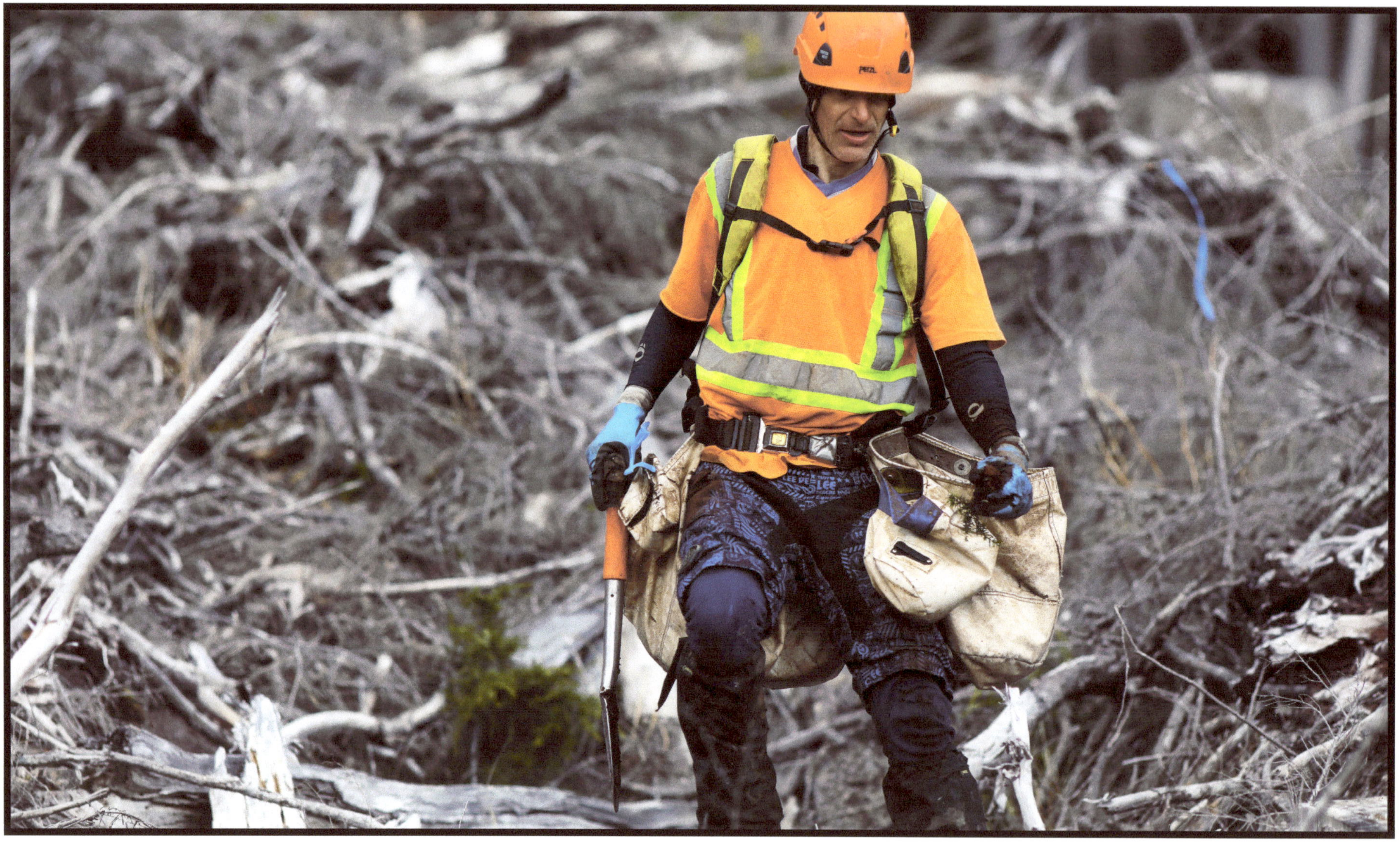

Climbing helmets have become much more popular than hard hats over the past decade. They're lighter and more comfortable, and the small side vents are very beneficial during hot weather.

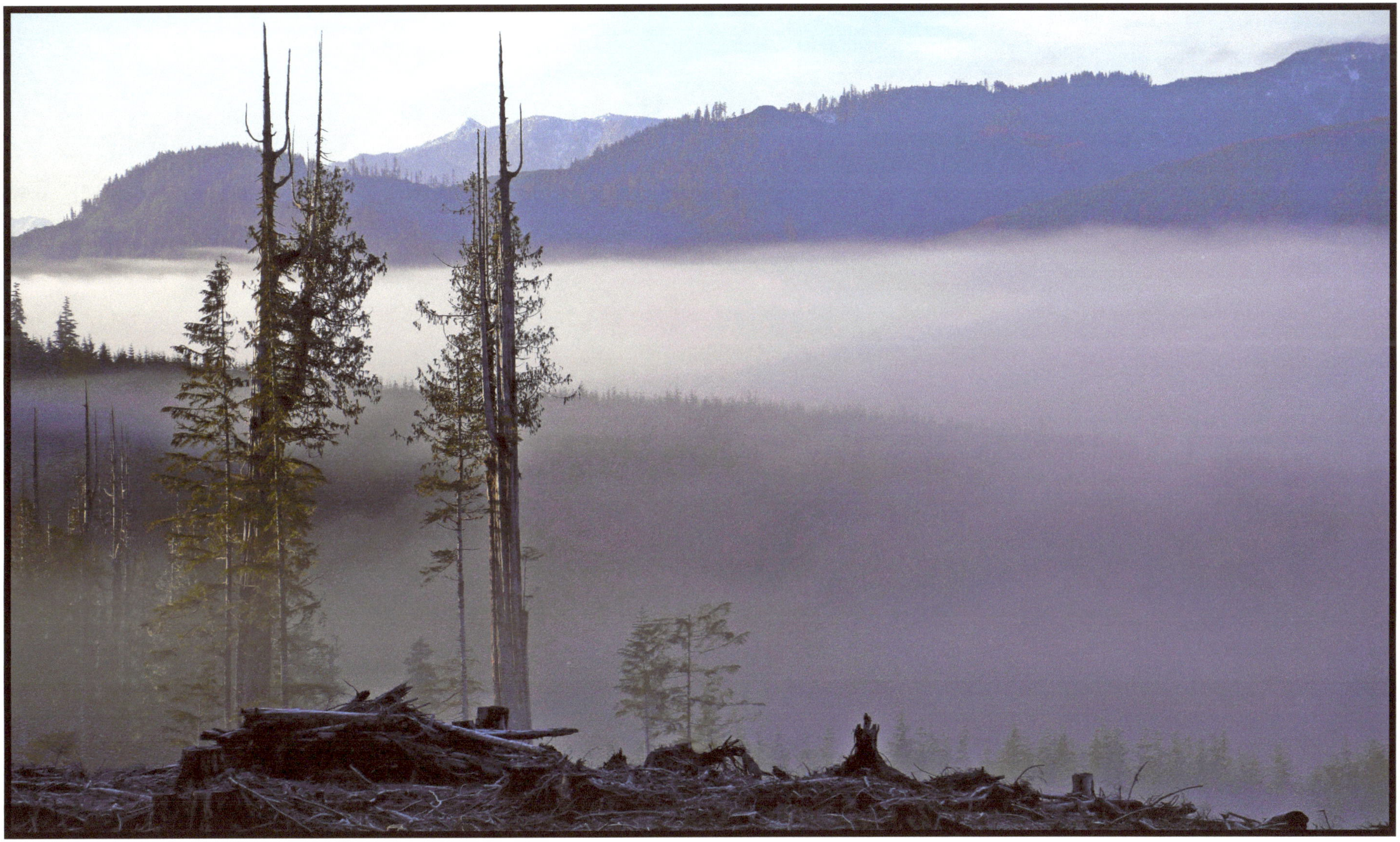

Coastal mornings usually mean lots of fog and mist.

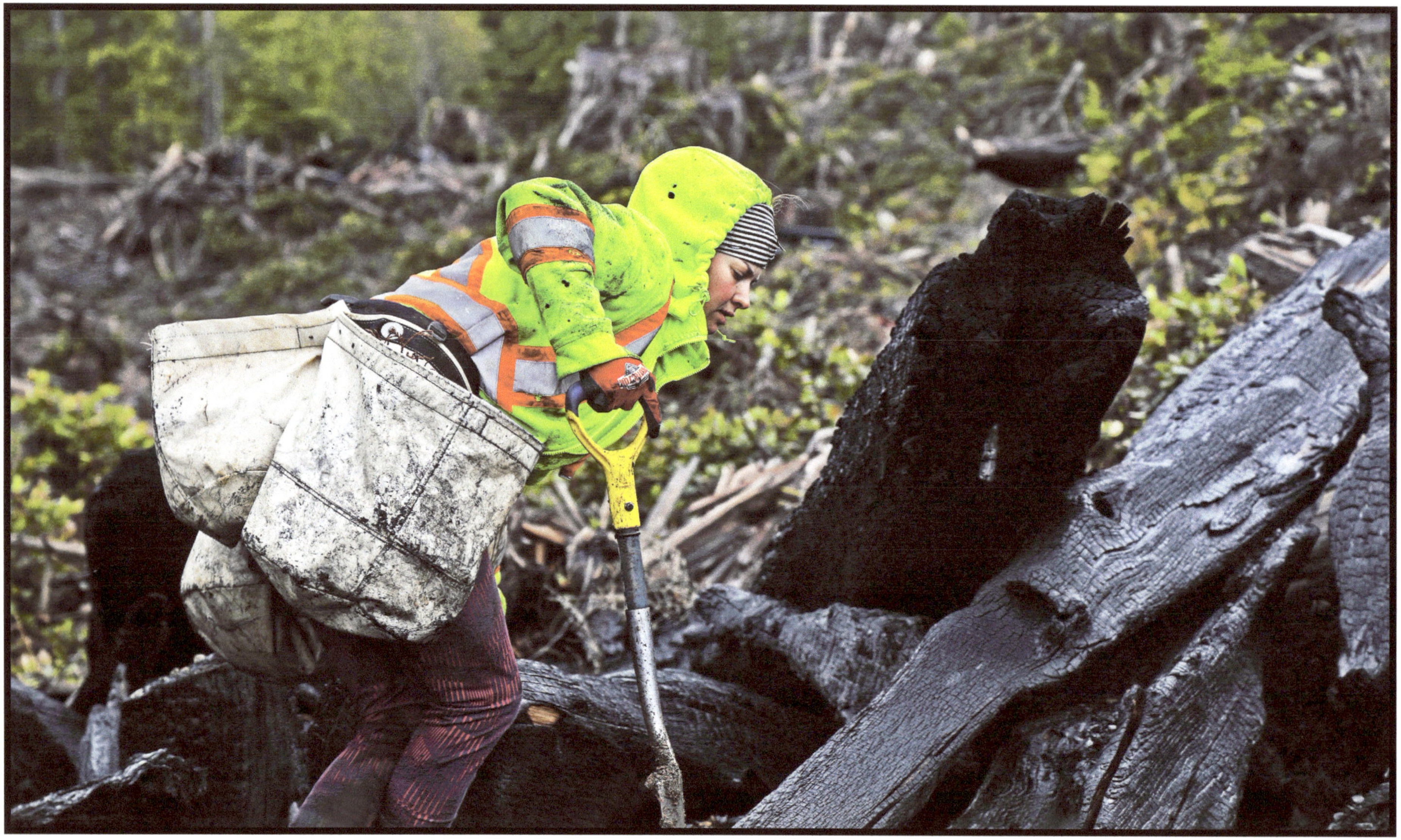

Crawling across the slash is always a good workout.

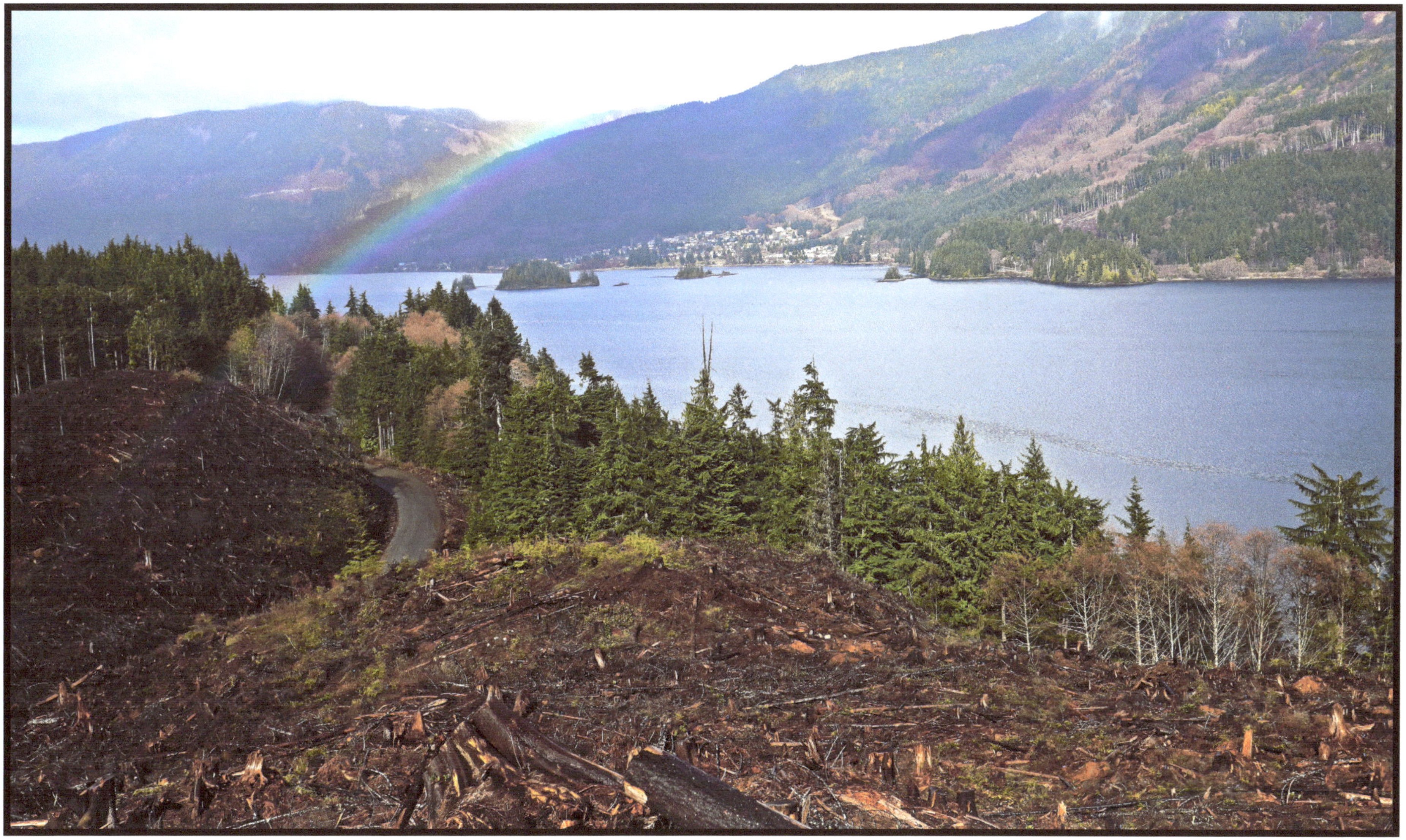

Rainbows are inevitable any time that the sun peeks through the overcast skies. That's Port Alice in the distance.

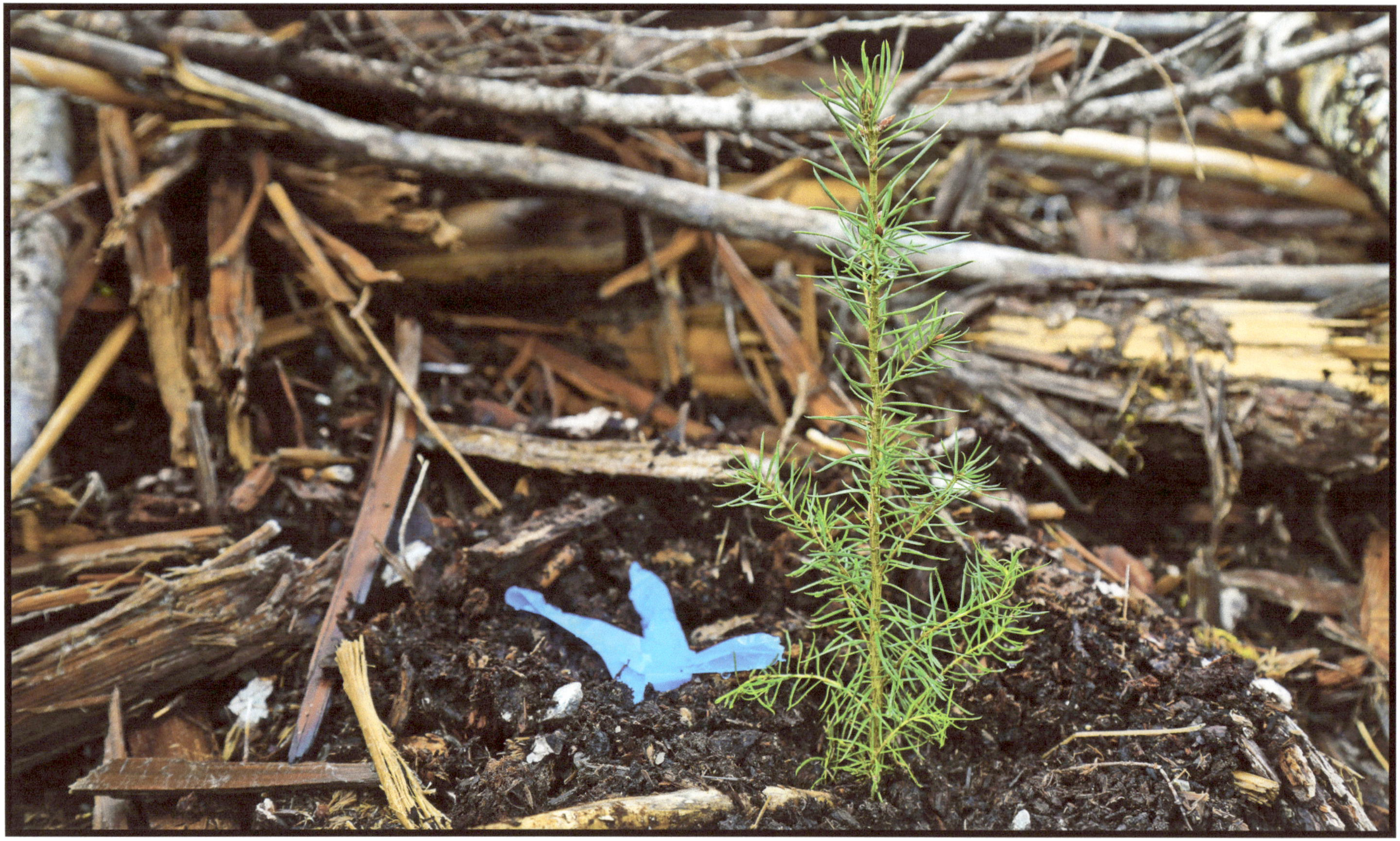

This is a Douglas Fir seedling. Mature trees of this species can eventually grow to more than a hundred meters in height in coastal regions (that's over three hundred feet tall, for American readers).

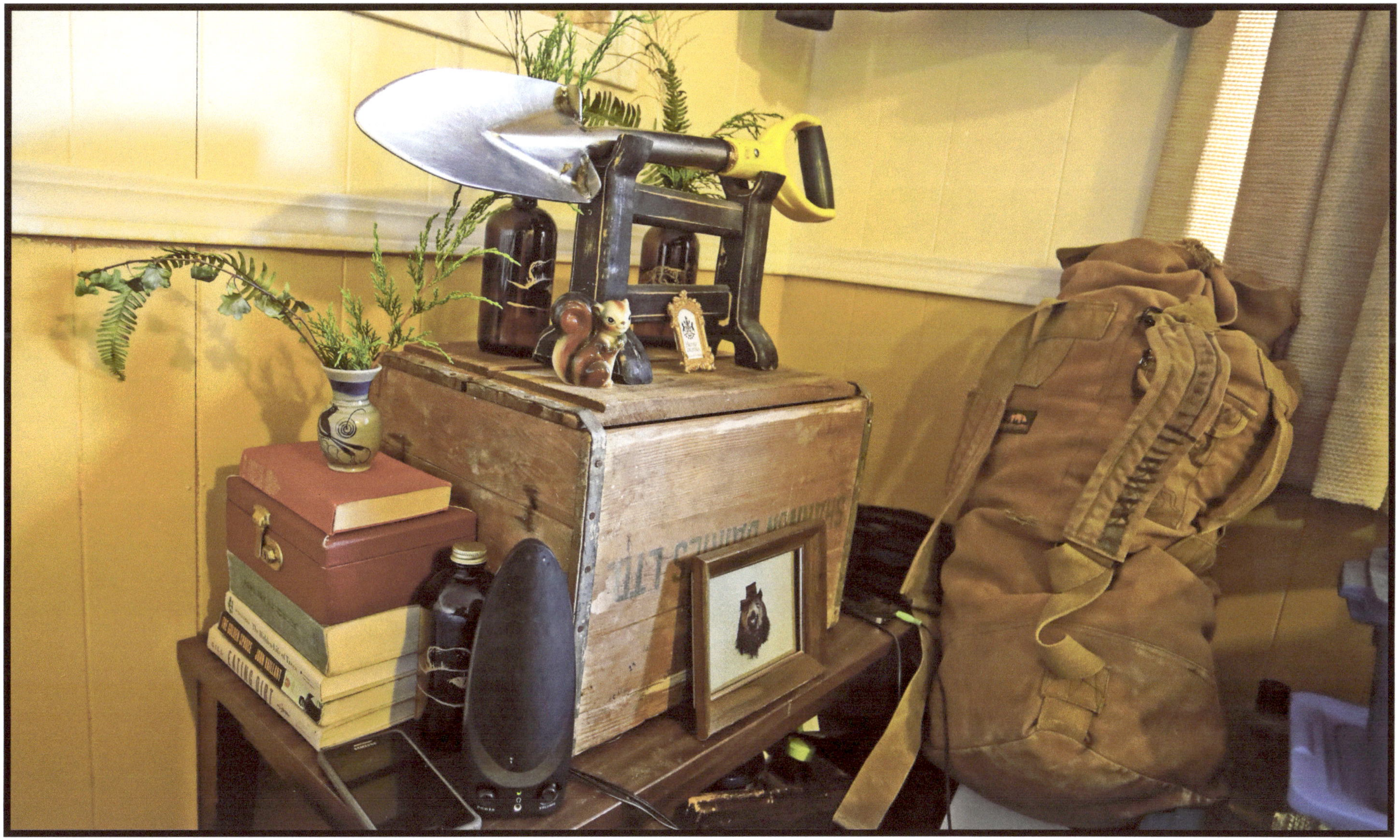

It is almost unheard of for coastal planters to work in tent camps. The weather is just too harsh. Motel rooms provide kitchenettes and hot showers, and facilitate setting up shrines to the tree gods.

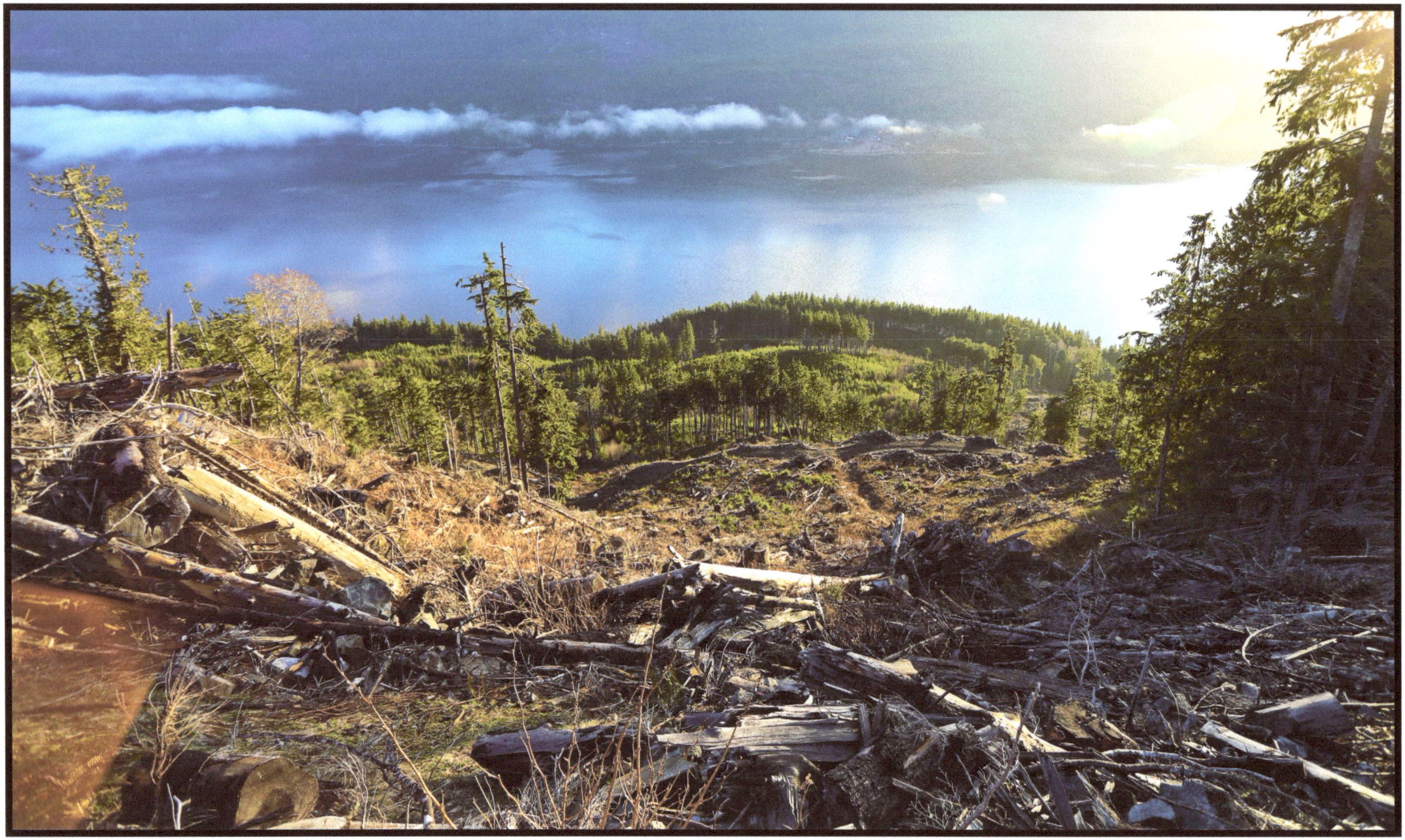

Going down a steep slope is just as challenging as going up one. For perspective, keep in mind that the mature trees at the bottom of this piece are more than a hundred feet high.

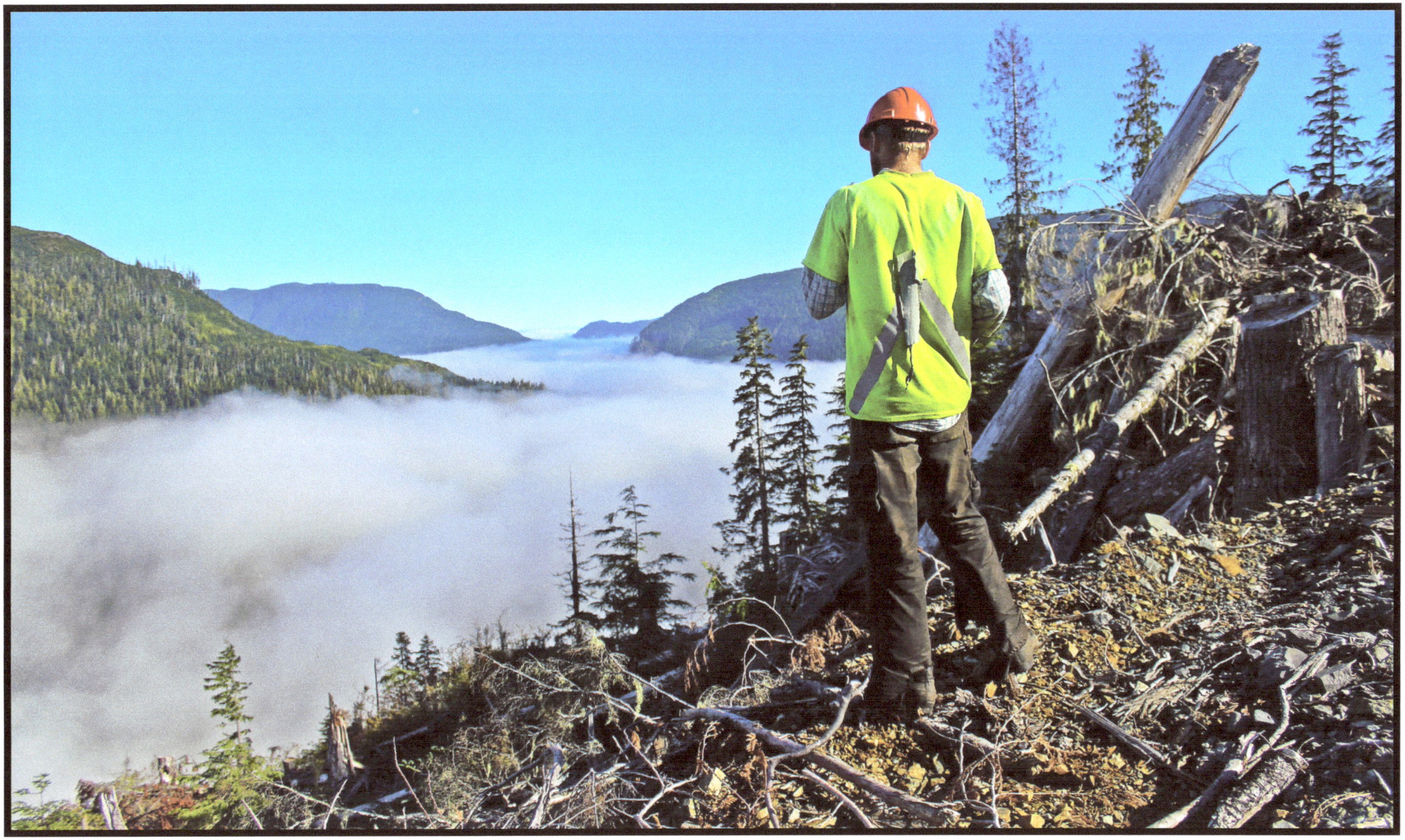

On a sunny day, fog in the valley bottoms will usually burn off within a few hours. However, it leads to some great early morning views.

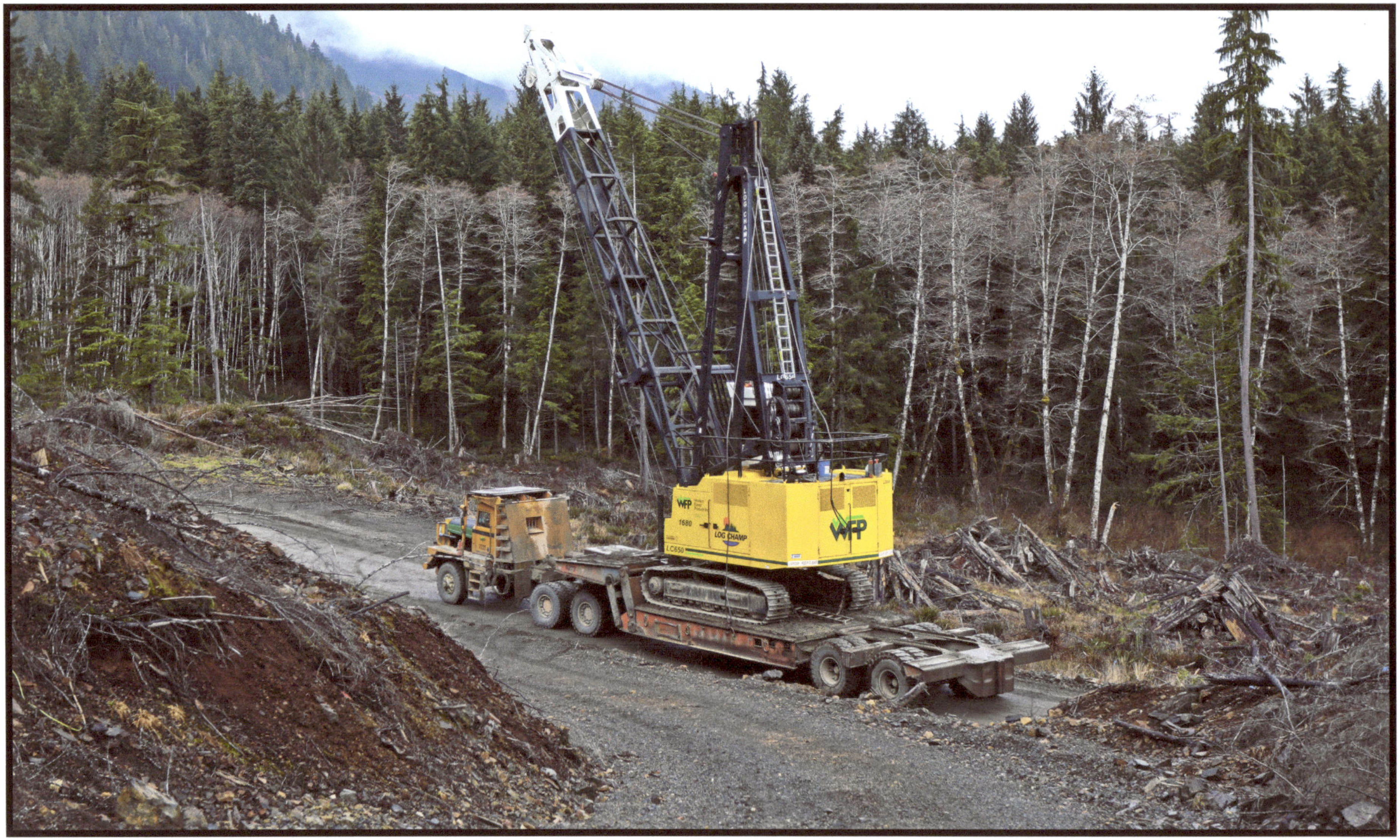

This yarder machine is being moved to a new block. The yarder is used to pull harvested trees out of the blocks and onto the roadsides, so logging trucks can pick them up and take them to the mills.

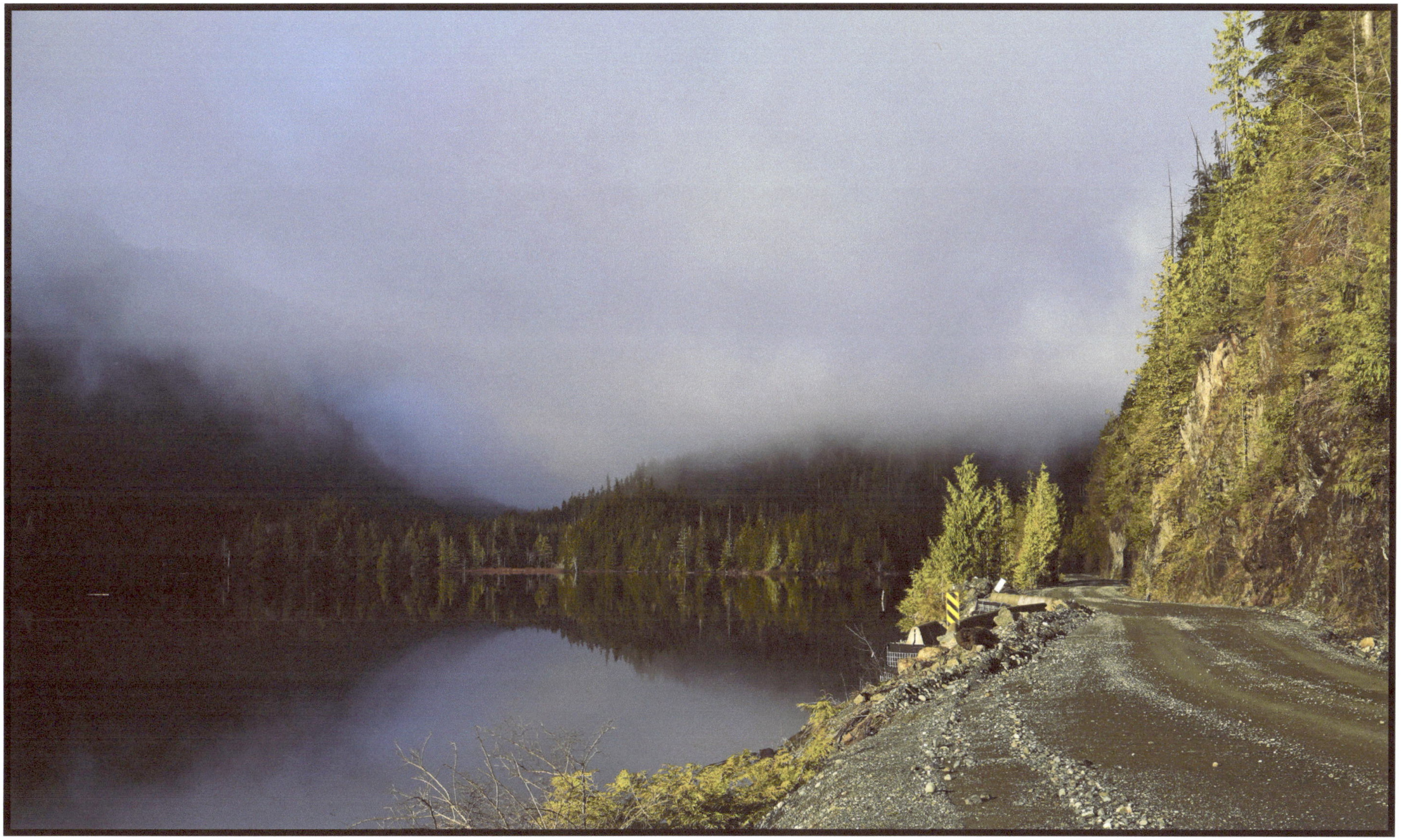

Here's a pleasant early-morning drive to the block.

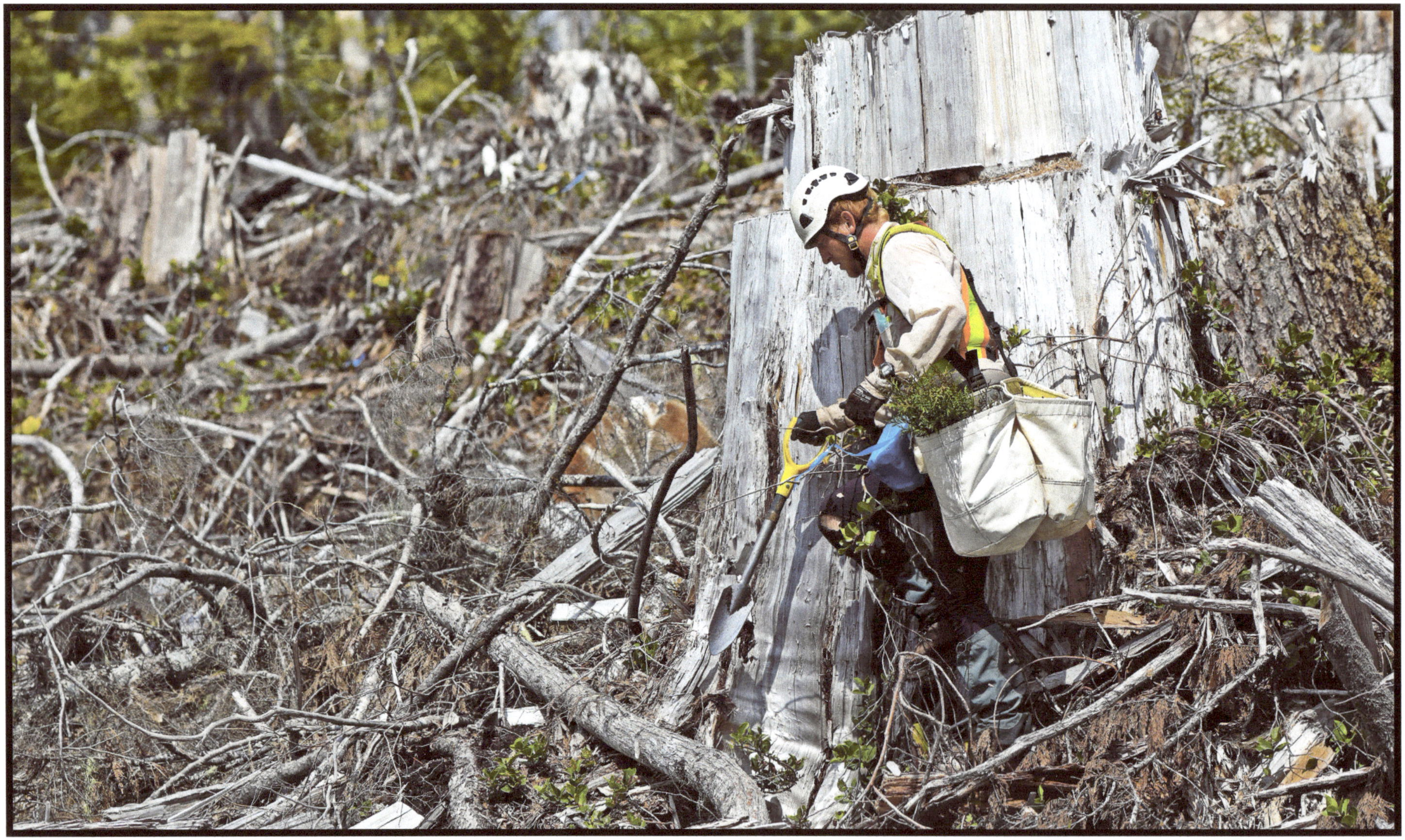

Willem is a civil engineer, but he loves planting. This photo was taken on day 38 of a season during which he planted more than 200 days, working for several different companies in two countries. A person has to be in outstanding physical and mental condition in order to get through such a long season. Many planters elsewhere in Canada burn out after only forty to sixty days.

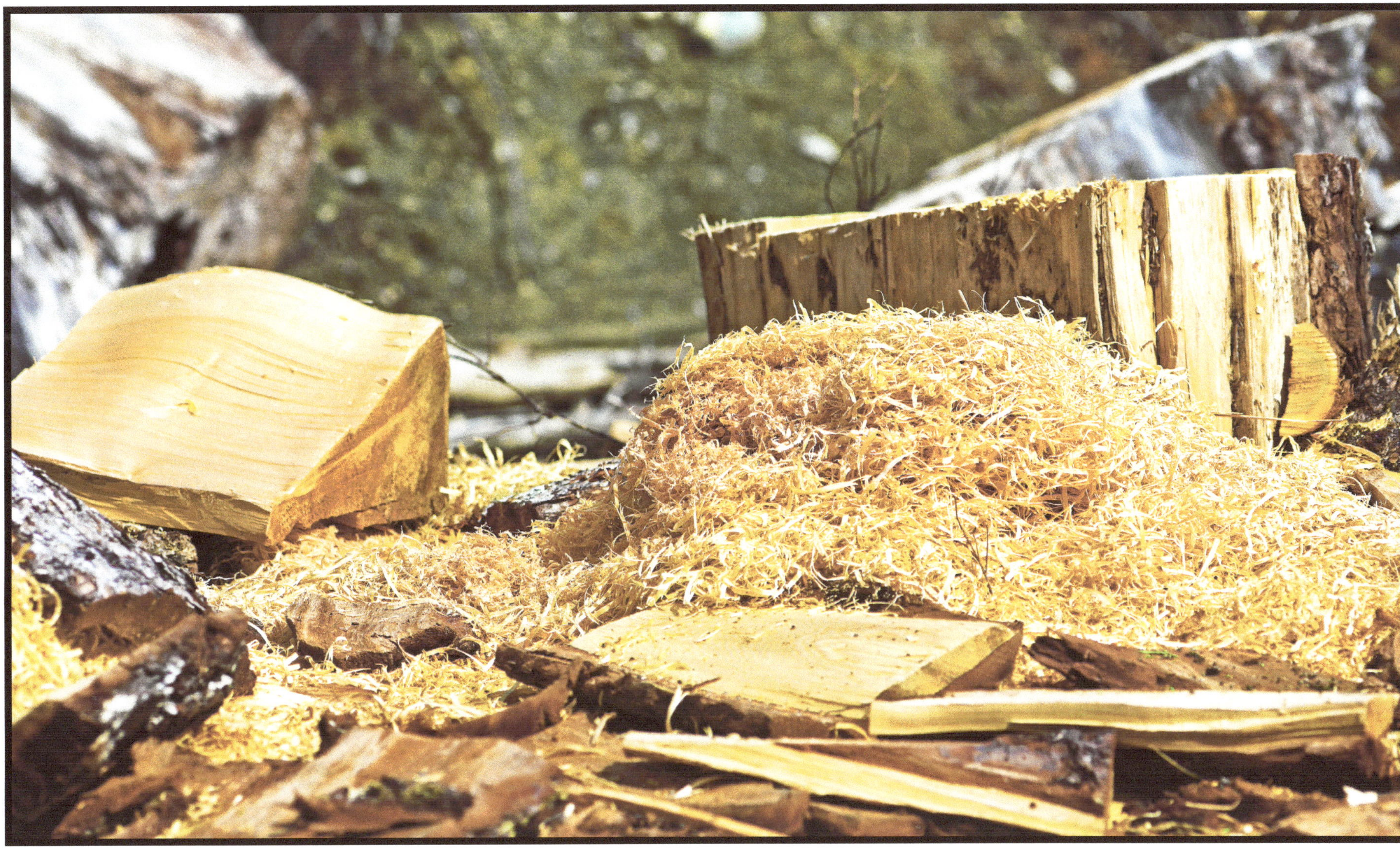

Two types of cedar are typically planted on coastal contracts. Western (red) cedar is probably more common, and suited for lower elevations down to sea level. Yellow cedar, shown above, prefers slightly higher elevations. It is quite easy to distinguish between these two species based upon the colour of their wood.

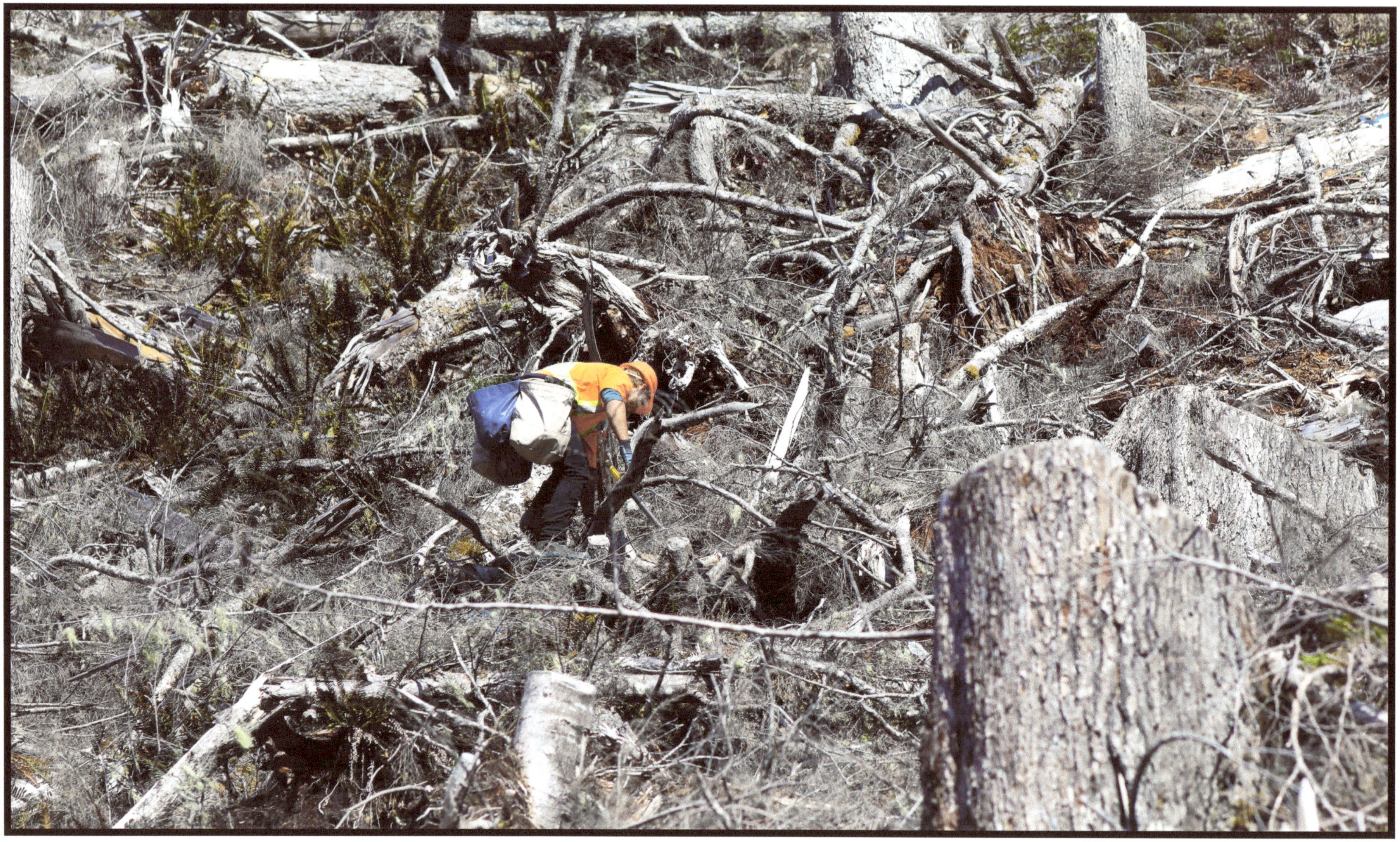

Although some people will probably look at this and be amazed by how much slash there is, most coastal planters would look at this photo and think that it's a pretty clean block.

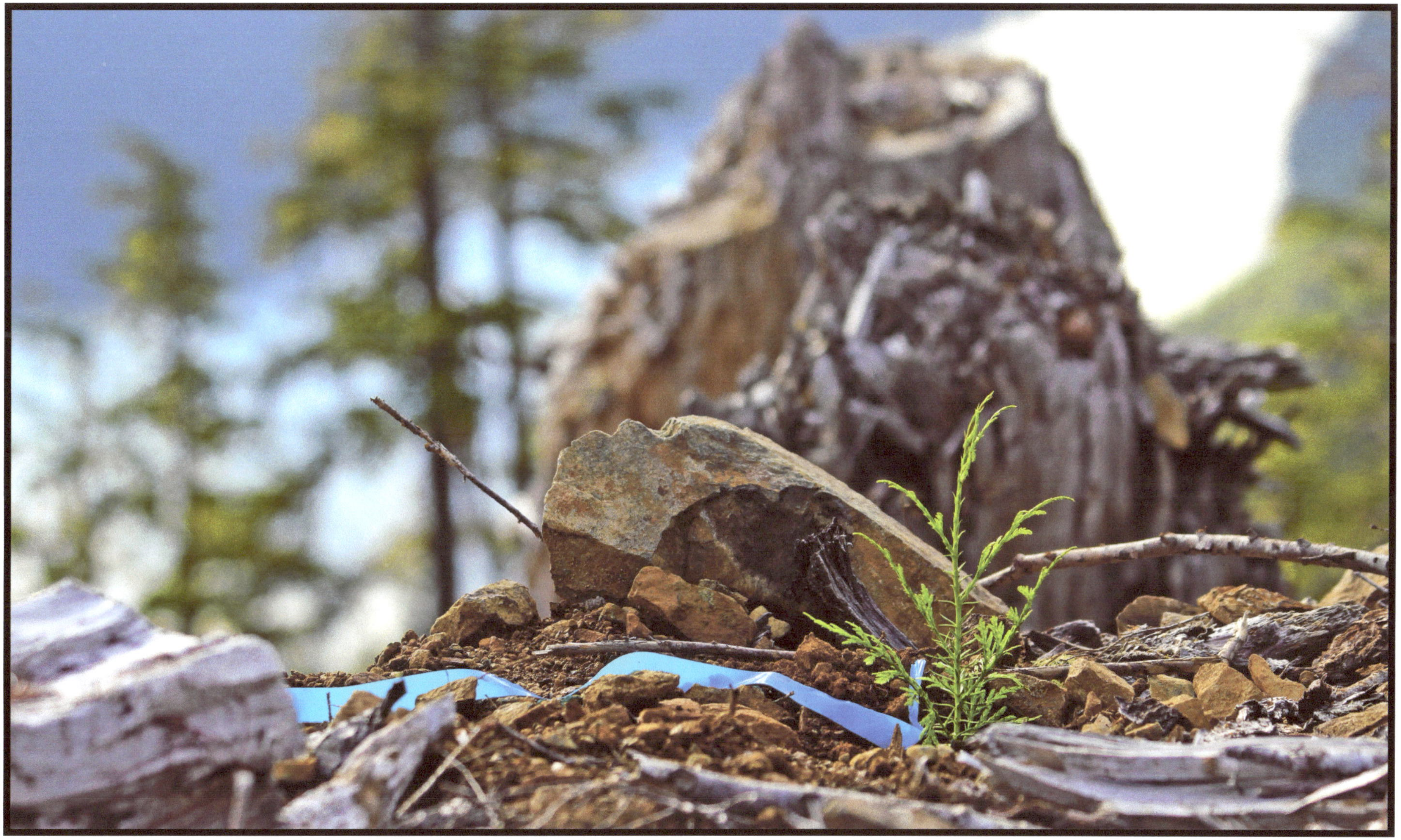

Here's a freshly planted western red cedar seedling.

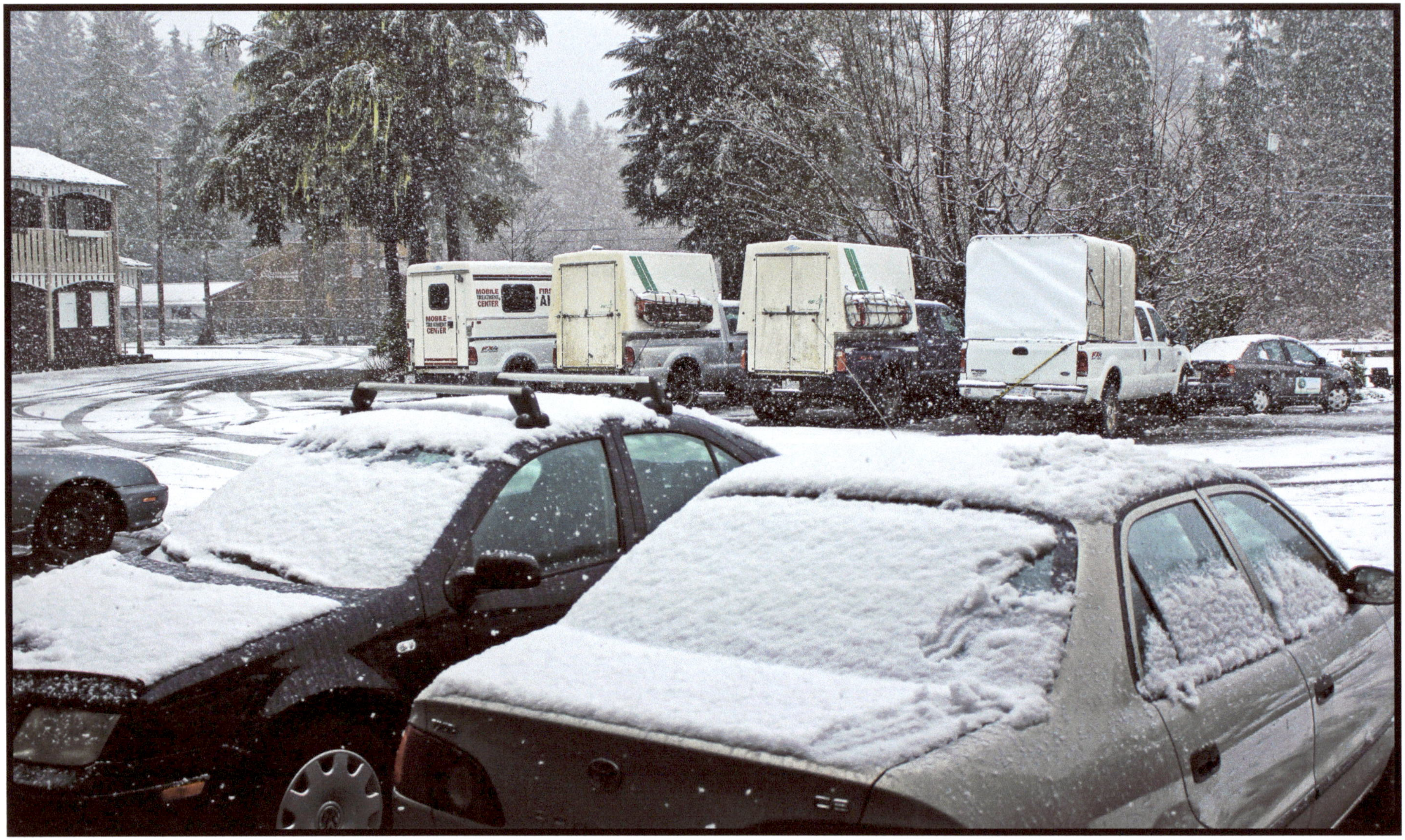

This looks like a snow day. Take advantage of the downtime when you can.

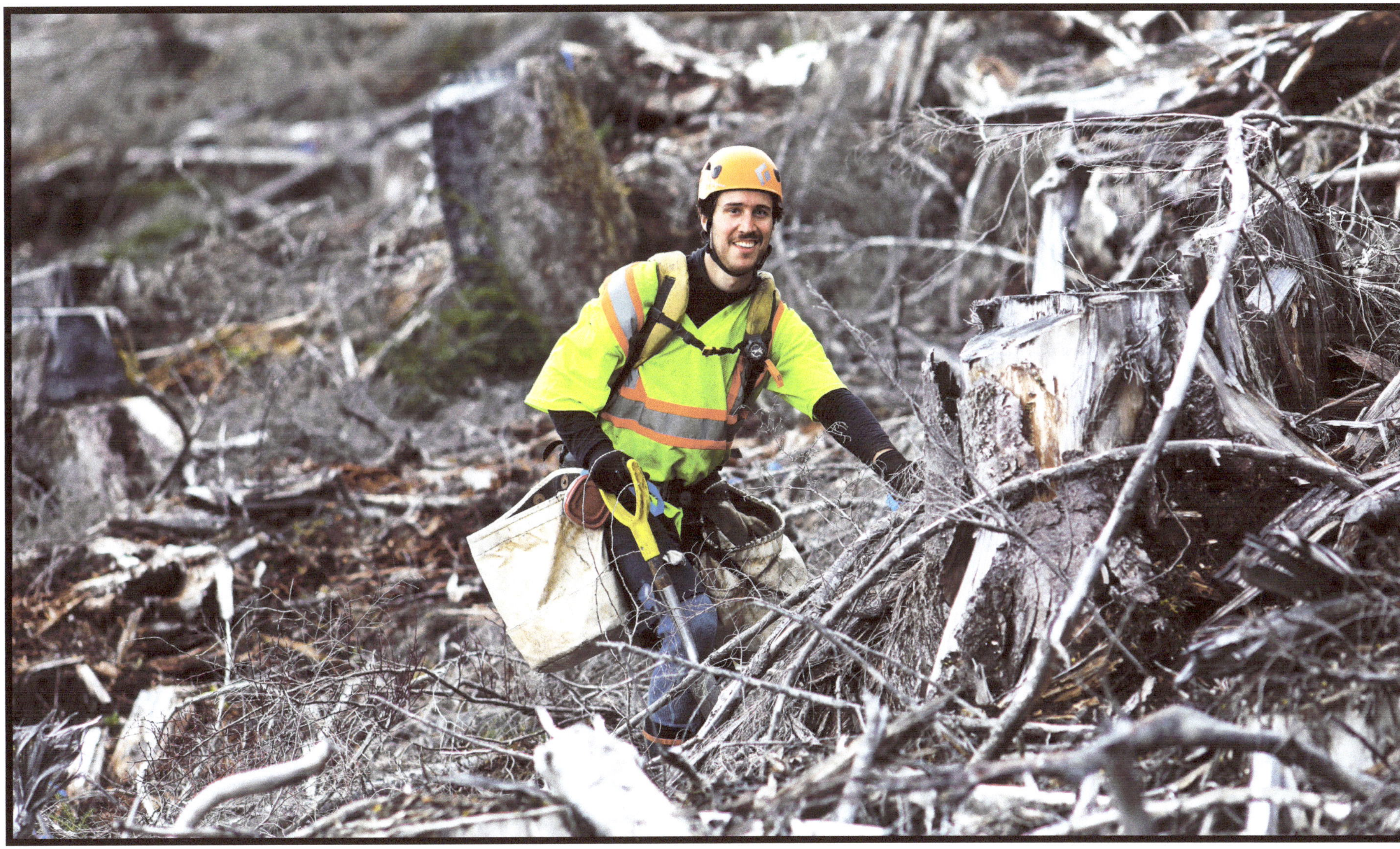

Stuart is from Ontario, but he loves planting in British Columbia. When he's not planting, his interests include hiking, biking, snowboarding, music, and cooking. Most coastal planters lead very healthy and active lifestyles.

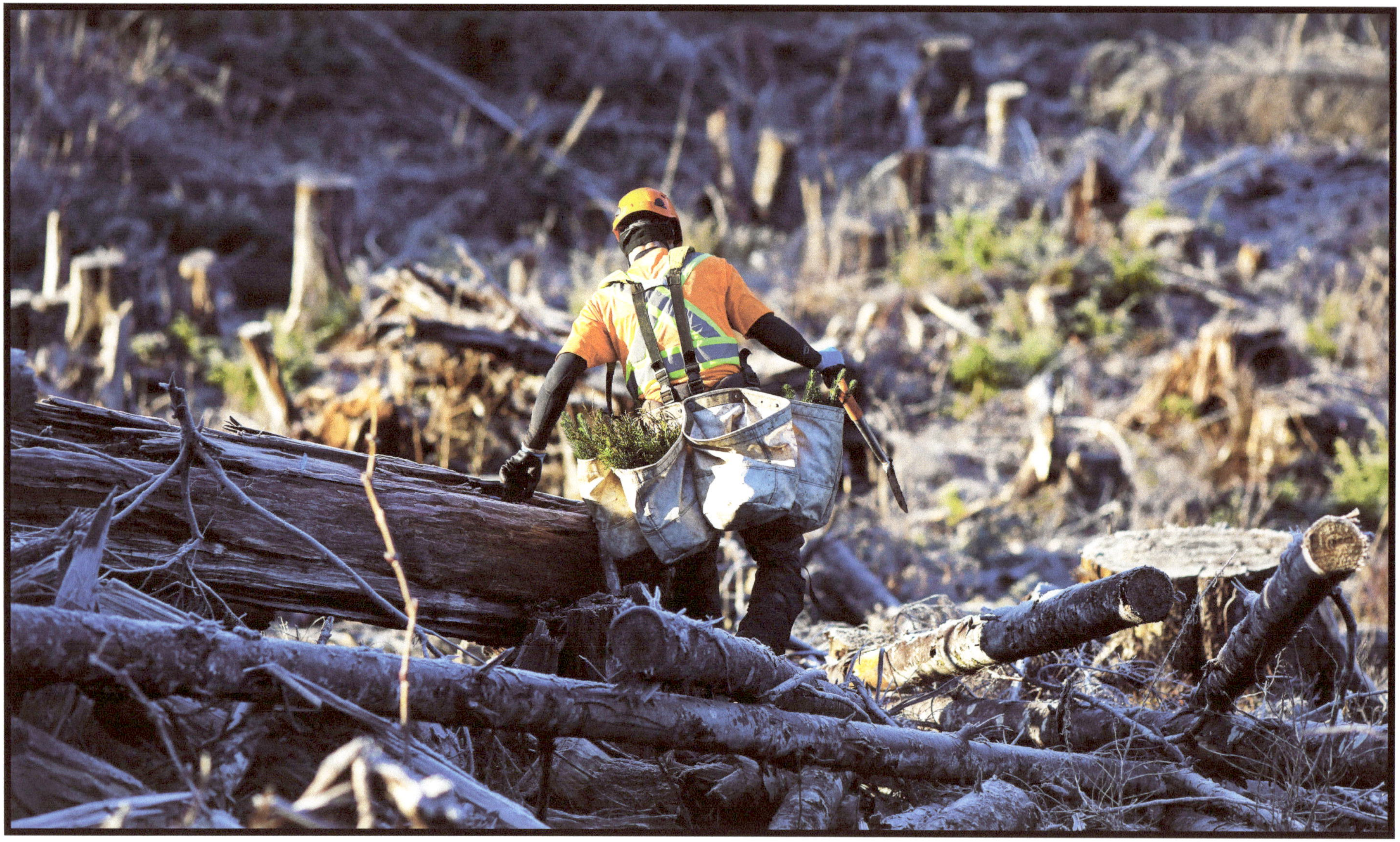

The first run of the morning is especially treacherous if there's frost on the slash.

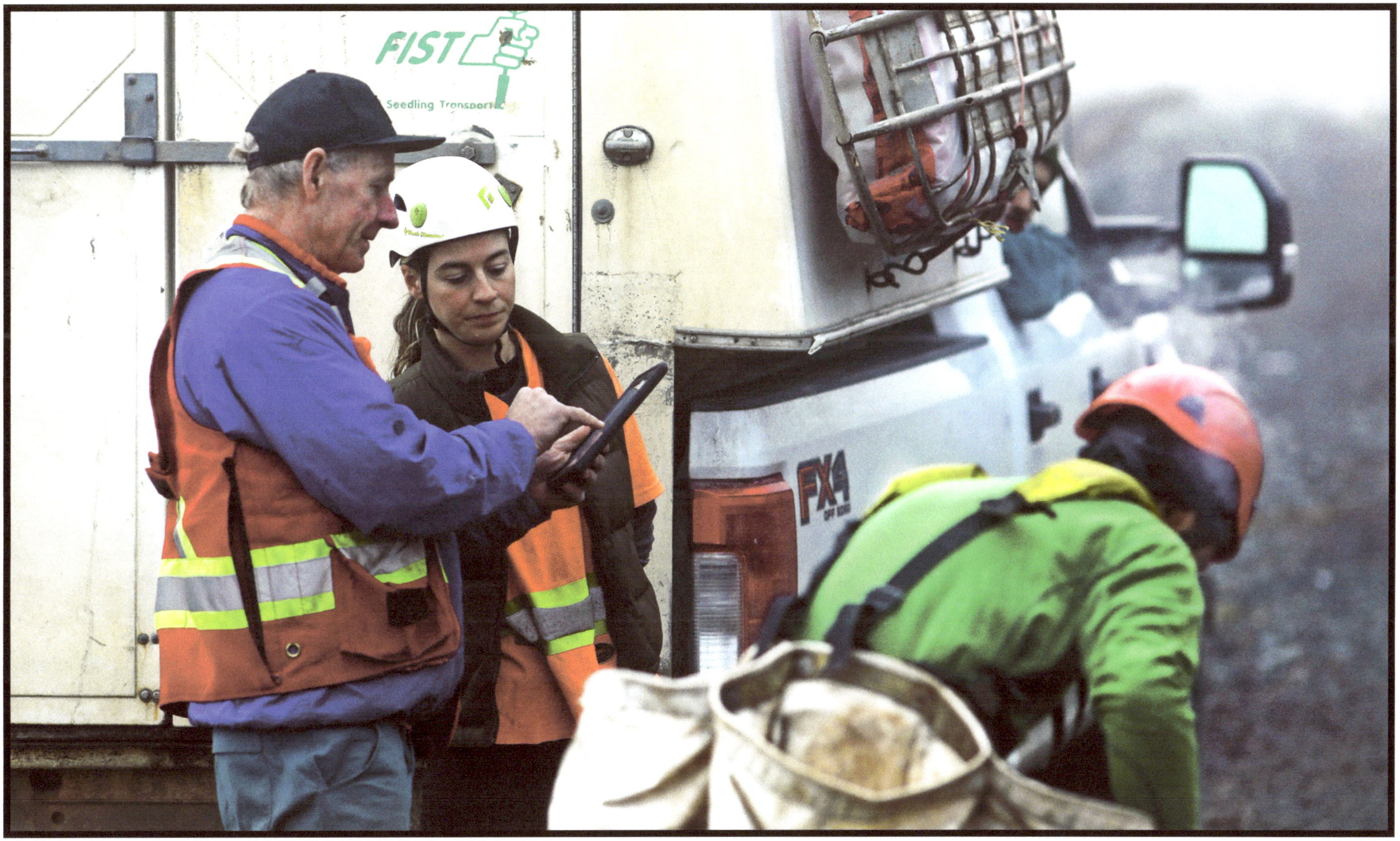

The planting industry has welcomed the advent of technology. The use of digital mapping has exploded over the past several years.

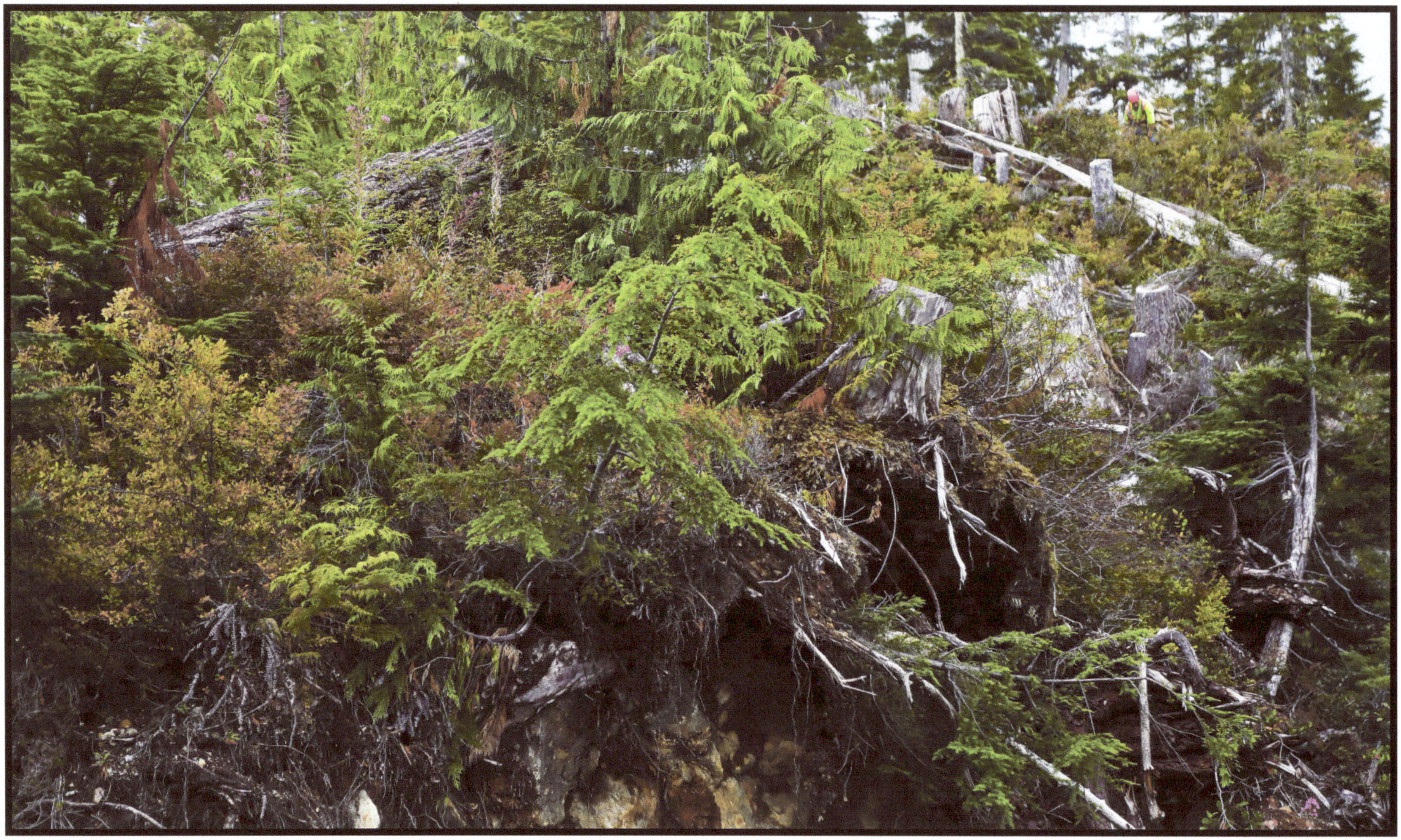

We should play a game of "find the planter" as we go through some of these photos. Here's an easy one to get you started. Yes, this is a block.

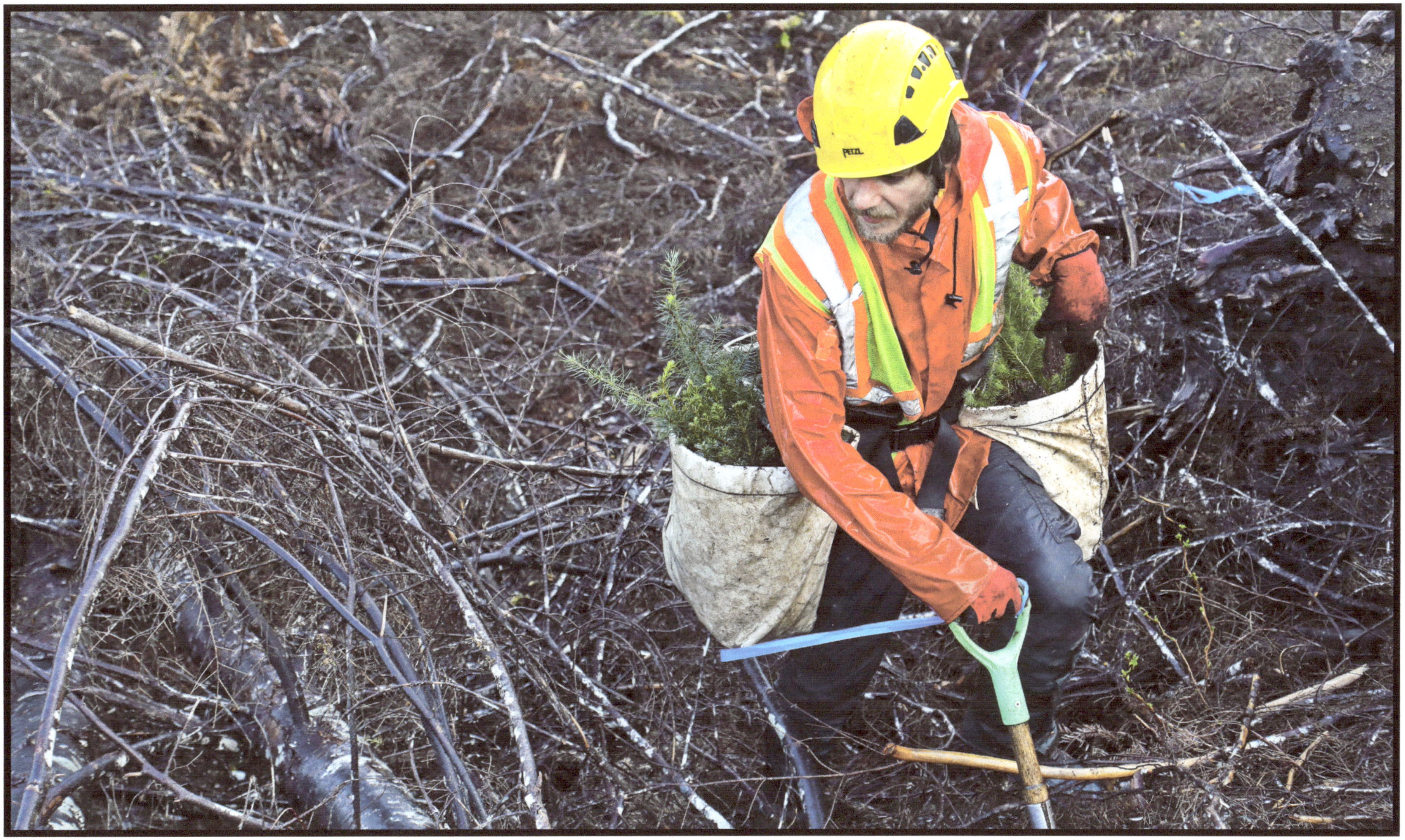

Thankfully, not all of the spring blocks are as lush and green as the block in the previous photo.

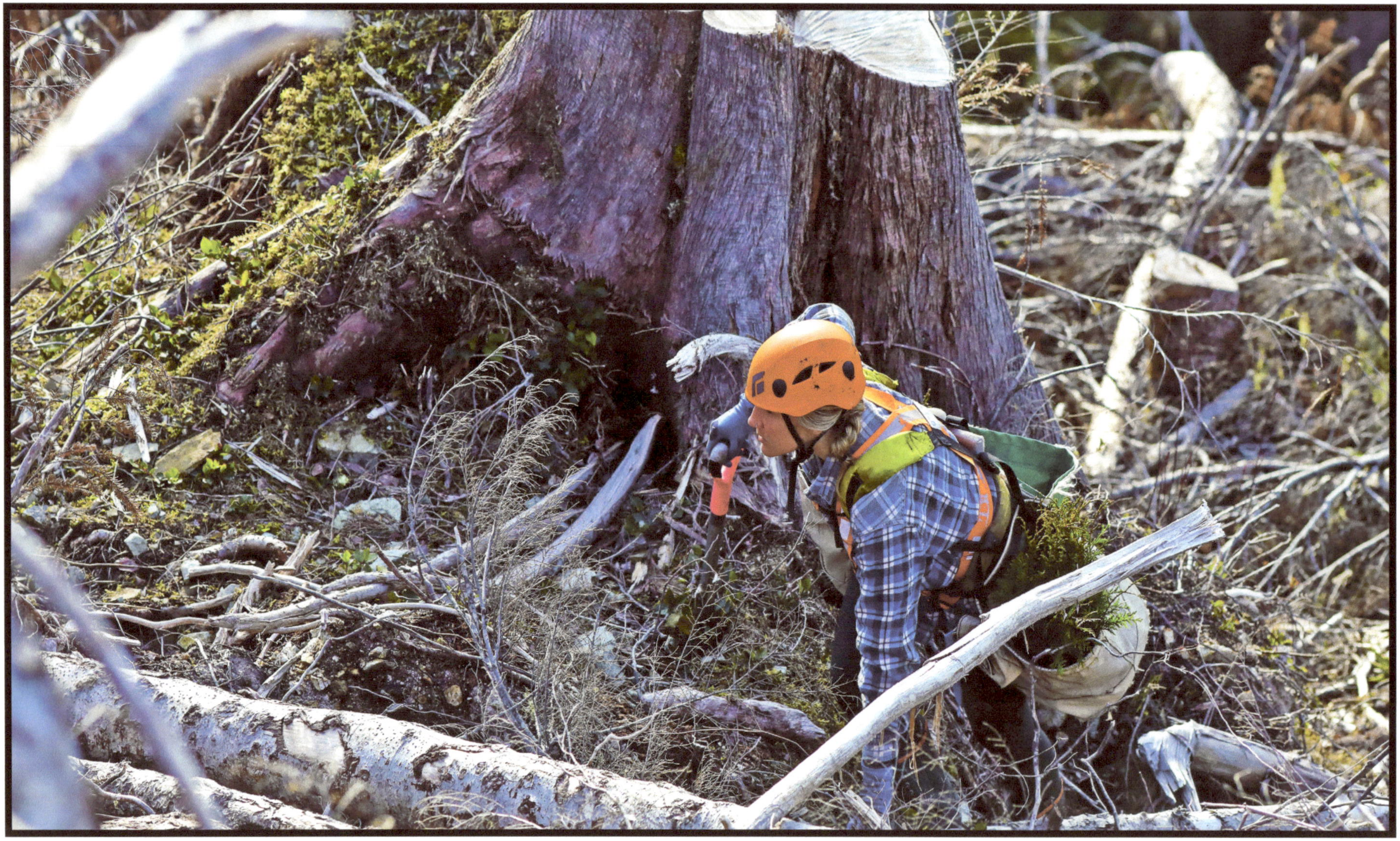

On some blocks, planters spend almost as much time crawling as they do walking.

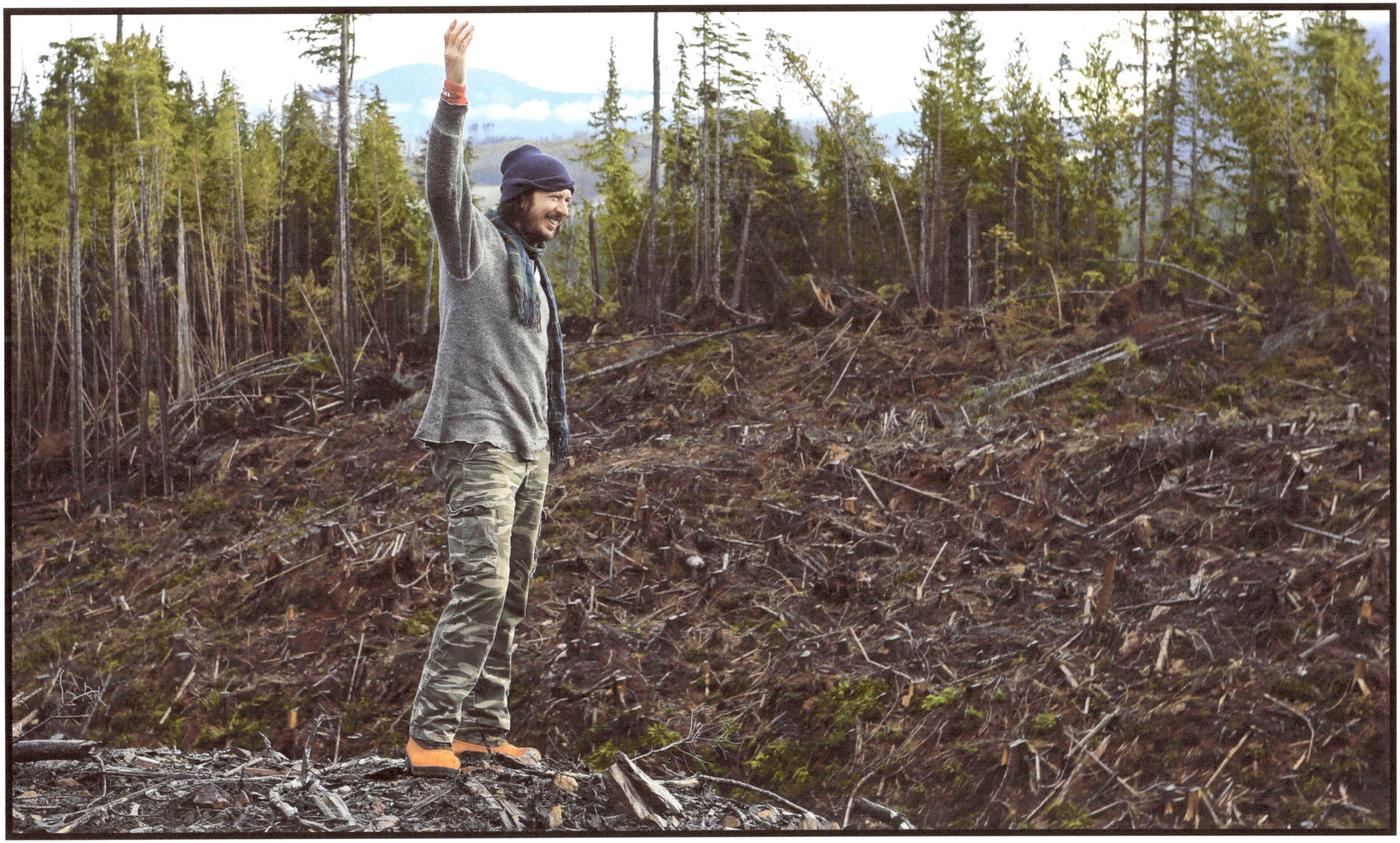

I appear to be pretty happy here. It must be time to head home for dinner.

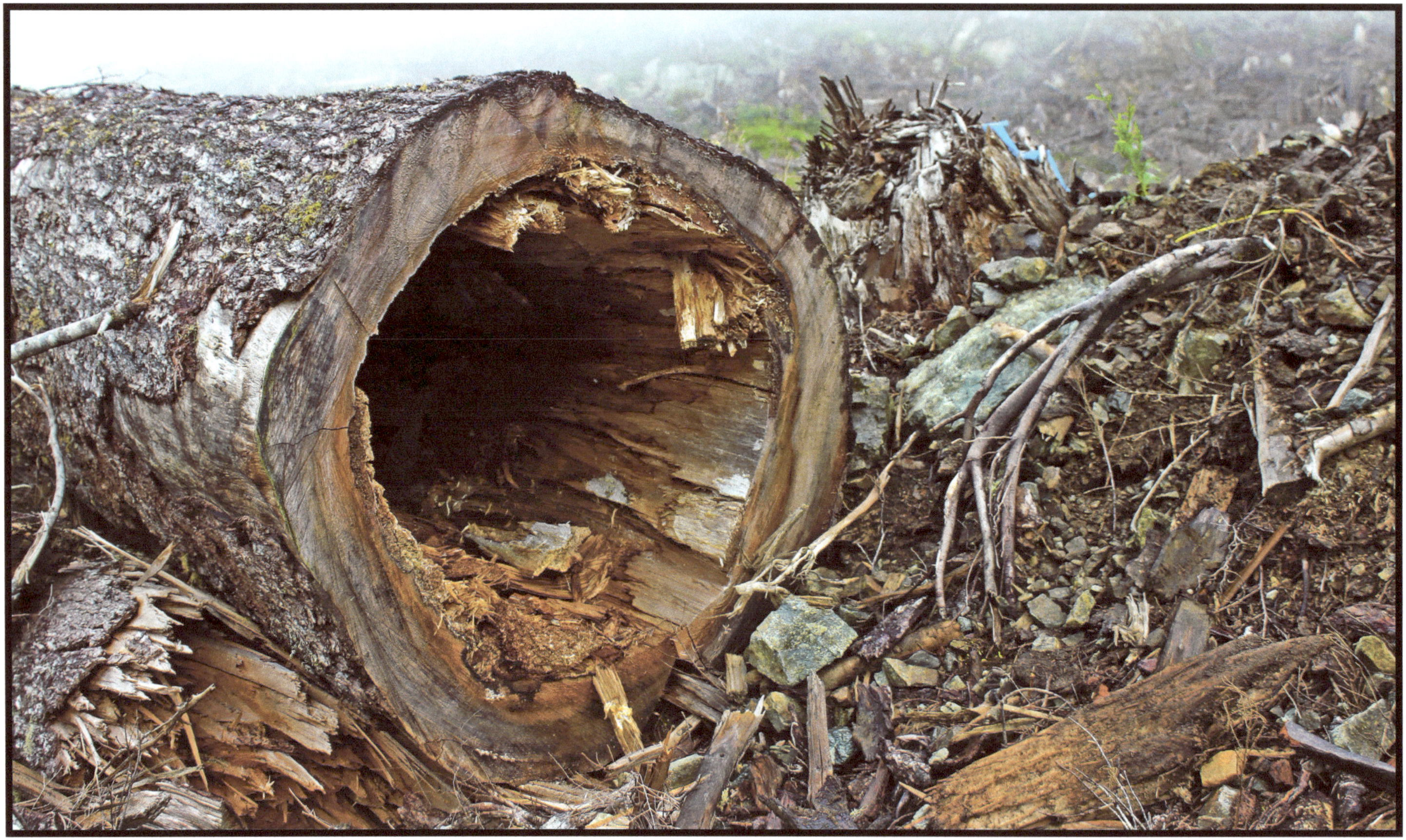

Many readers will be dismayed at seeing how much slash is left behind on most of the coastal blocks. Often, the commercial value is not high enough to justify the transportation costs of moving scrap wood to the mill.

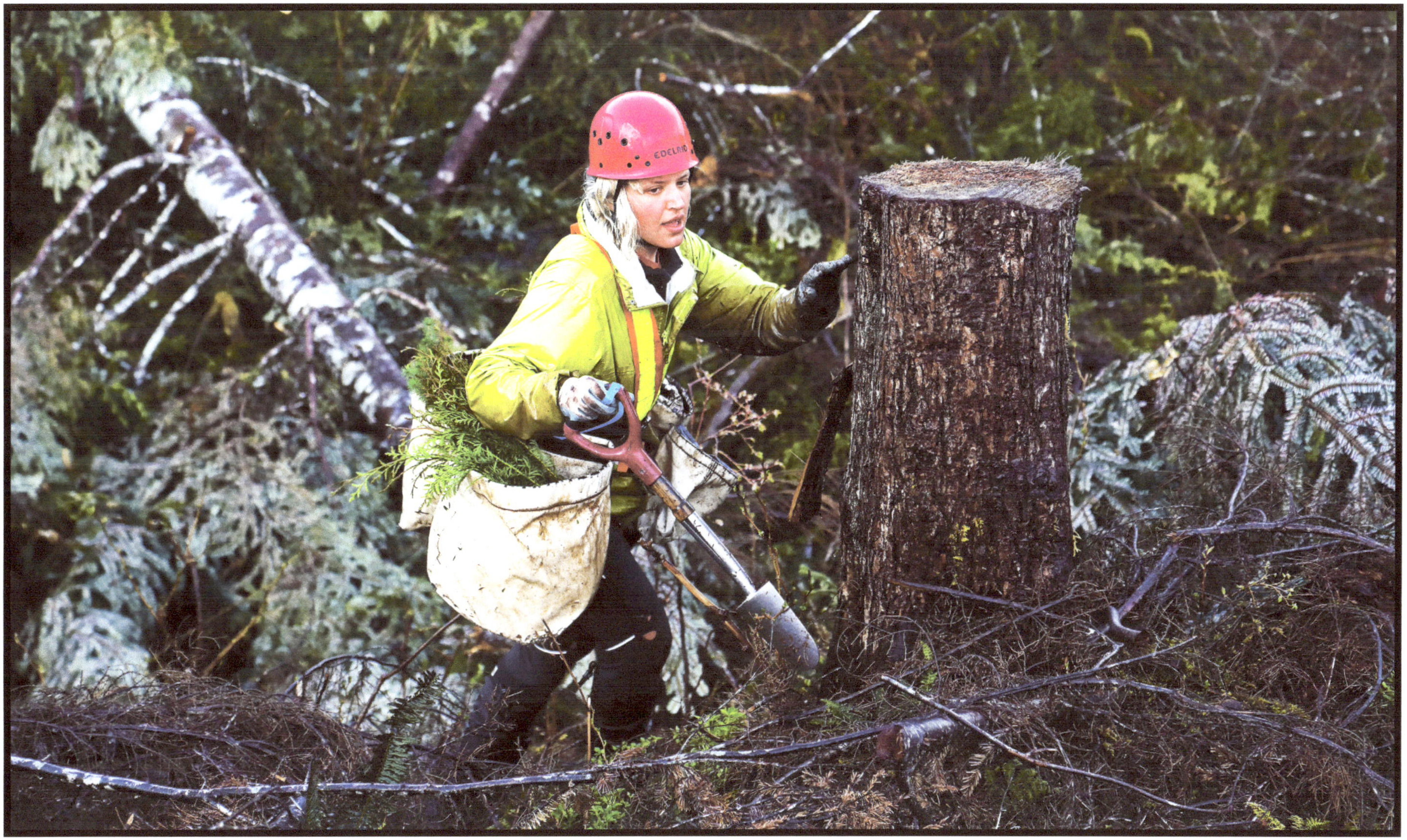

Most coastal planters invest in layers of wool and high-quality rain gear. Of course, rain gear doesn't keep us dry, because we sweat so much inside. However, the point of wool and rain gear is to keep us warm on days when the temperature is a few degrees above zero, and when we're working in freezing rain all day. On the first day of this particular spring season, the daily forecast was for 100 millimeters of rain, and 100 km/hr winds. We worked.

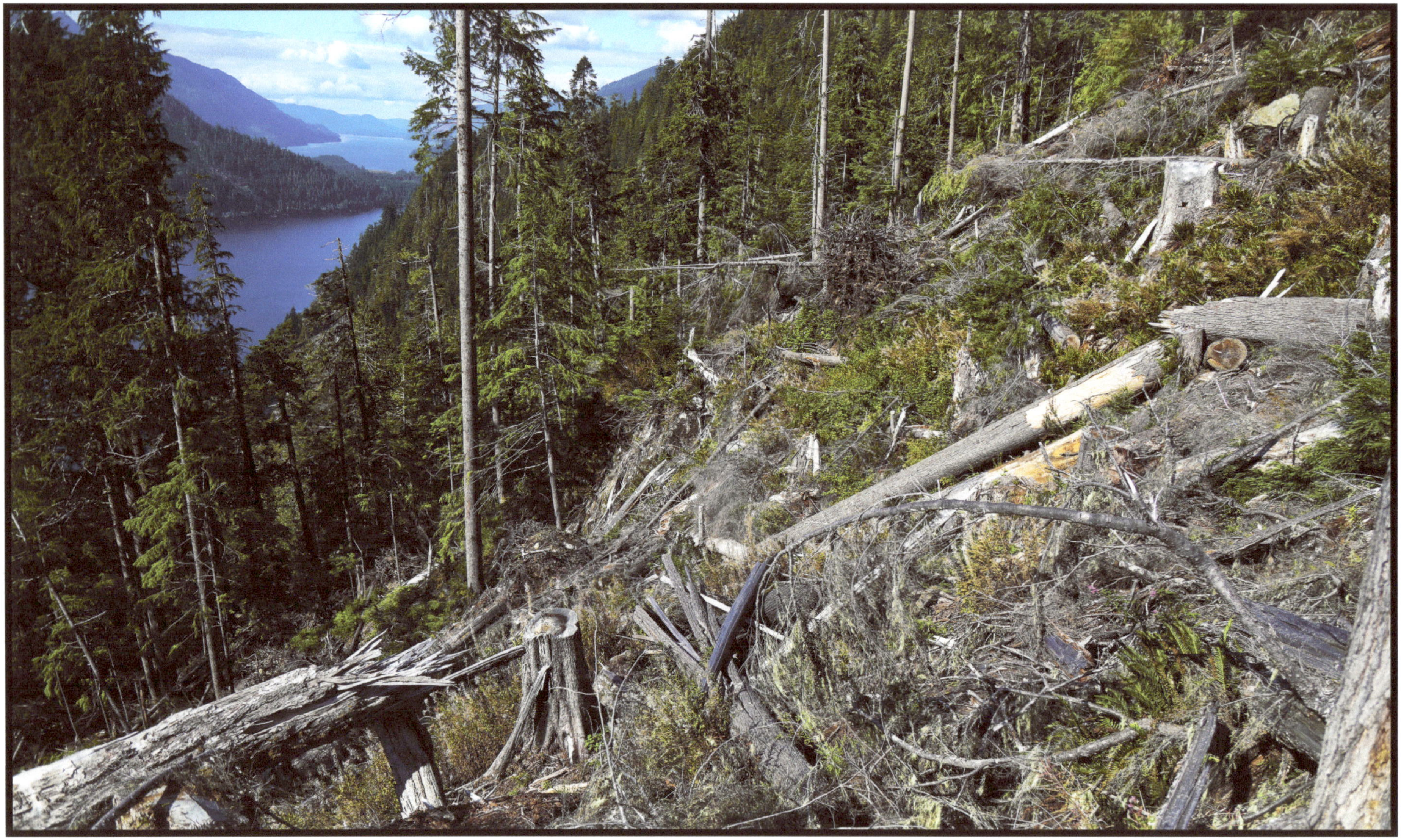

I would call this a fairly typical north island coastal block.

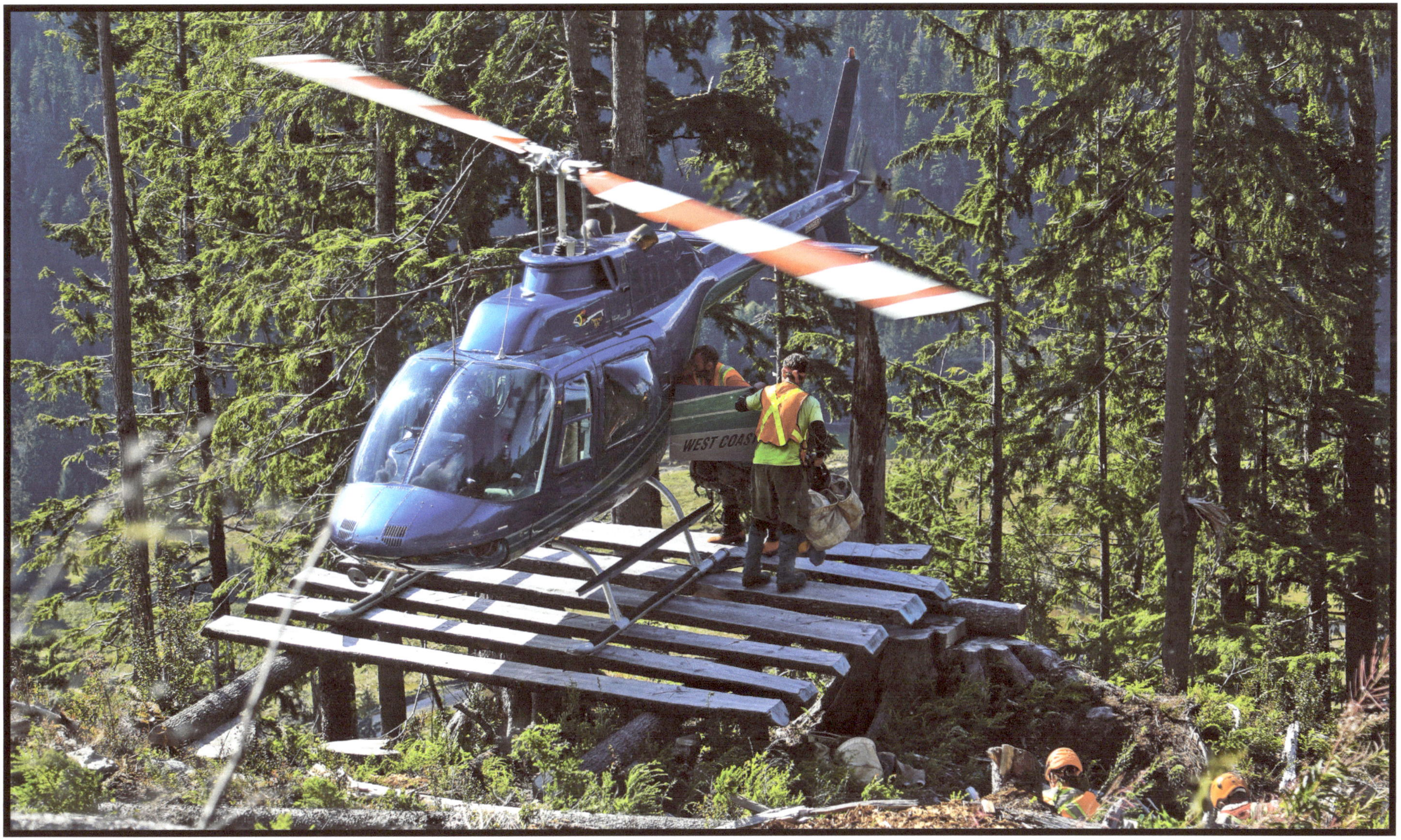

There is no place for helicopters to land on most coastal blocks, so the fallers typically walk in with chain saws and build heli-pads. Coastal fallers, like coastal planters, are the true professionals of their industry.

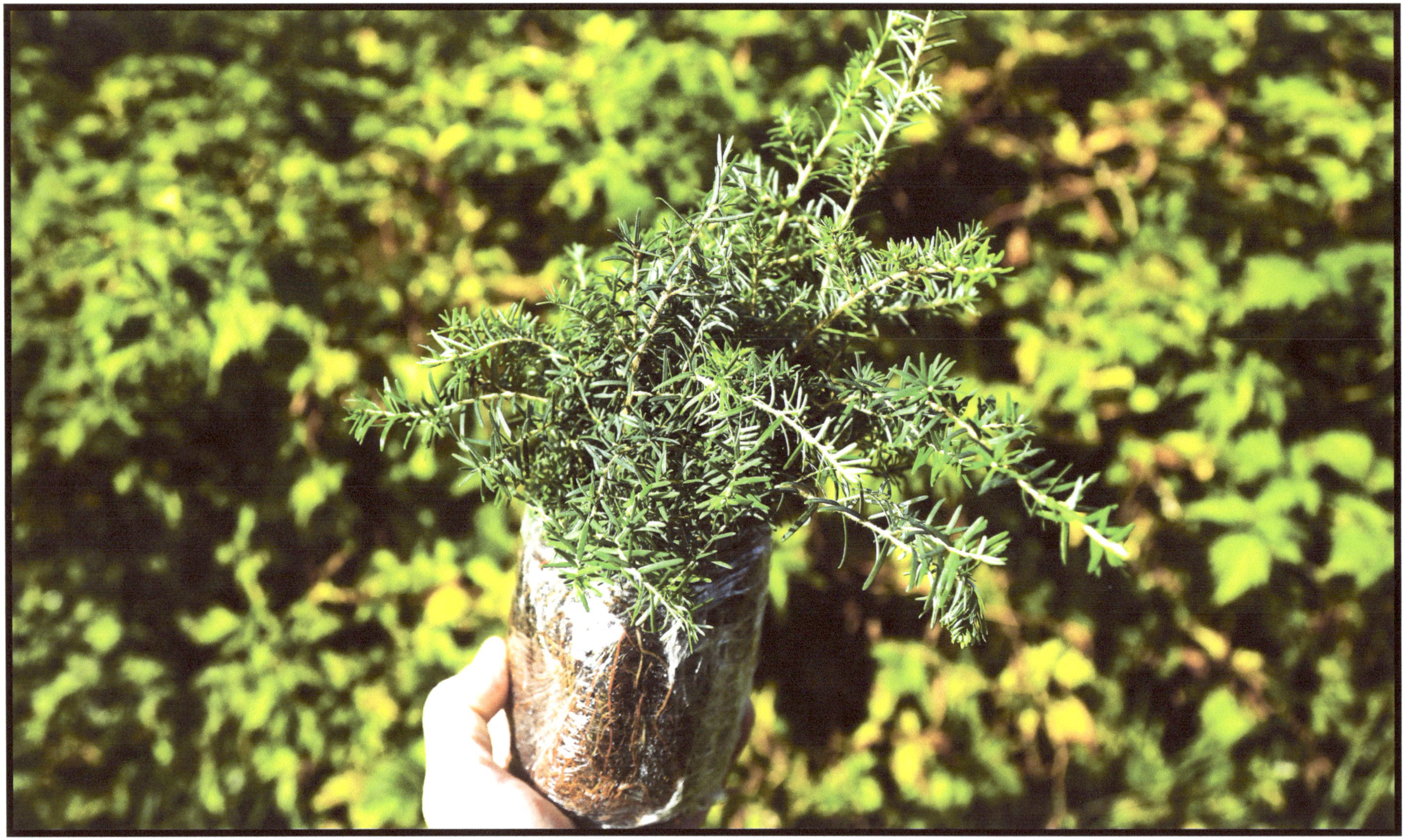

This is a bundle of mountain hemlock seedlings. Mature mountain hemlock trees often reach heights of forty to sixty meters, and typically grow on cold, snowy, subalpine slopes. Mountain hemlocks can sometimes reach an age of more than 800 years old.

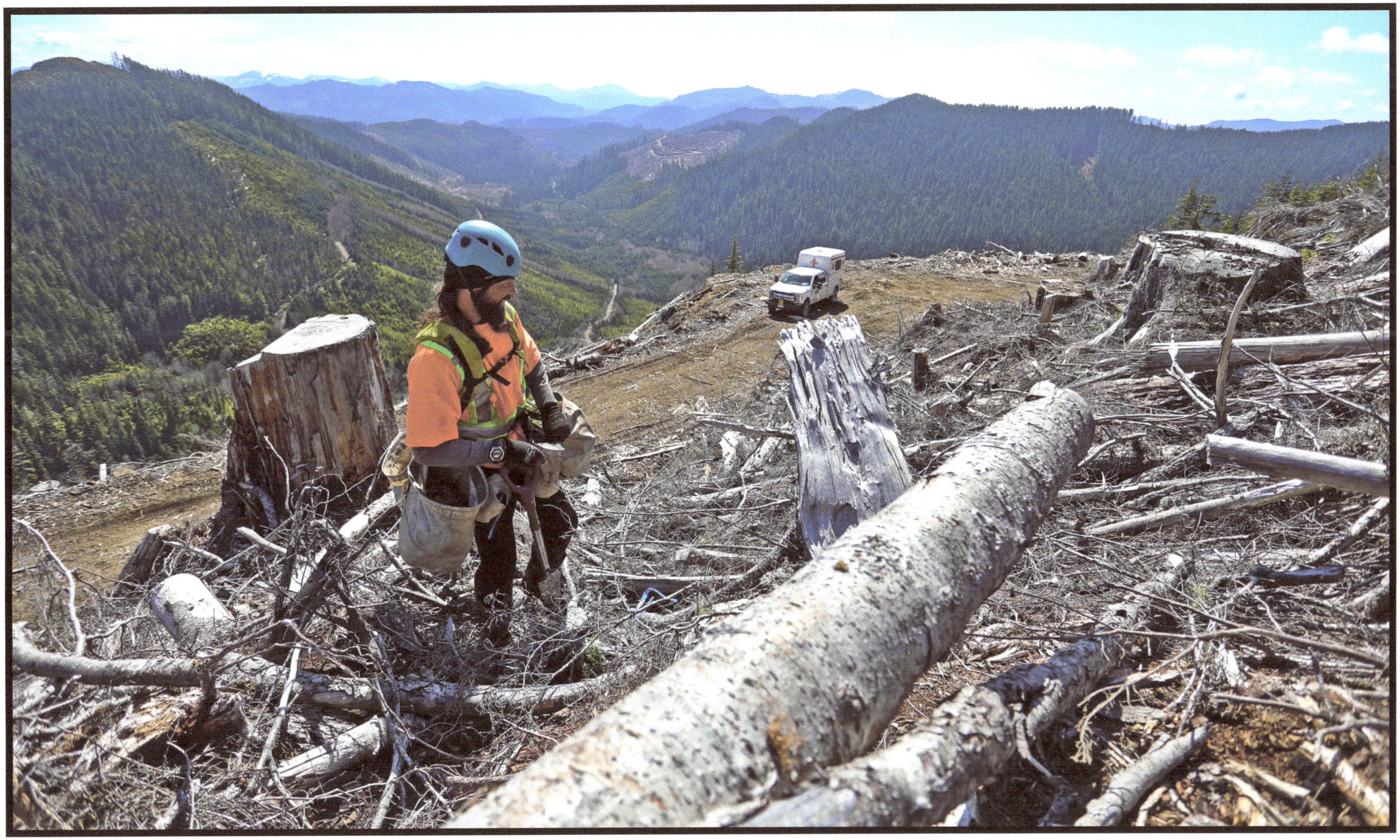

Notola is from Quebec. Although he studied finance, he hopes to work in the outdoors for his entire life. He is in love with nature, and spends a lot of time surfing and snowboarding.

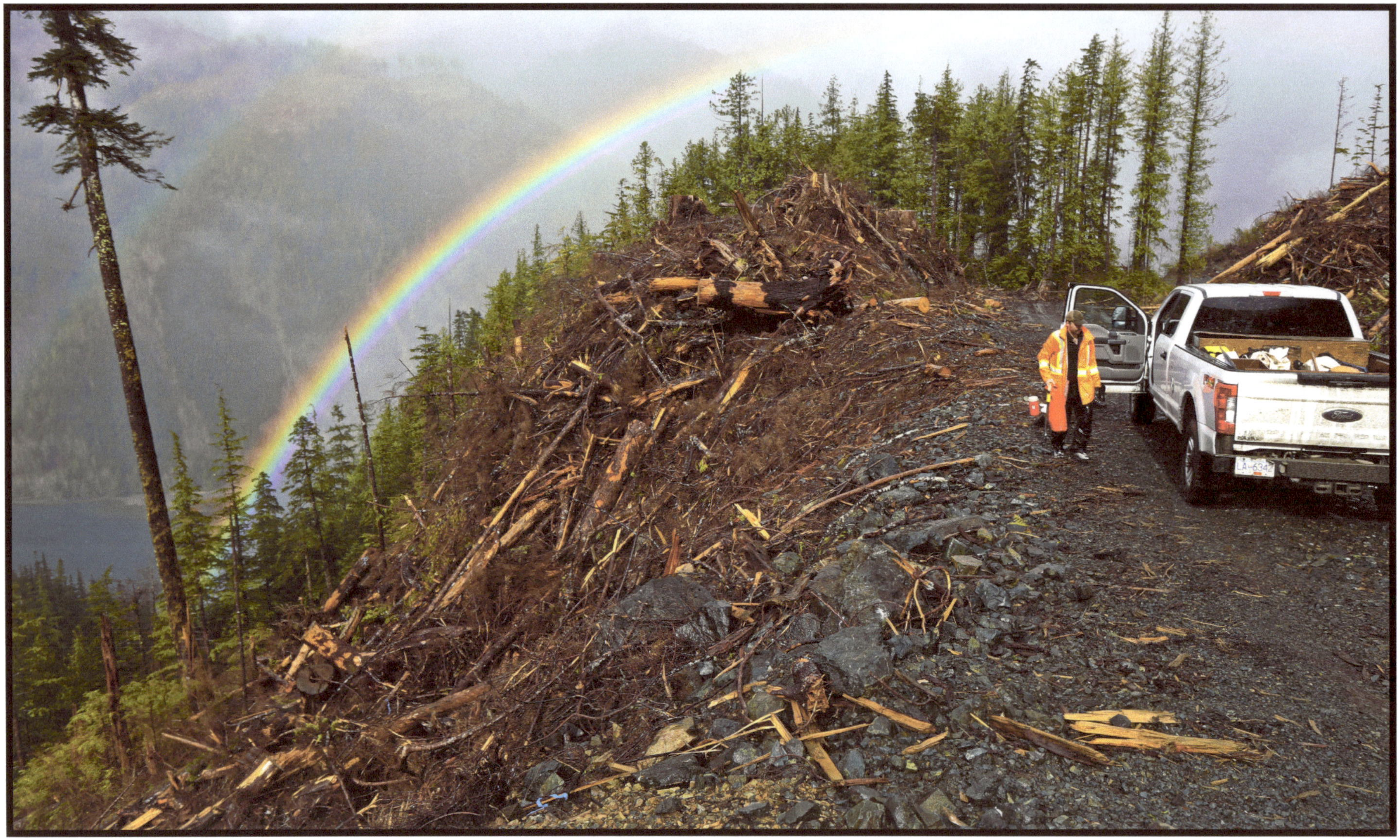

This rainbow is nice, but the block was a bit steep.

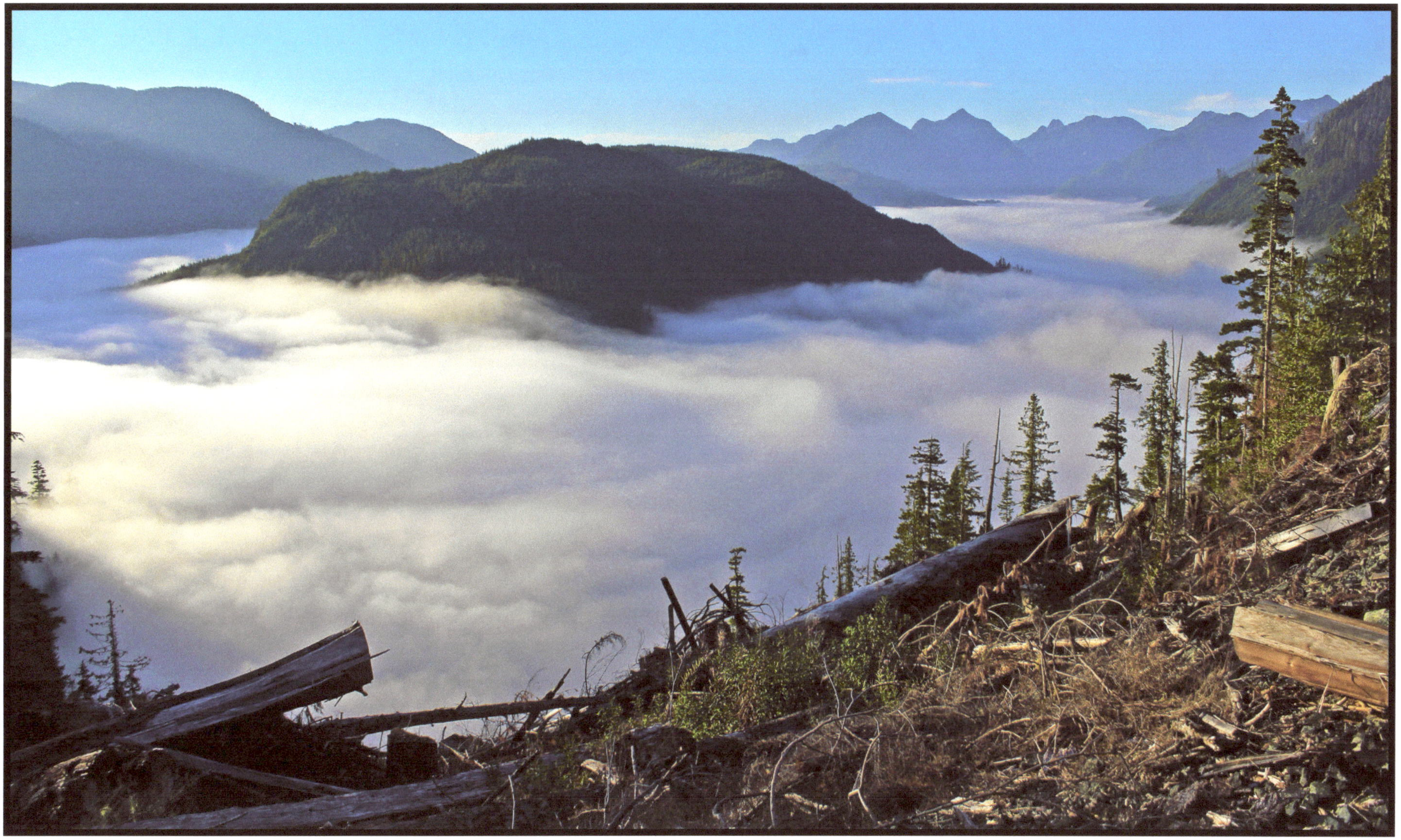

It's hard to ignore the beauty of a hilltop surrounded by a sea of fog.

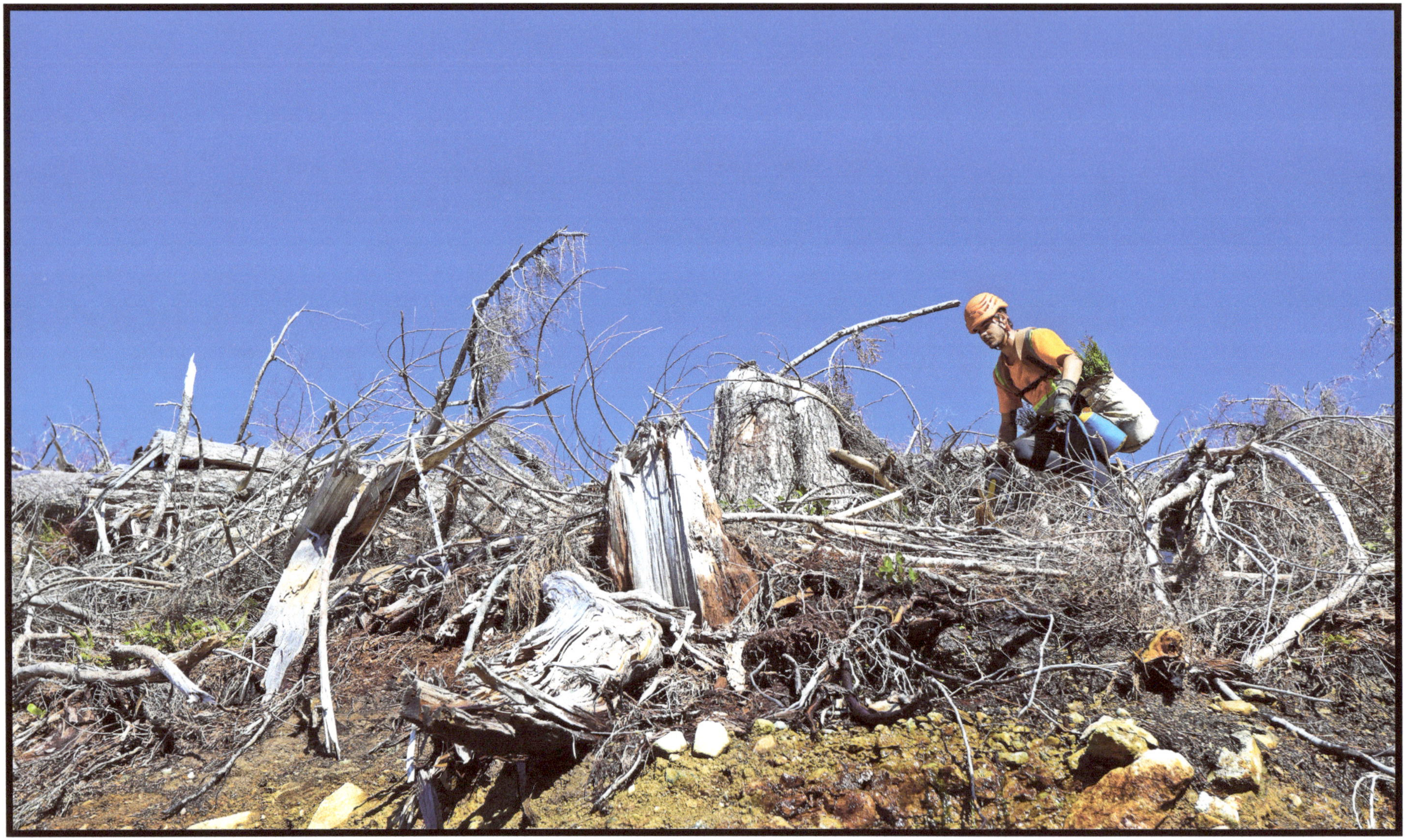

We rarely see blue skies in the spring season, but some companies have a shorter fall planting season in September and early October. There's usually a 50/50 chance of blue skies on any given day during the fall plant. However, the fall blocks are usually at higher elevations, and have steeper ground.

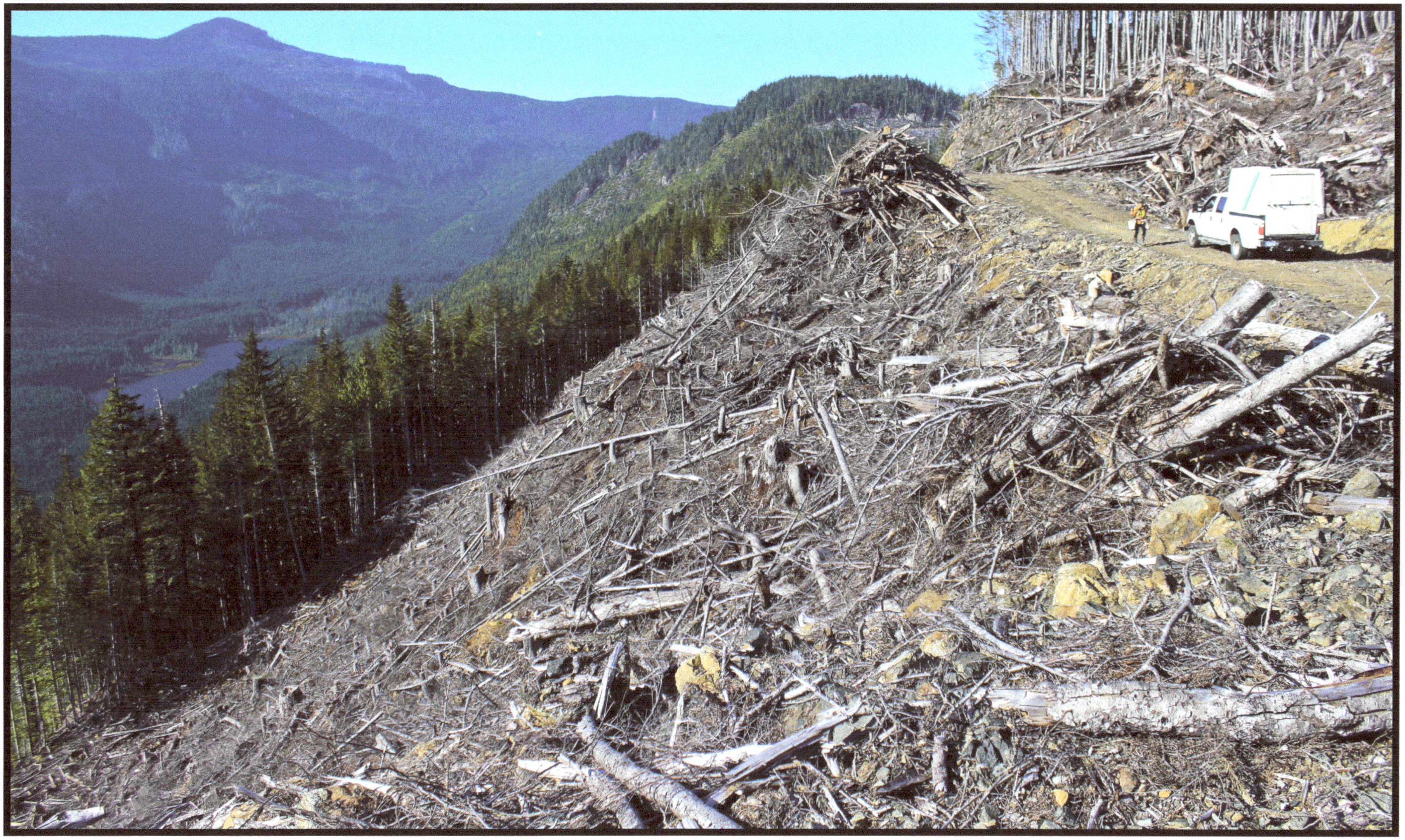

I like a block that isn't too green.

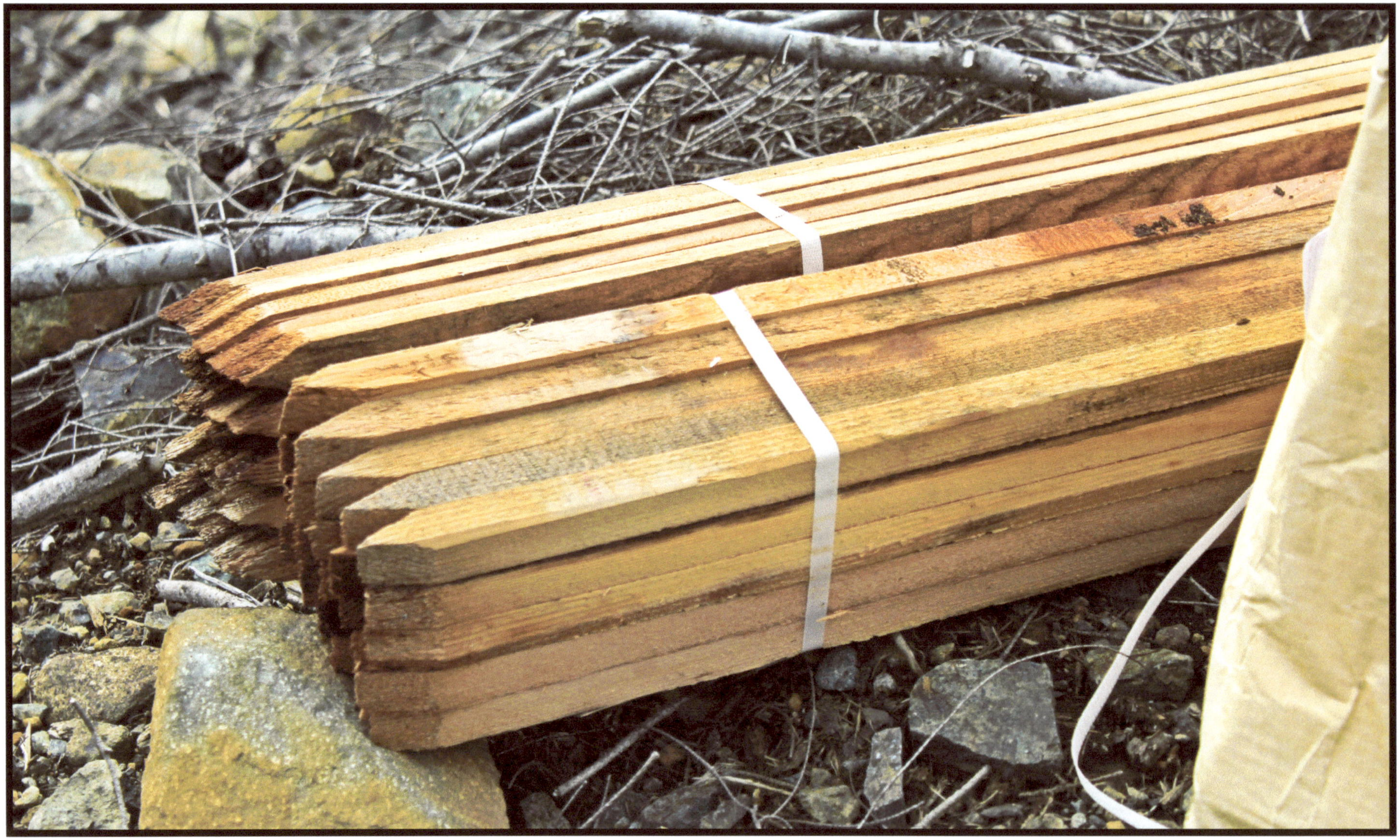

On some contracts, planters are required to install browse protectors, so deer and elk don't eat the freshly planted seedlings. These animals seem to love the taste of young cedar trees. To protect the seedlings, we'll pound a four-foot high stake into the ground beside each seedling, then cover the seedling and the stake with a browse protection device (usually referred to as a cone). The cones are then fastened to the stakes with zip-ties. Cones are left in place for about three or four years, until the trees have grown enough that they are no longer appealing to the deer and elk. At that point, other planters go through the blocks to remove the cones.

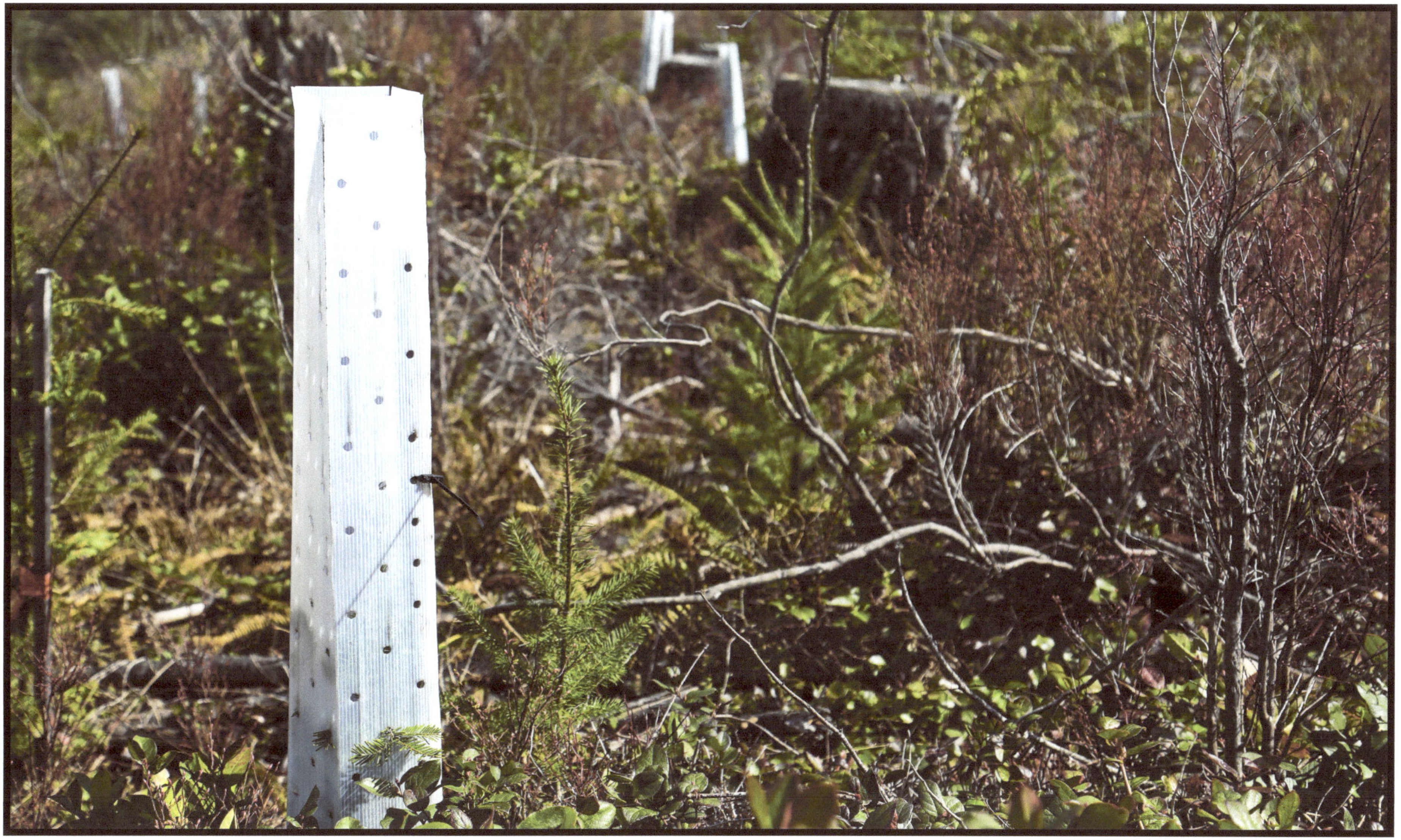

Here's a close-up of a cone. You can partly see the outline of the cedar tree inside the base of the cone. The cones are translucent, to help let sunlight through for photosynthesis.

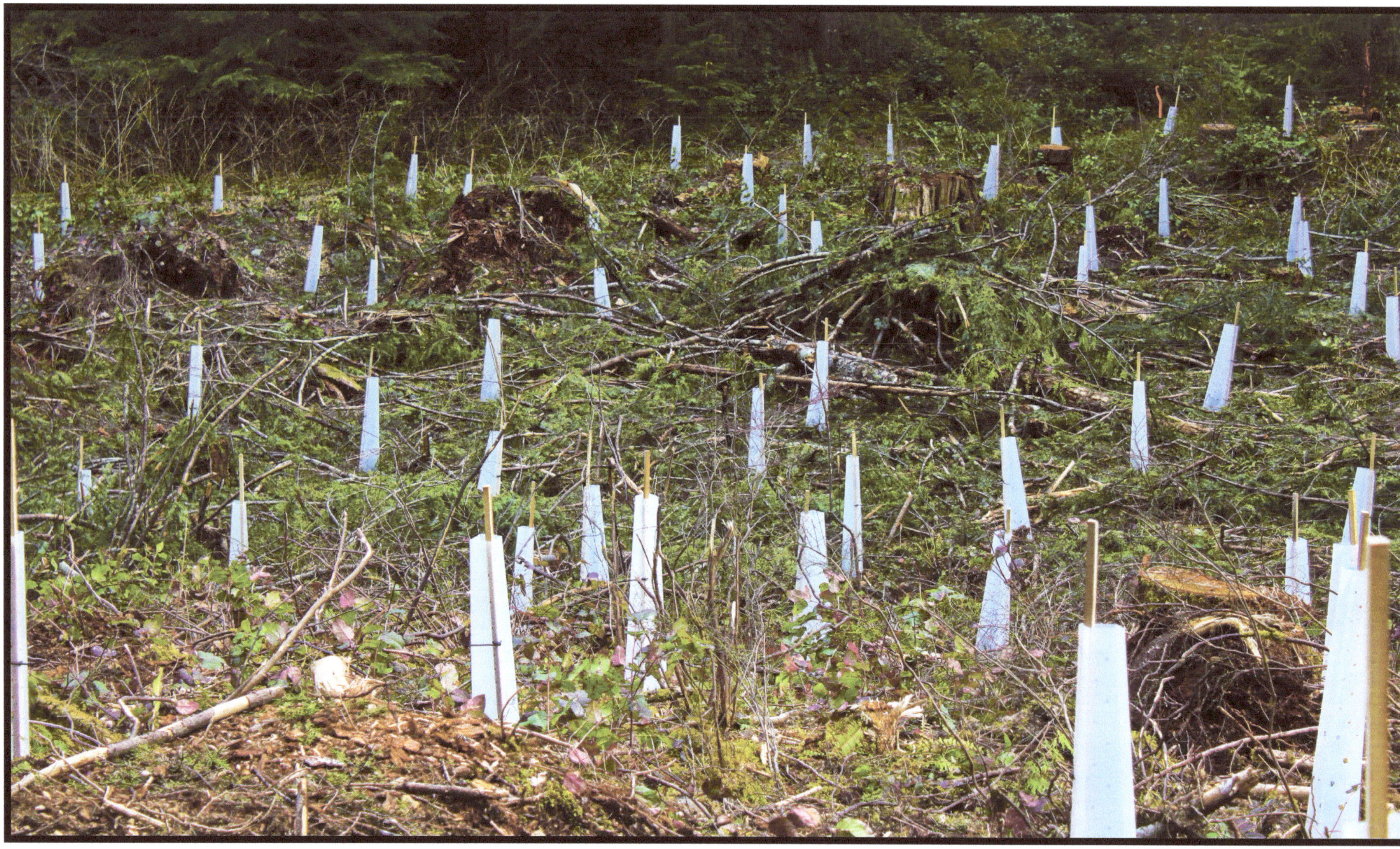

The cedar seedlings on this block are well protected from the deer and elk.

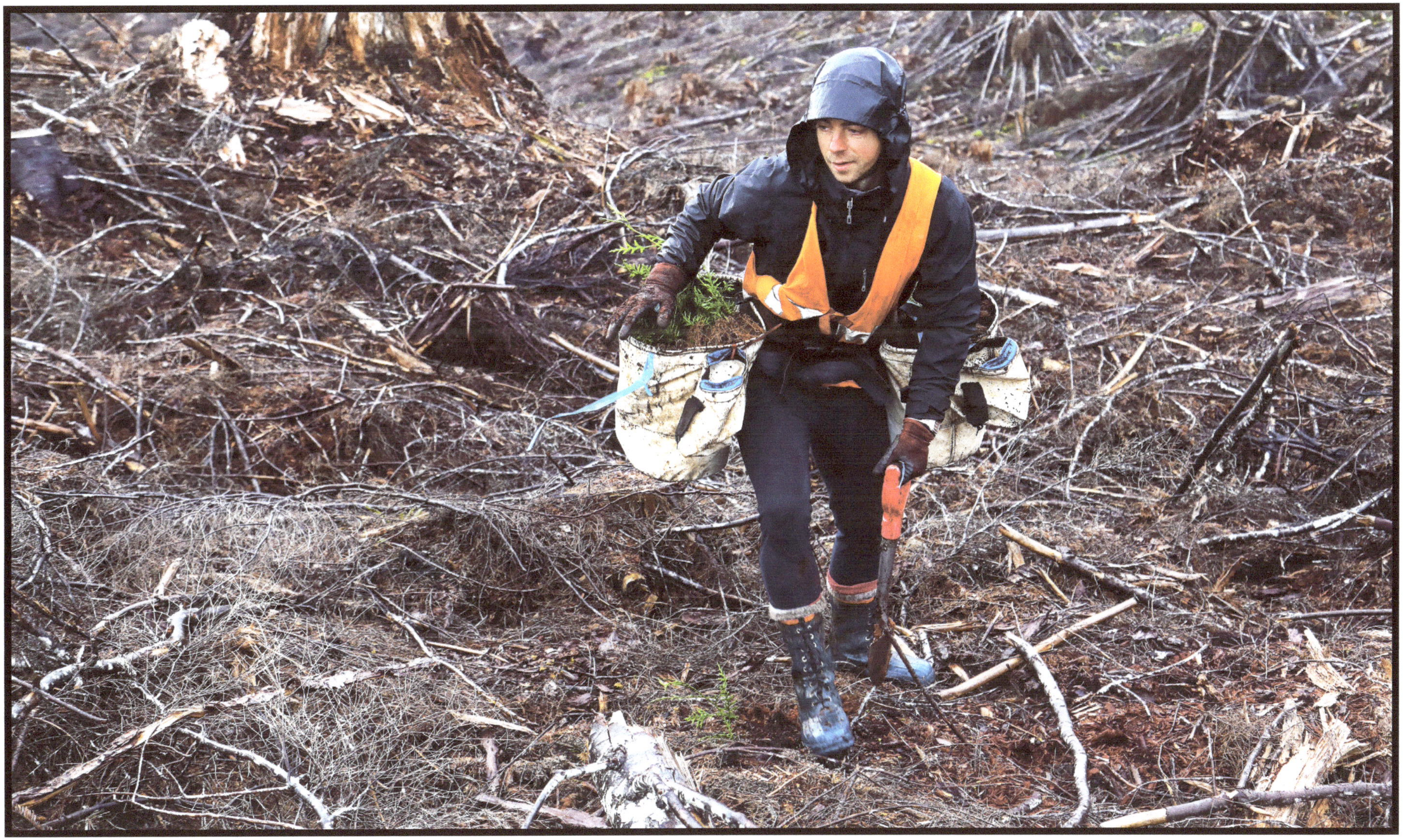

Lukas was moving pretty quickly through this creamy block. When a block is easy and the price is fair, we say that it is creamy.

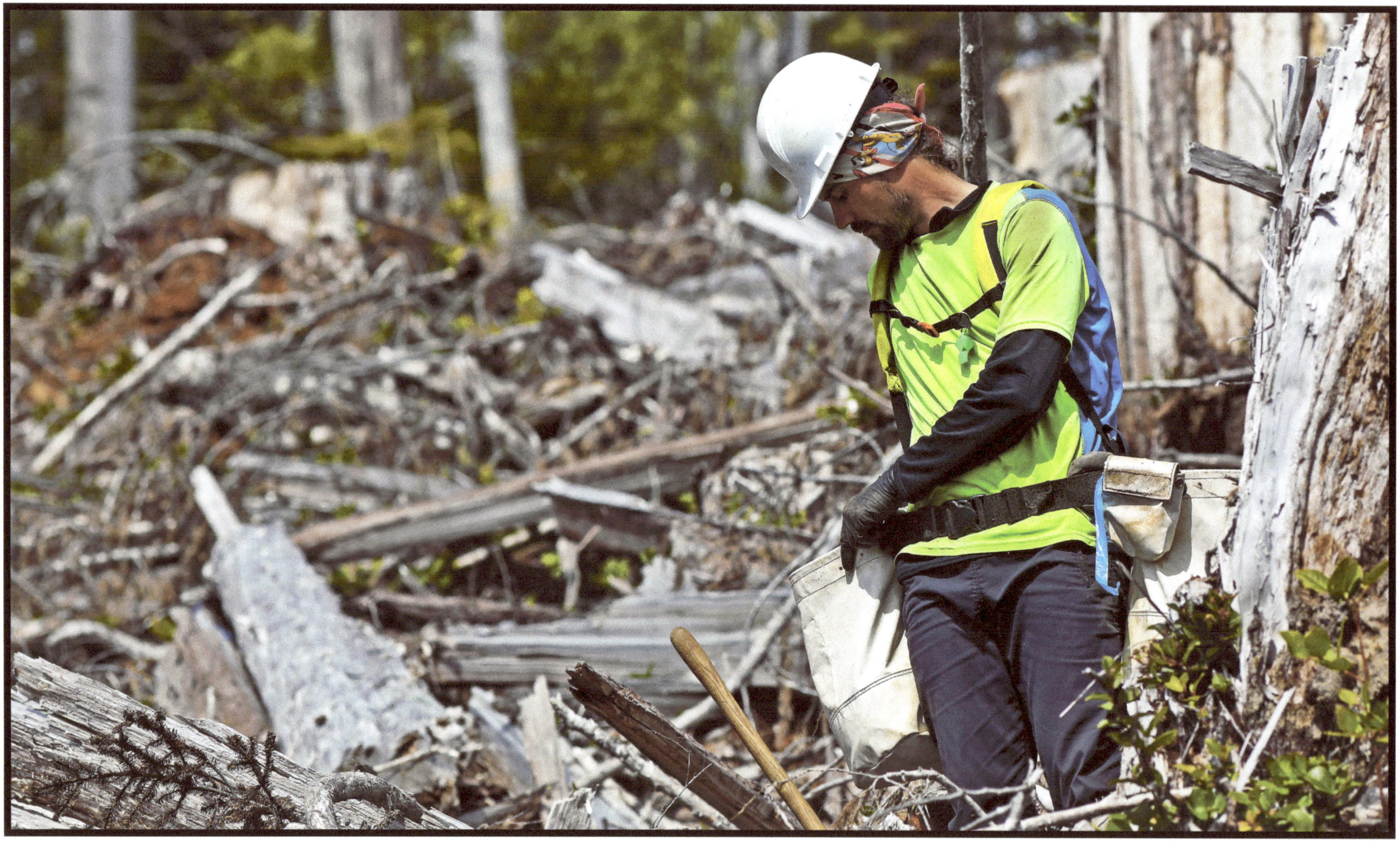

I see a staff shovel. Not very many planters in western Canada use staff shovels instead of D-handle shovels. D-handles are used by around ninety percent of coastal planters. That's not to say that there's anything wrong with using a staff shovel.

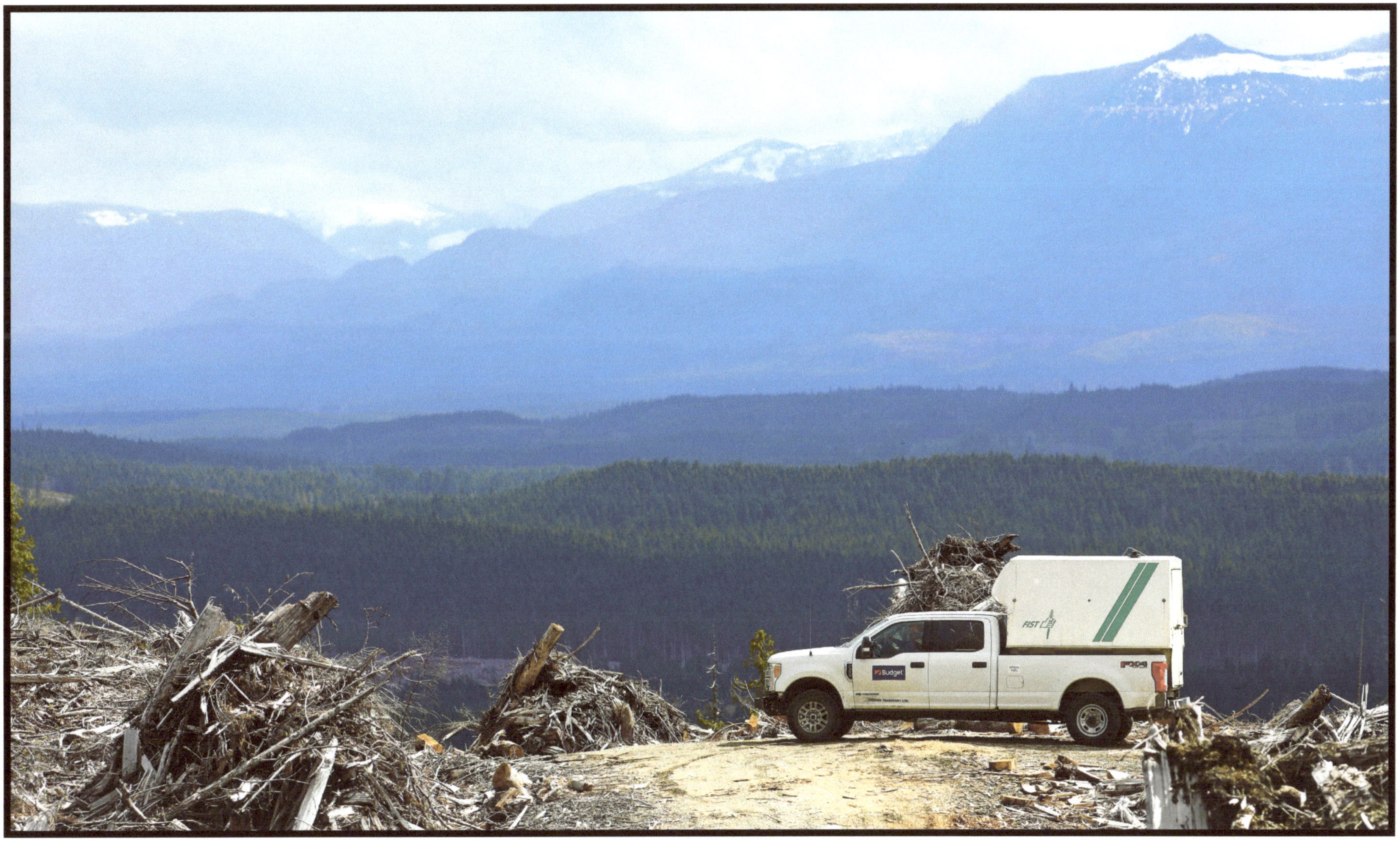

Many of the crew trucks have a FIST canopy on the back. FIST stands for Fibreglass Insulated Seedling Transport. These canopies are insulated to help keep boxes of trees cool on hot days. Hot days are not really a problem during the spring coastal season. In fact, I'd be more likely to refer to hot days as a fantasy.

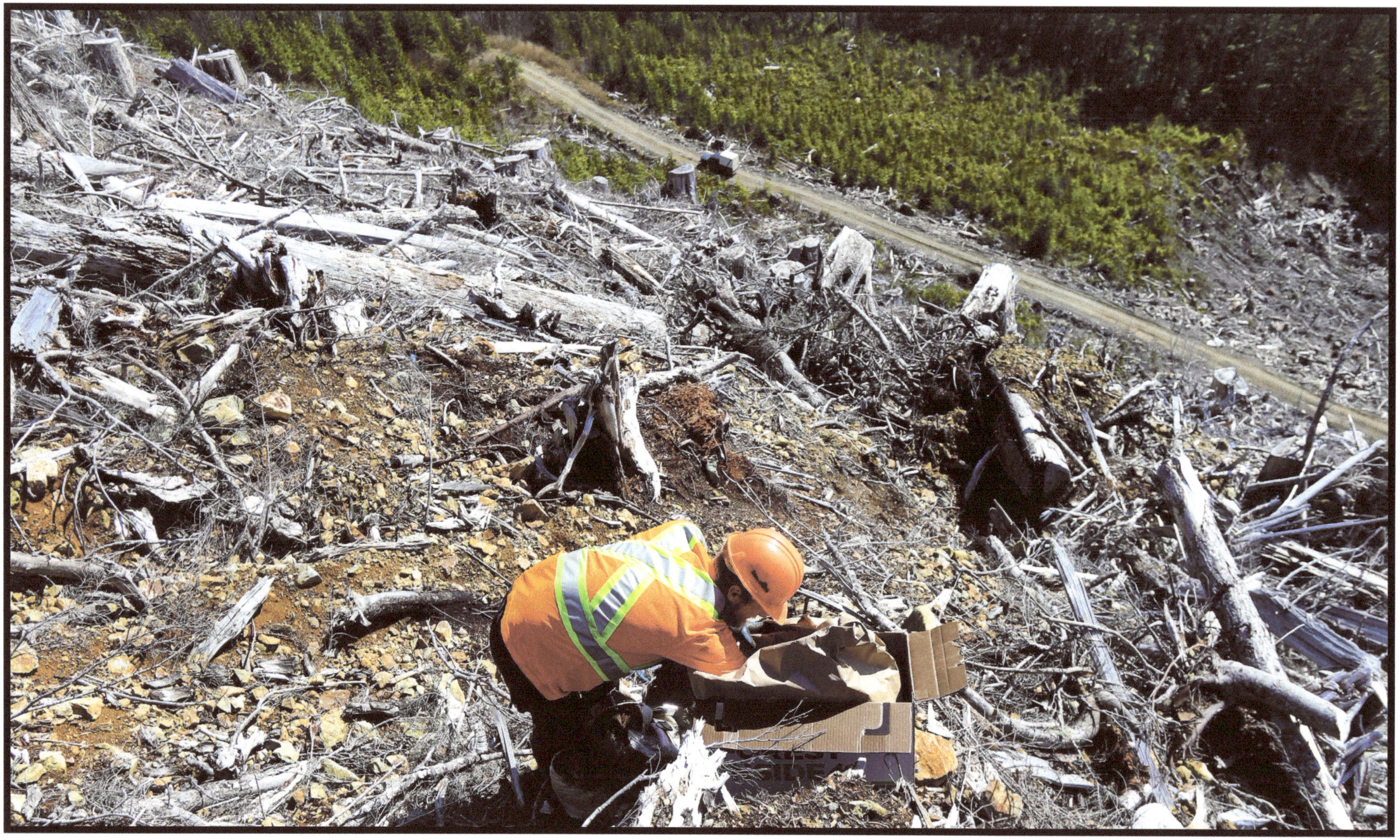

It's unfortunate that a camera can never truly capture the severity of the slopes that we work on. If you're thinking about the truck parked down on the road at the bottom of the block, and trying to understand the perspective, remember that those mature trees behind the new plantation across the road are probably over a hundred feet tall.

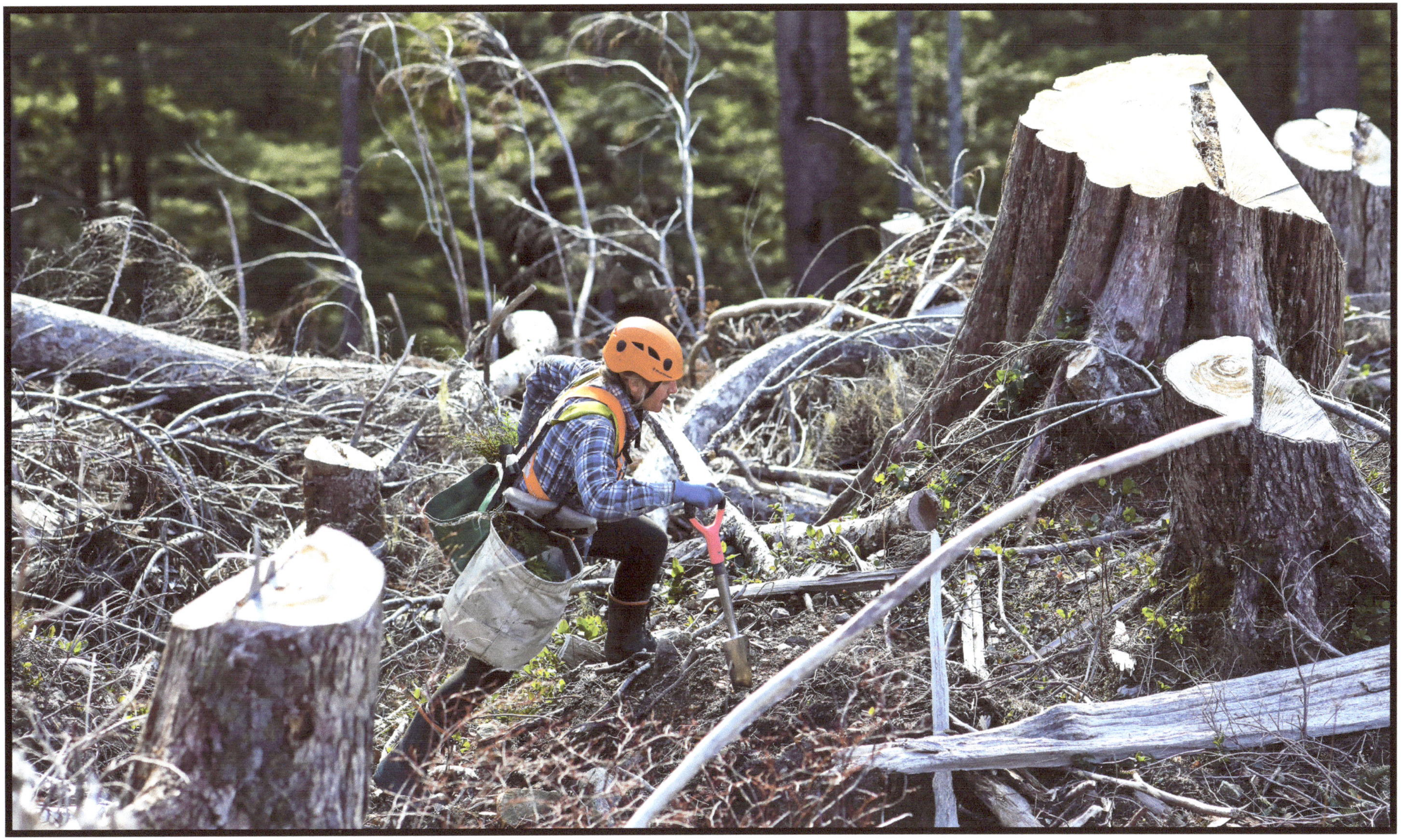

This looks like fast ground.

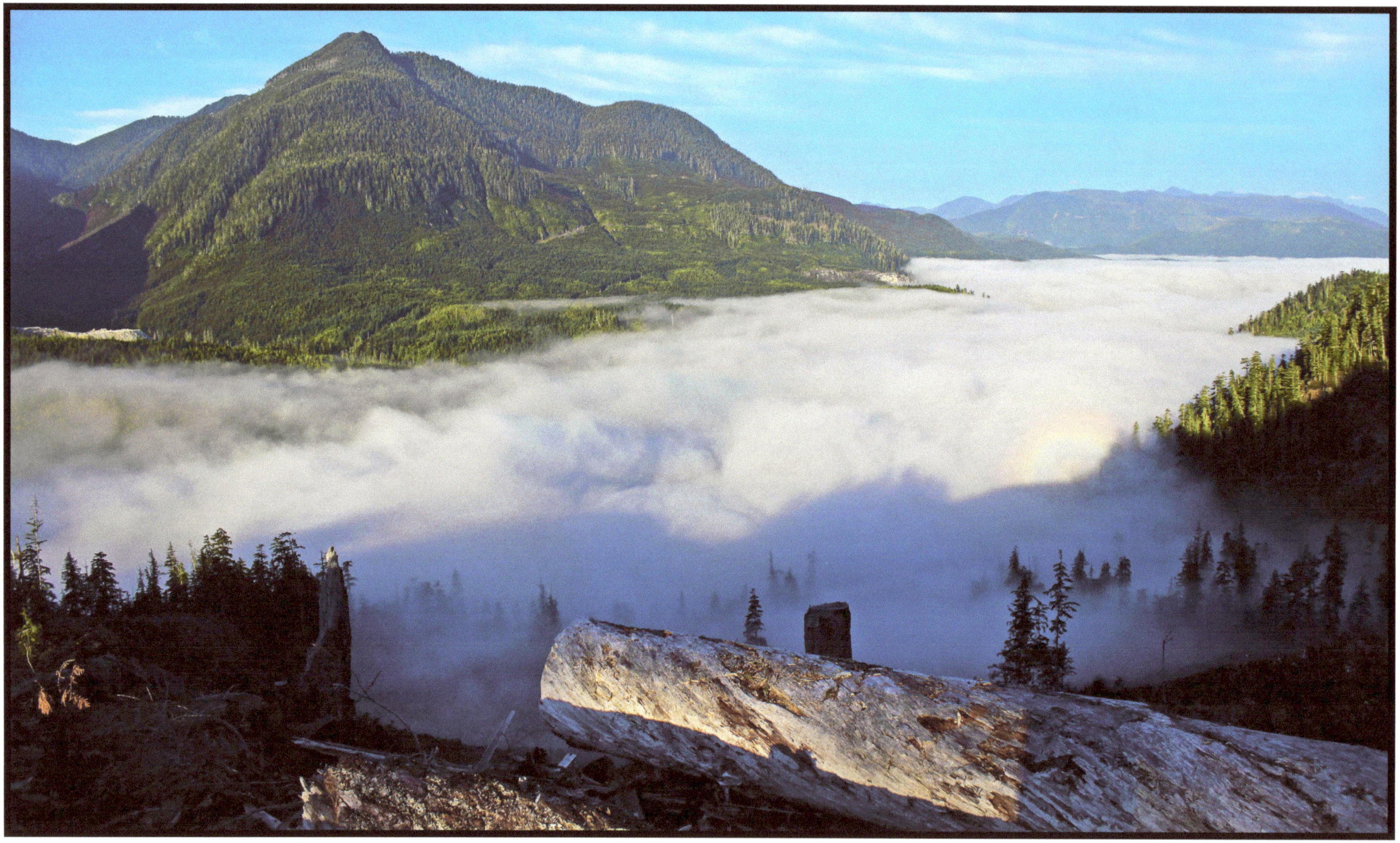

Another beautiful early morning in the late fall.

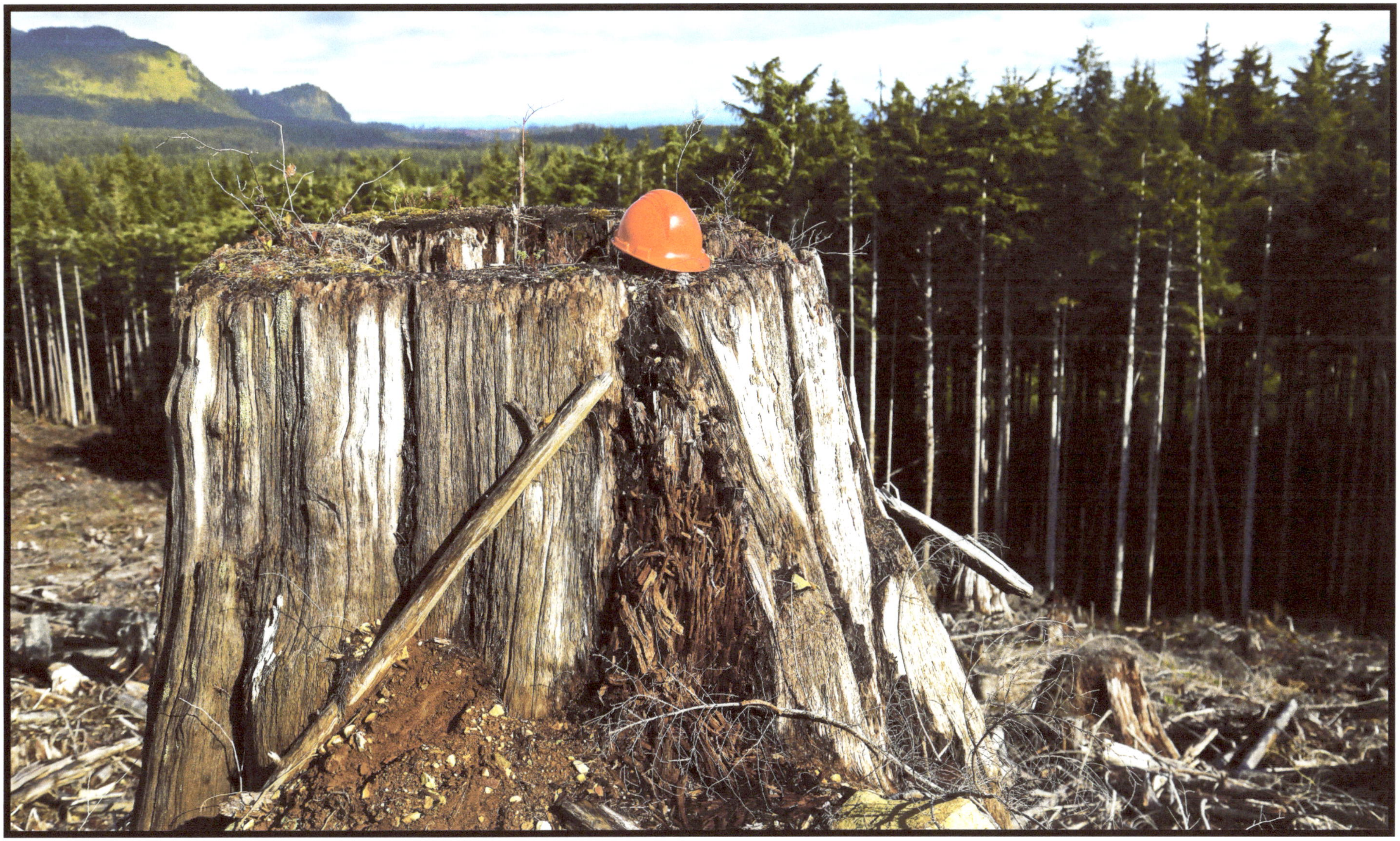

Don't worry, there's nobody in the stump. The hard hat is just for perspective.

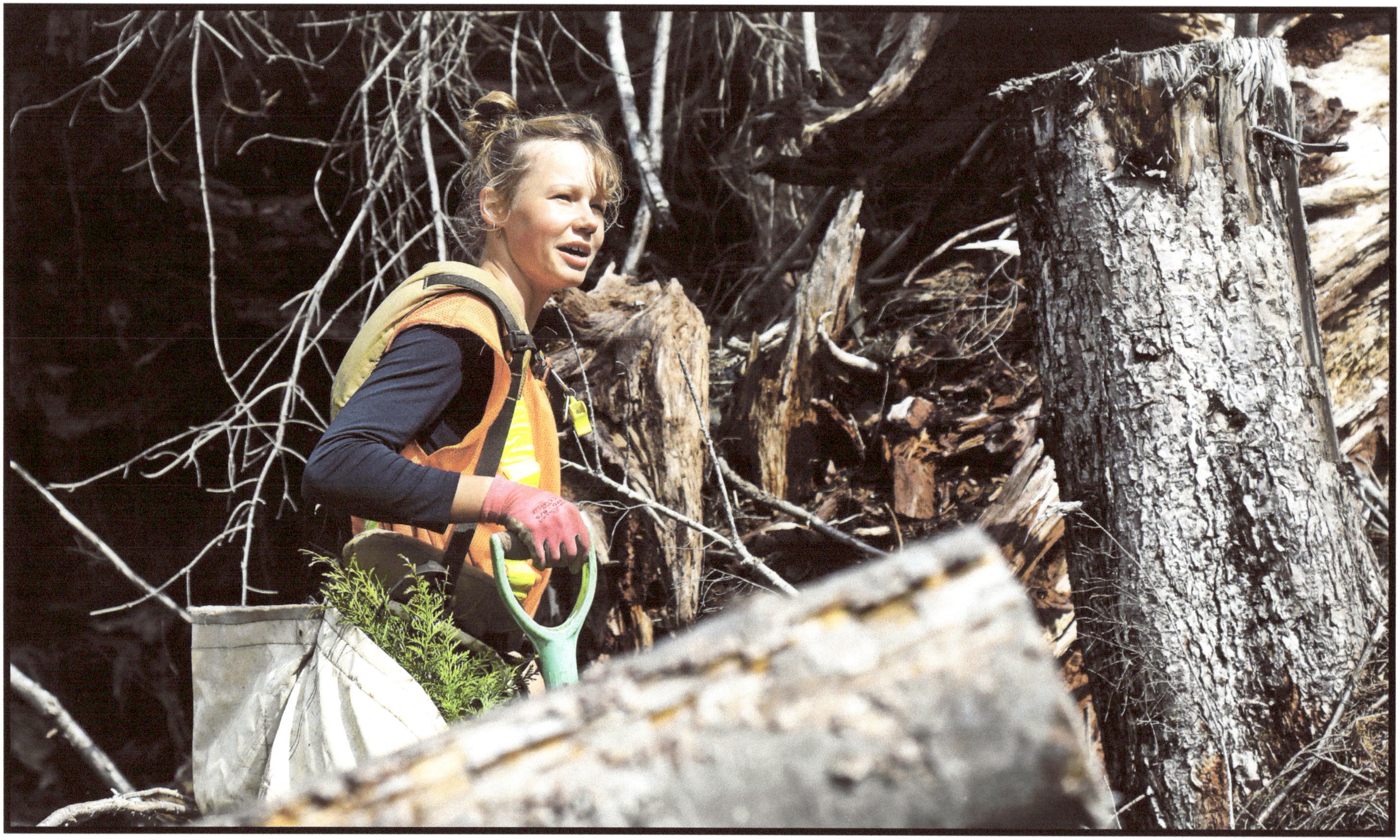

Any day without freezing rain is a good day.

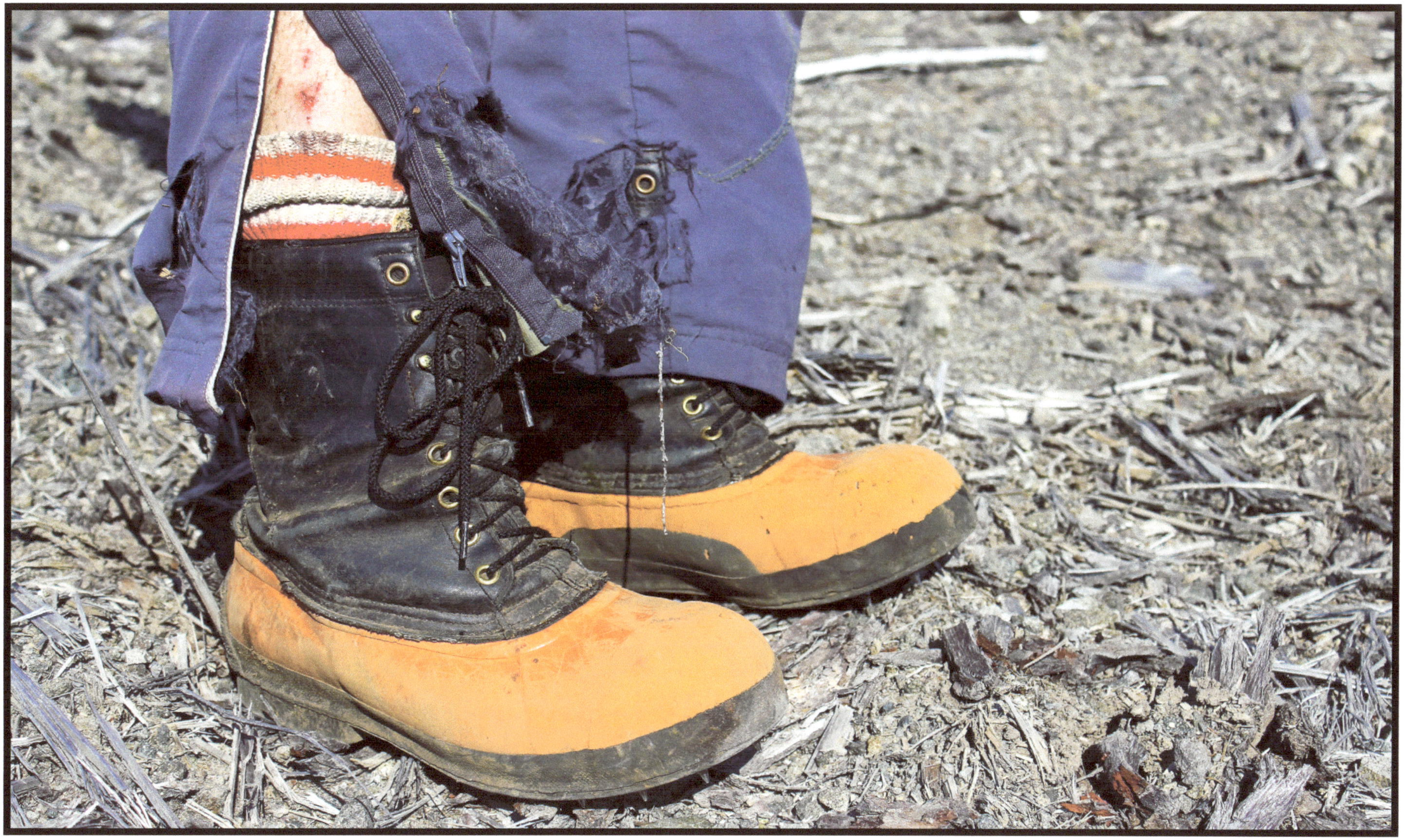

Everything that we wear gets shredded eventually. This type of caulk boot typically lasts about sixty planting days before I need to buy a new pair. I've had cheaper boots that have been destroyed in the space of a few shifts.

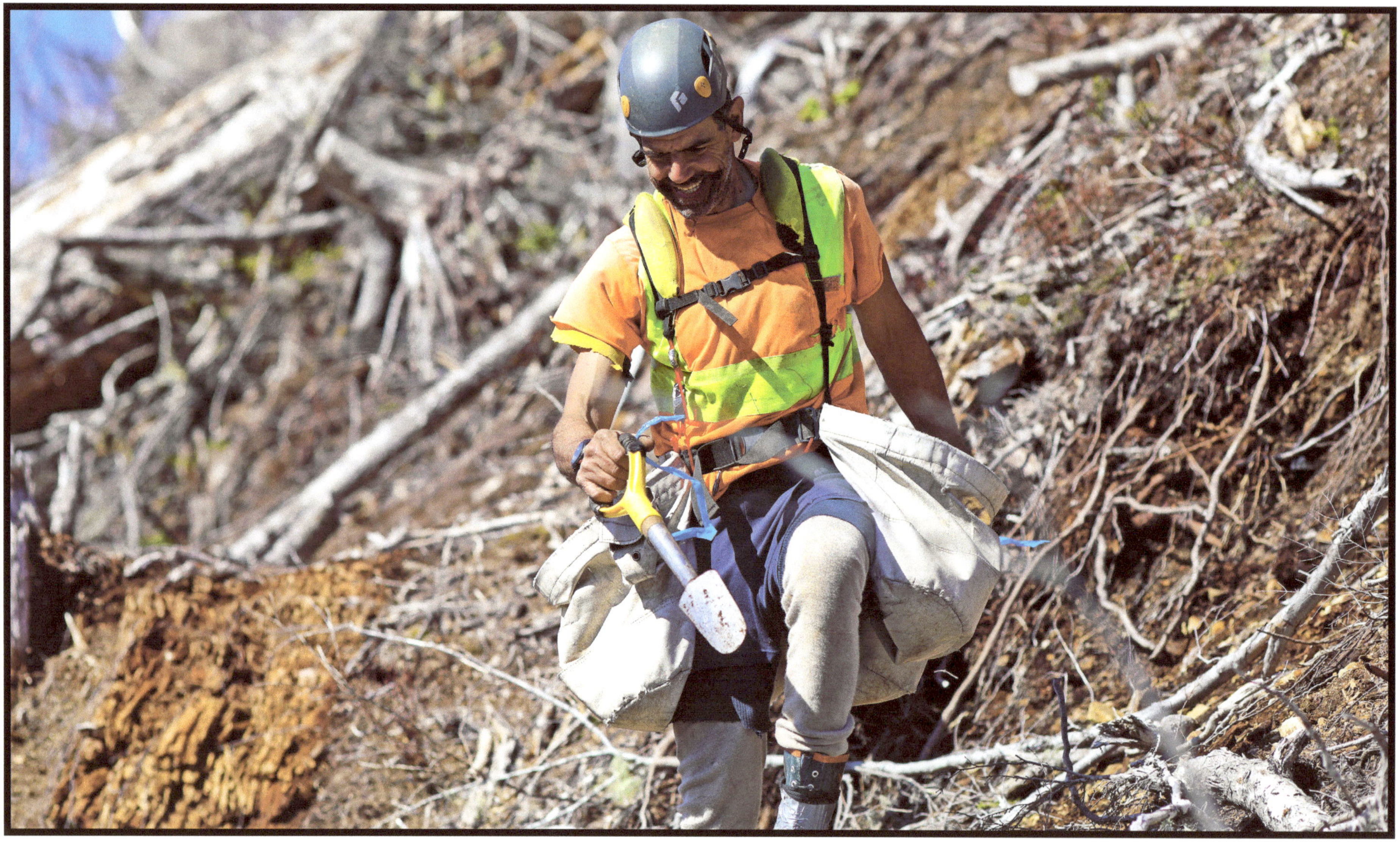

Simon is a planting machine. During his career, he has planted close to four million trees, and many of them have been on coastal contracts.

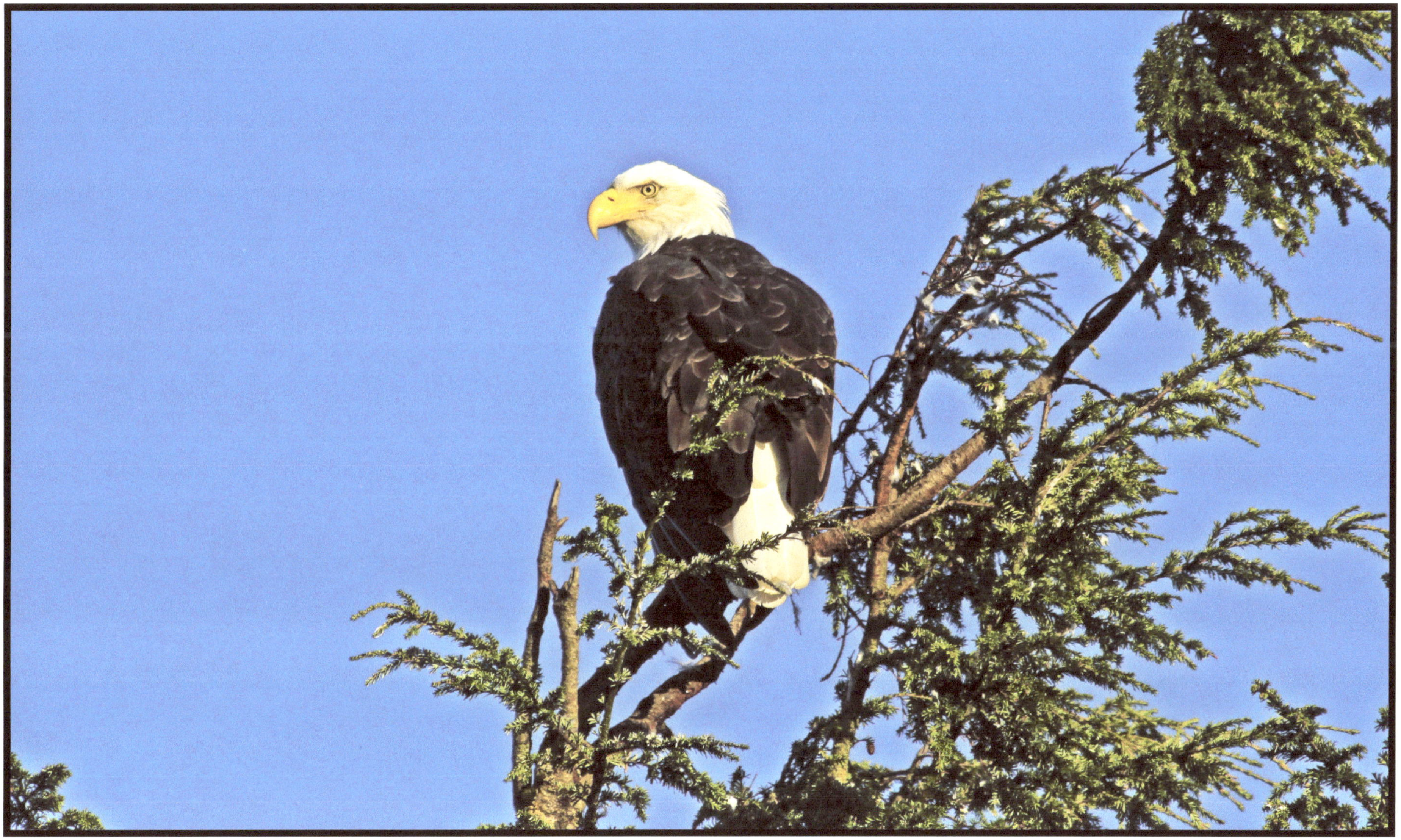

We see a lot of bald eagles up close.

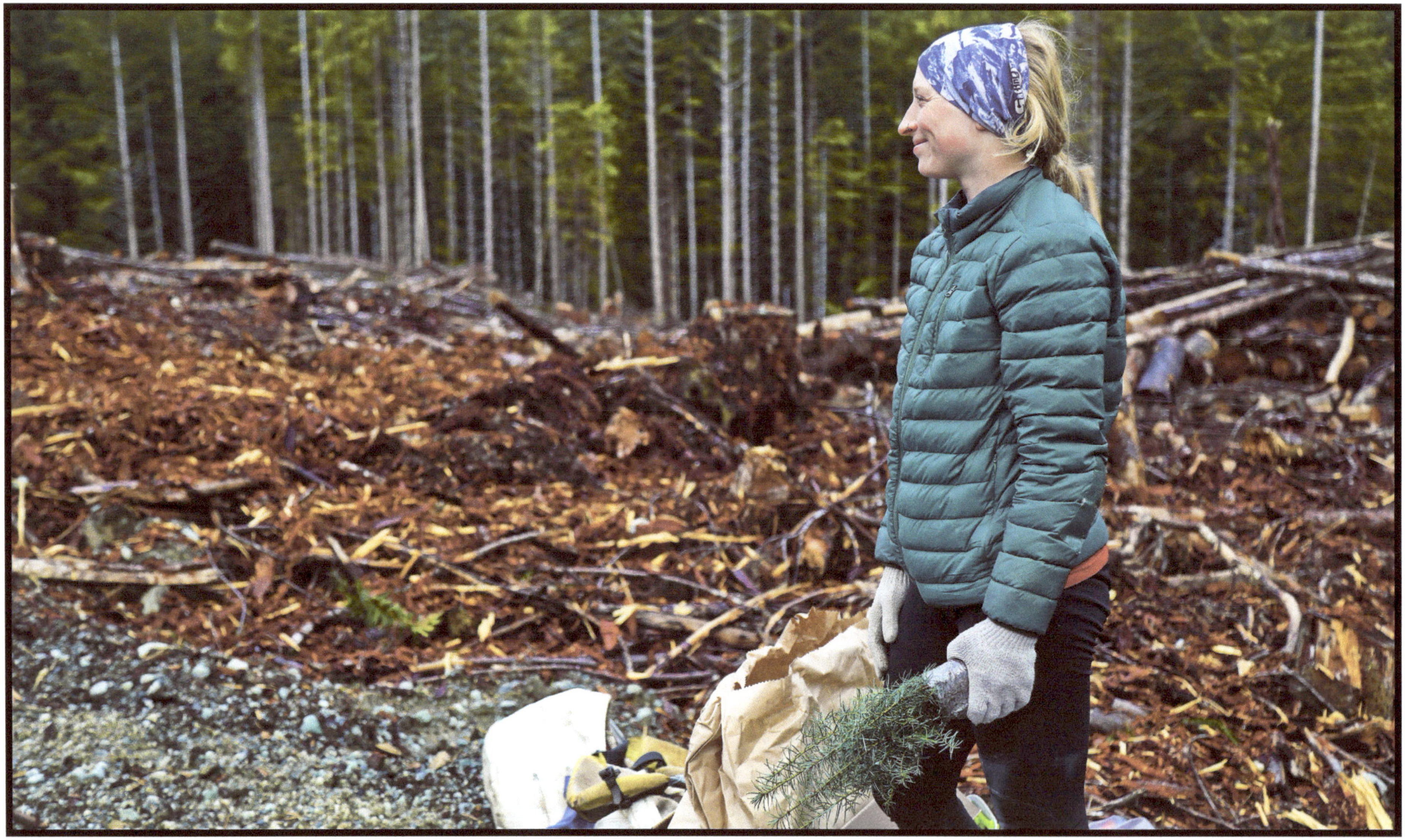

If you can't smile about the challenges ahead of you, then you shouldn't be planting on the coast.

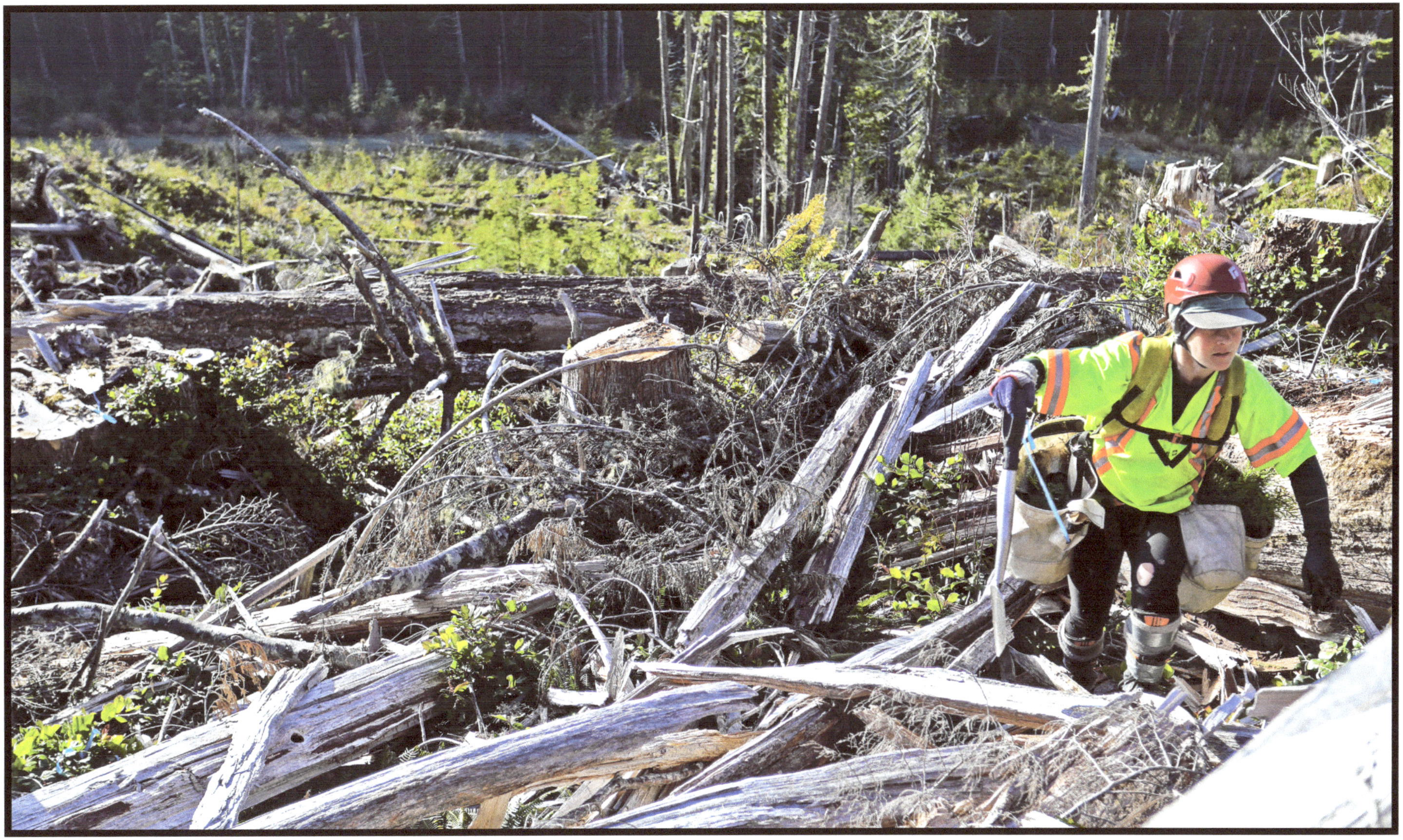

Sometimes, the worst parts of the block are the ten or twenty meters immediately below the road.

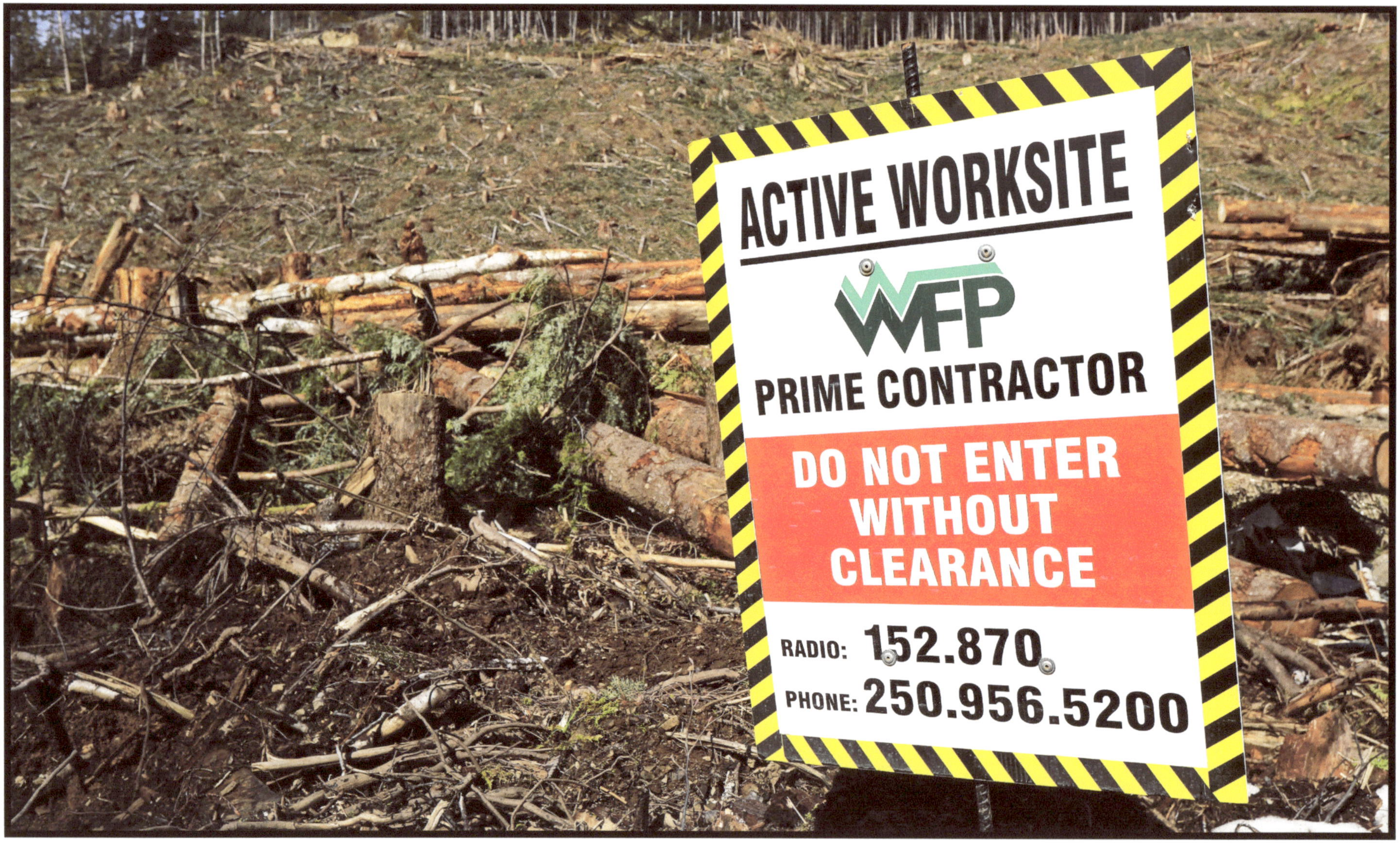

This block has a surprising lack of green slash on the ground for a fresh block that is still in the process of being logged.

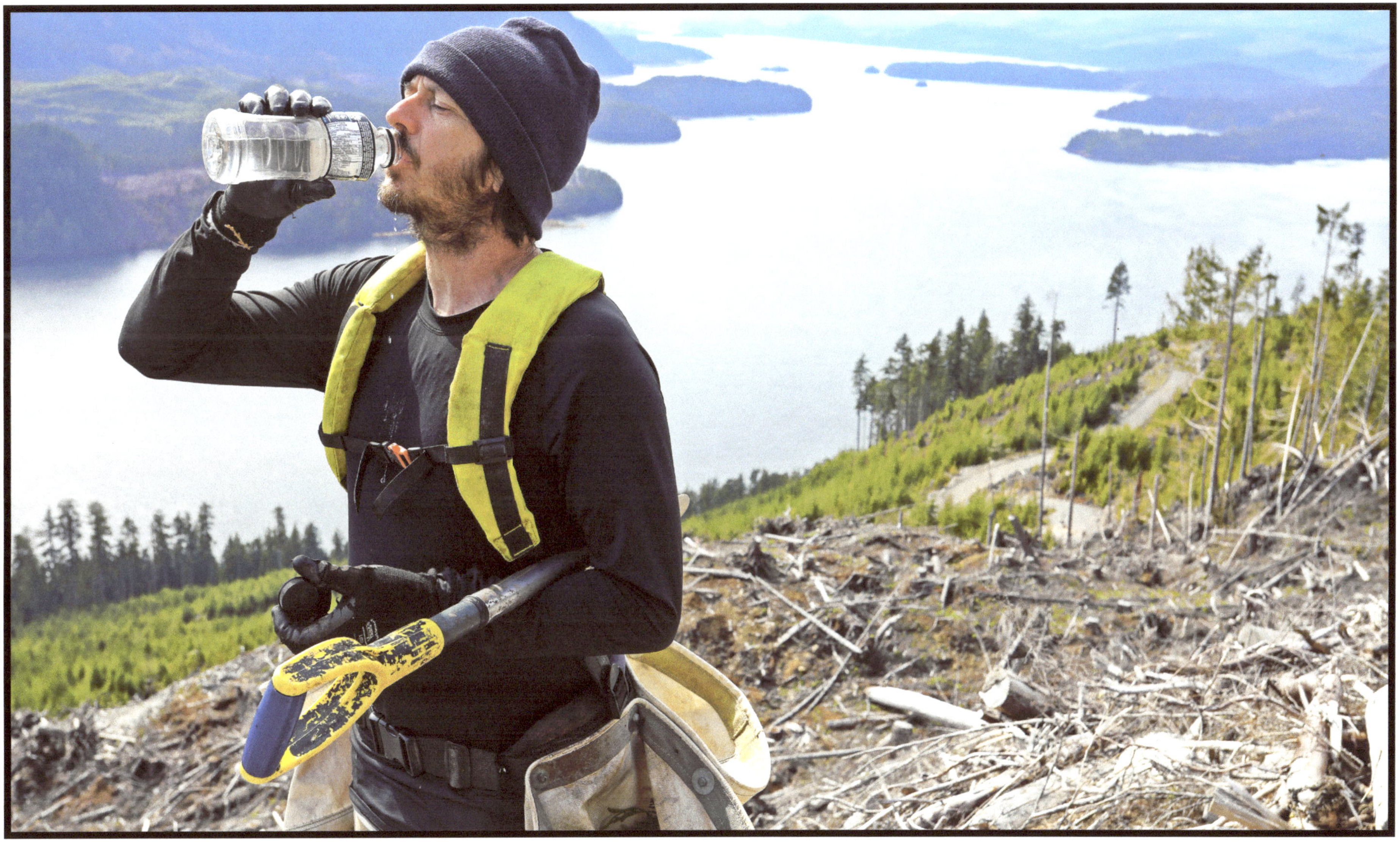

You can never drink too much water. A lot of coastal planters are guilty of dehydrating themselves. When you're soaking wet on a rainy day, and the temperature is just a couple degrees above freezing, you may not feel like drinking a lot of water. However, you're still sweating inside that rain gear, so you need to force yourself to drink as much as possible.

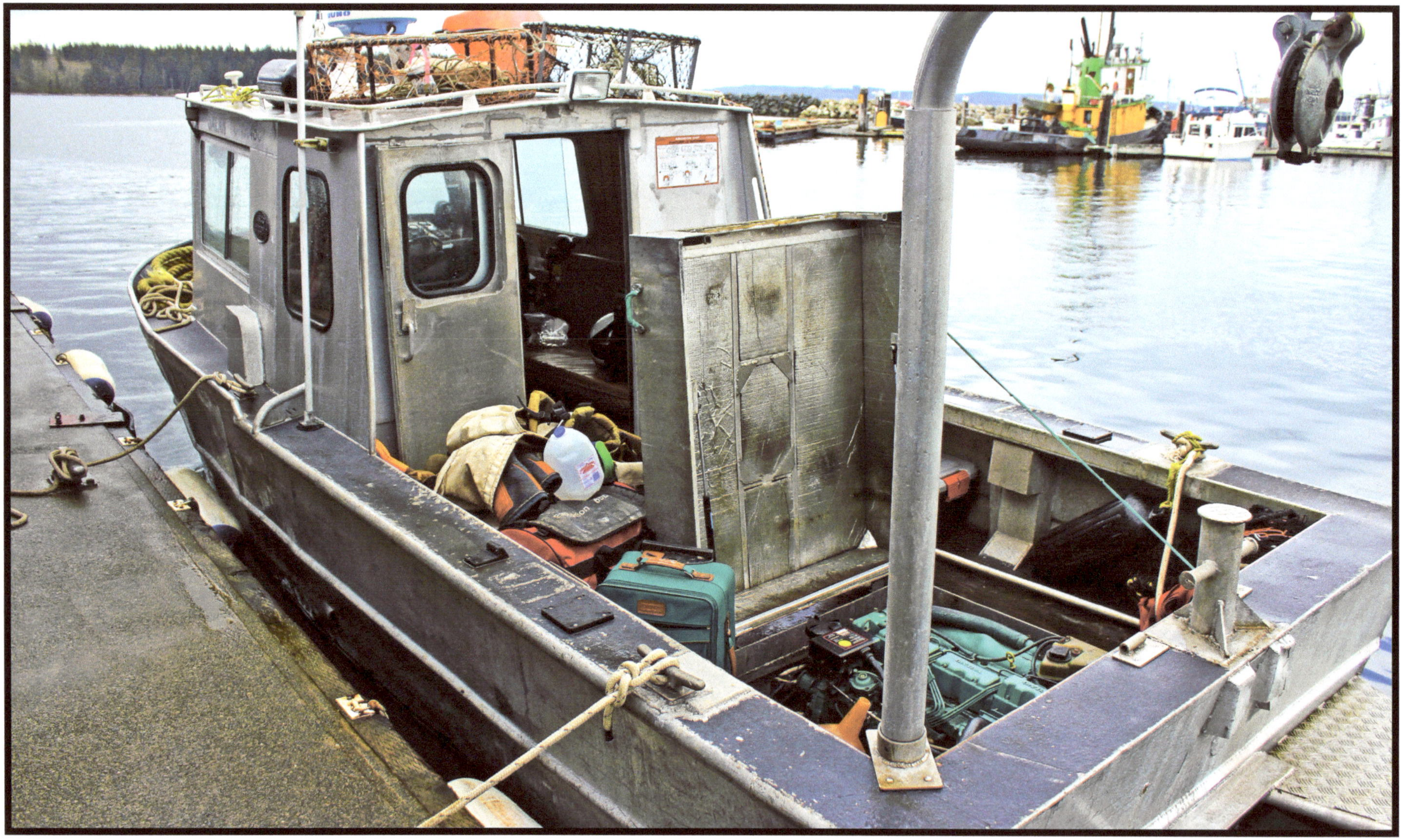

Here's a water taxi, sometimes used to access remote blocks.

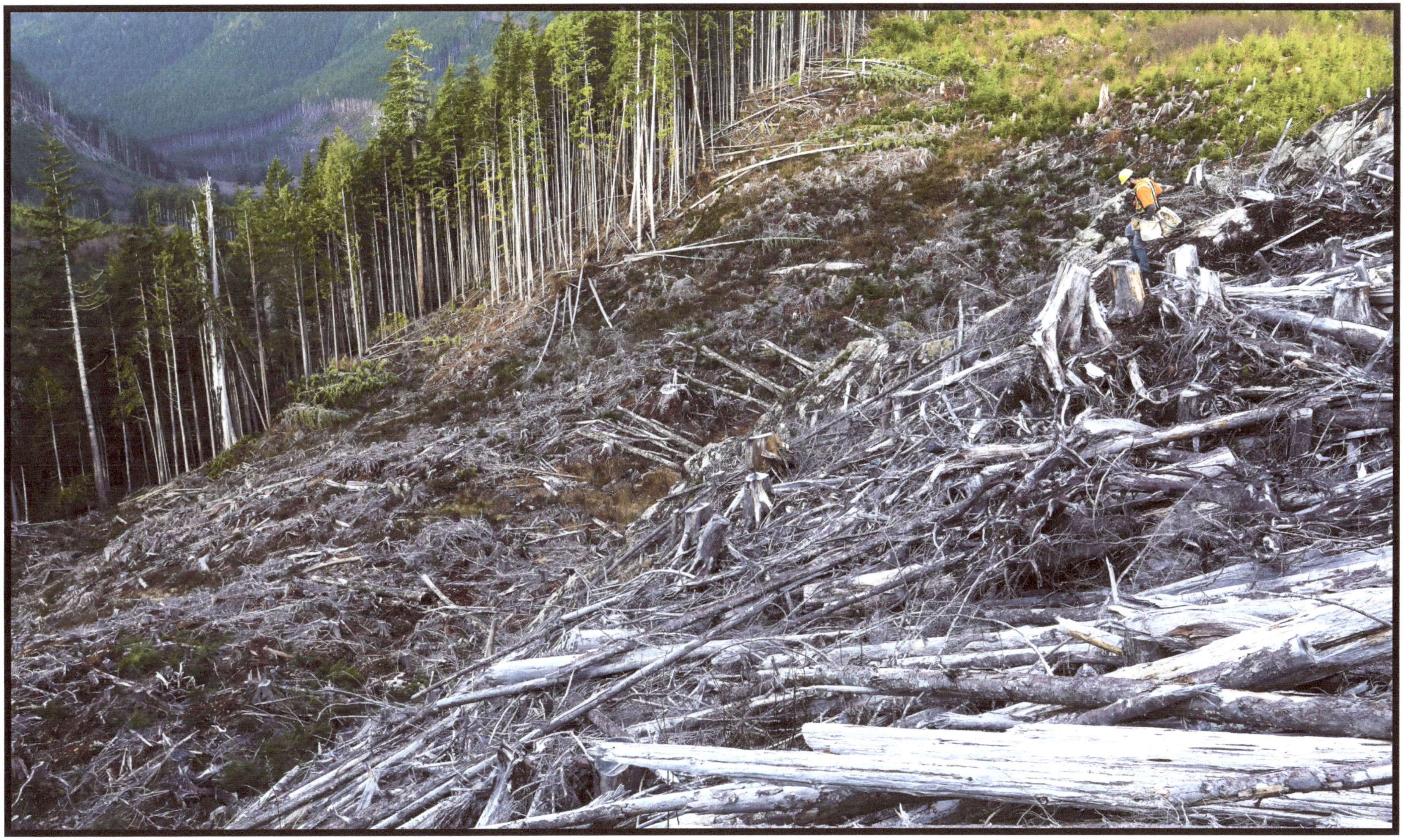

Kiernan is going to be hitting a steep section in just a few more trees.

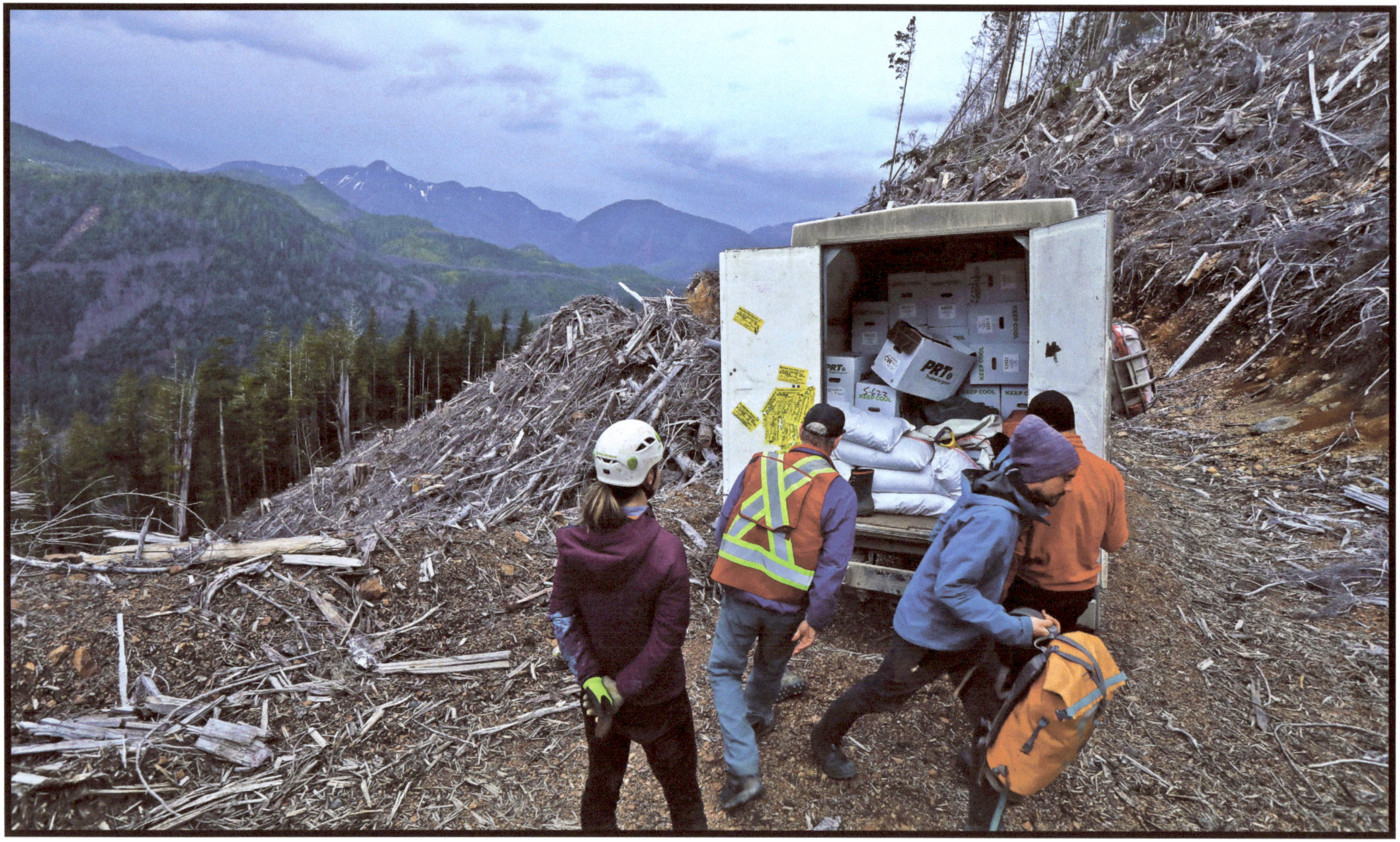

There's always a fast flurry of activity when planters arrive at the block in the morning. Coastal planters aren't the type to sit around at the cache.

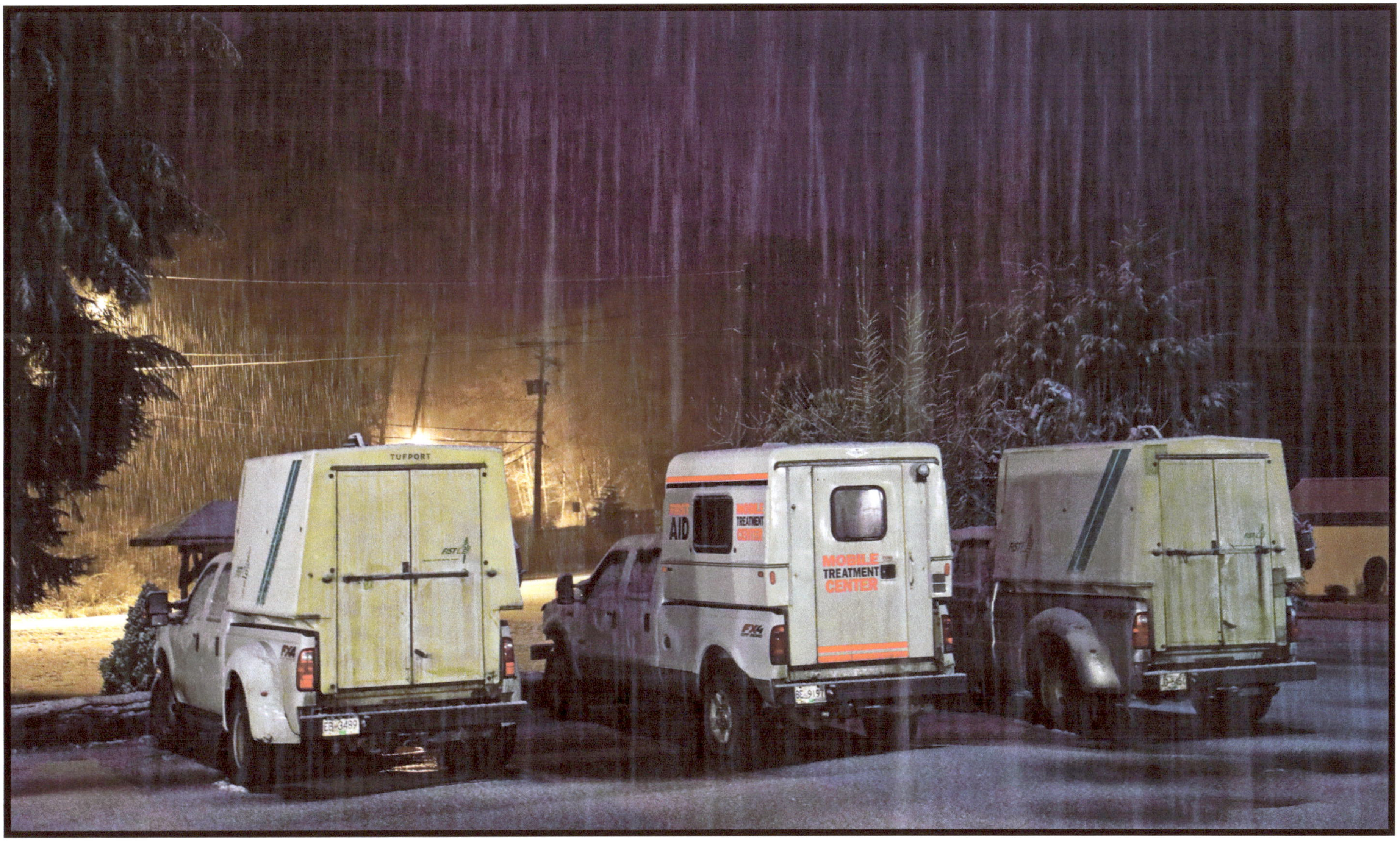

If snow is falling in town, you can pretty much guarantee that there's no point in driving out to the blocks.

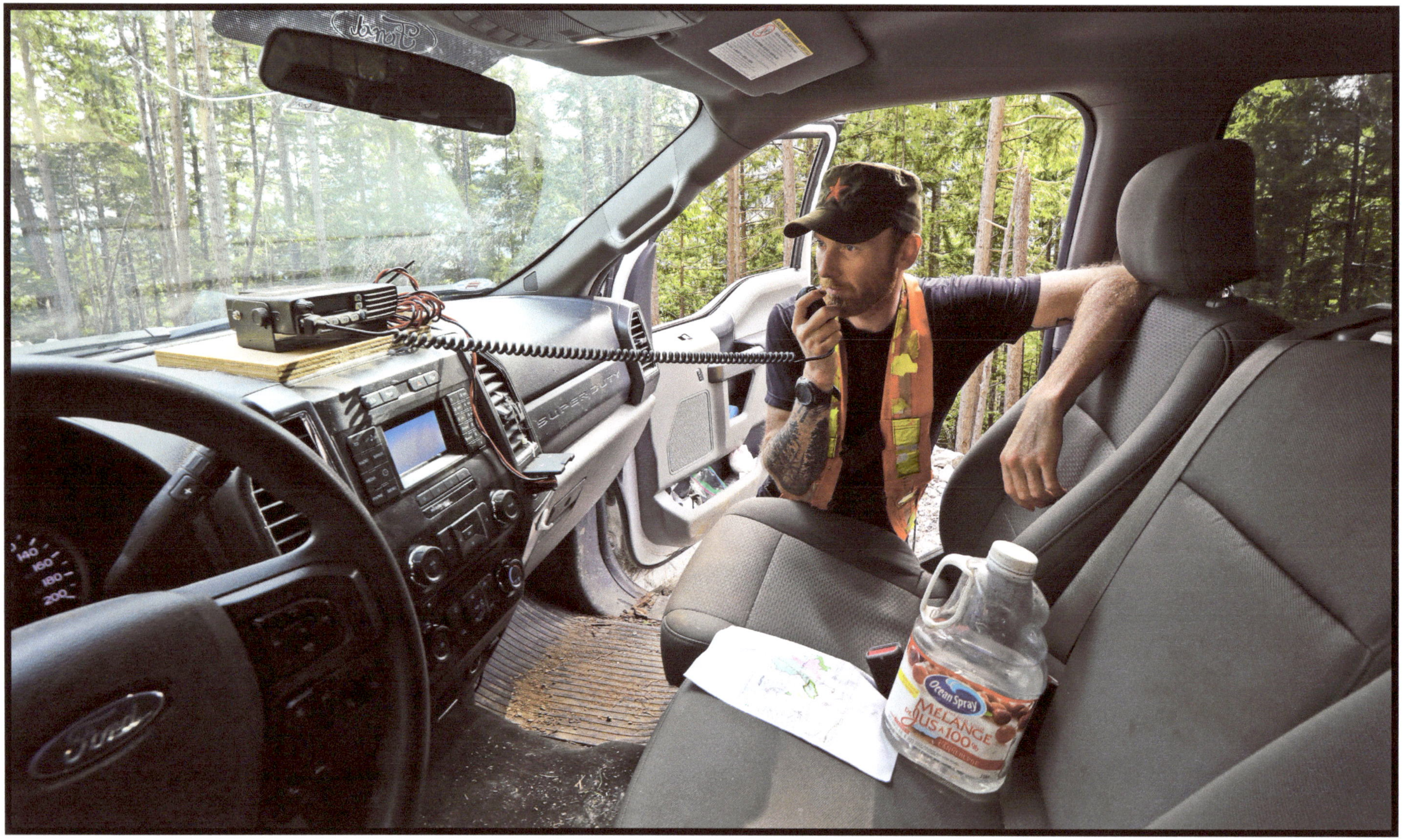

Andrew and I have planted together for more than a decade.

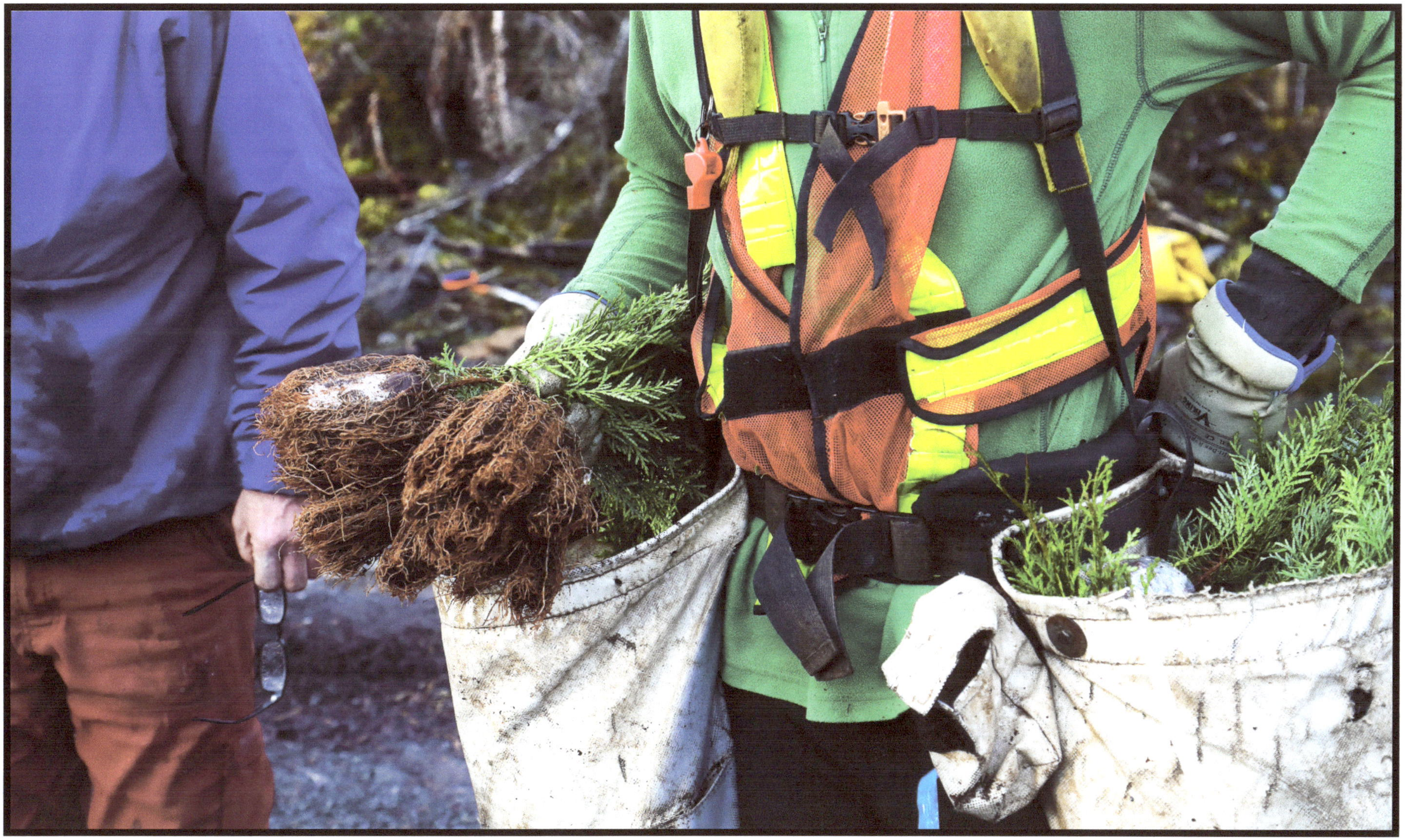

Some of these trees appear to be frozen together.

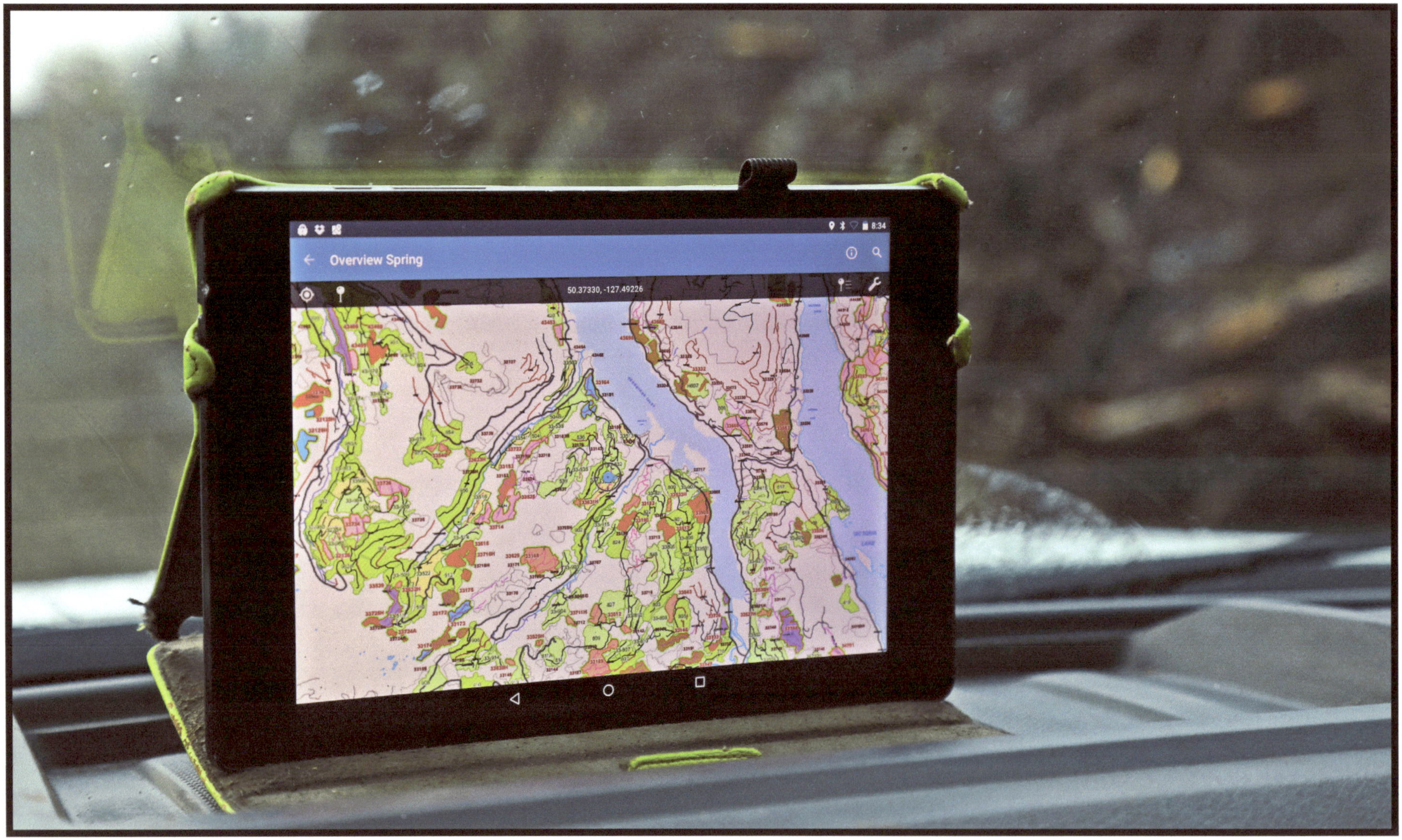

Each of the foremen has a tablet with digital maps and GPS positioning capabilities.

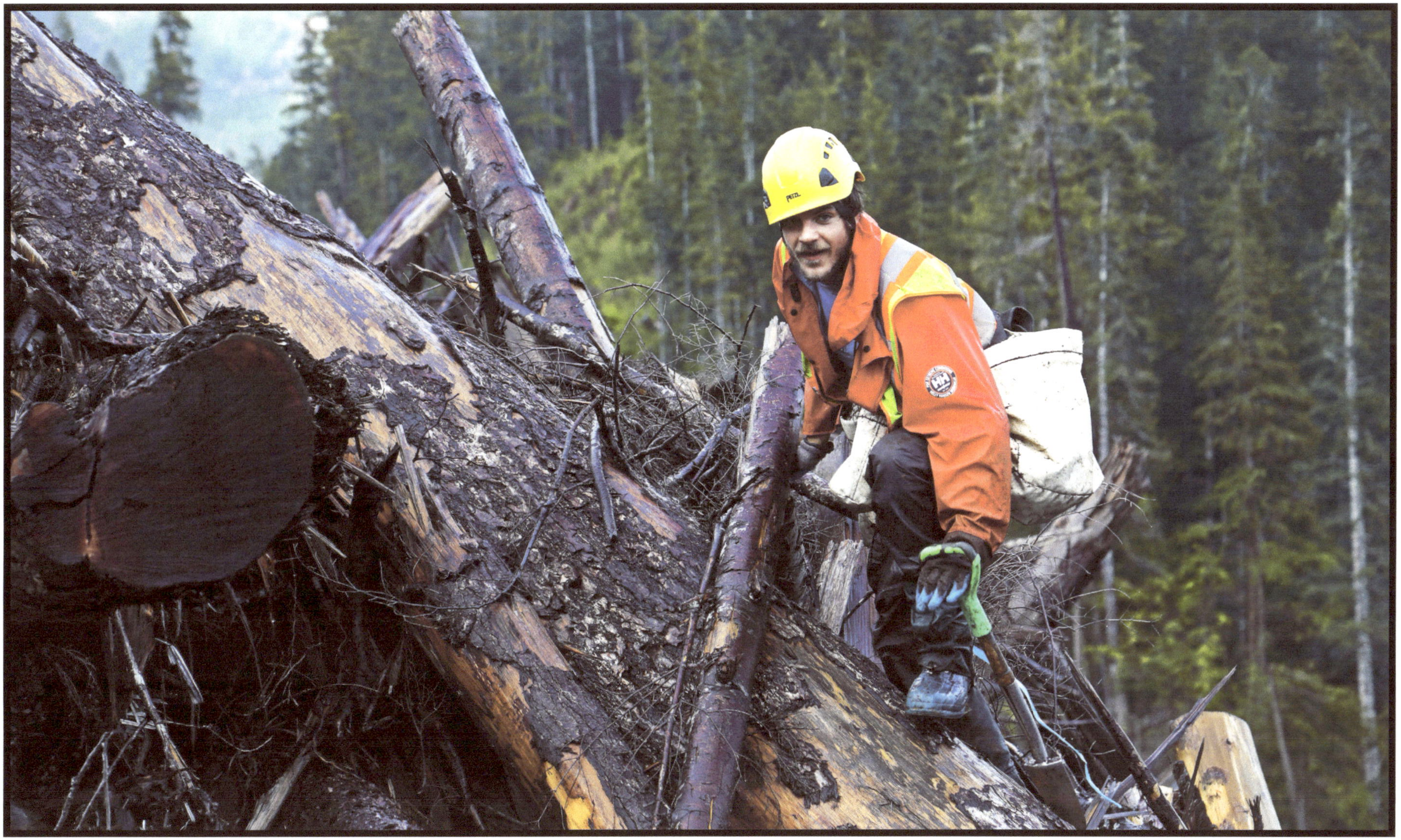

Vincent knows the value of keeping his caulks sharp.

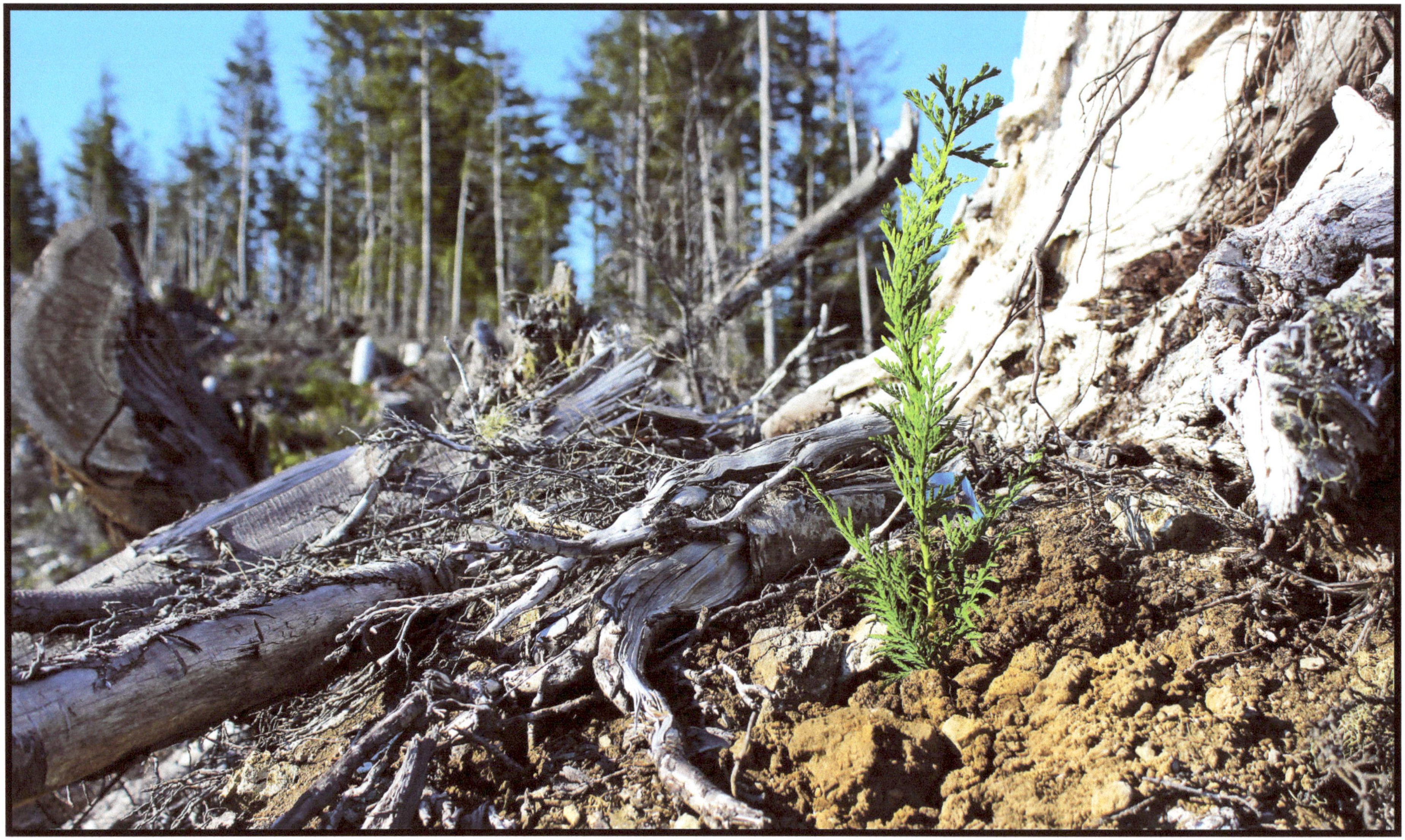

Another healthy-looking cedar seedling, freshly planted. I wonder if this one will grow to be a thousand years old. I like to hope that's a possibility.

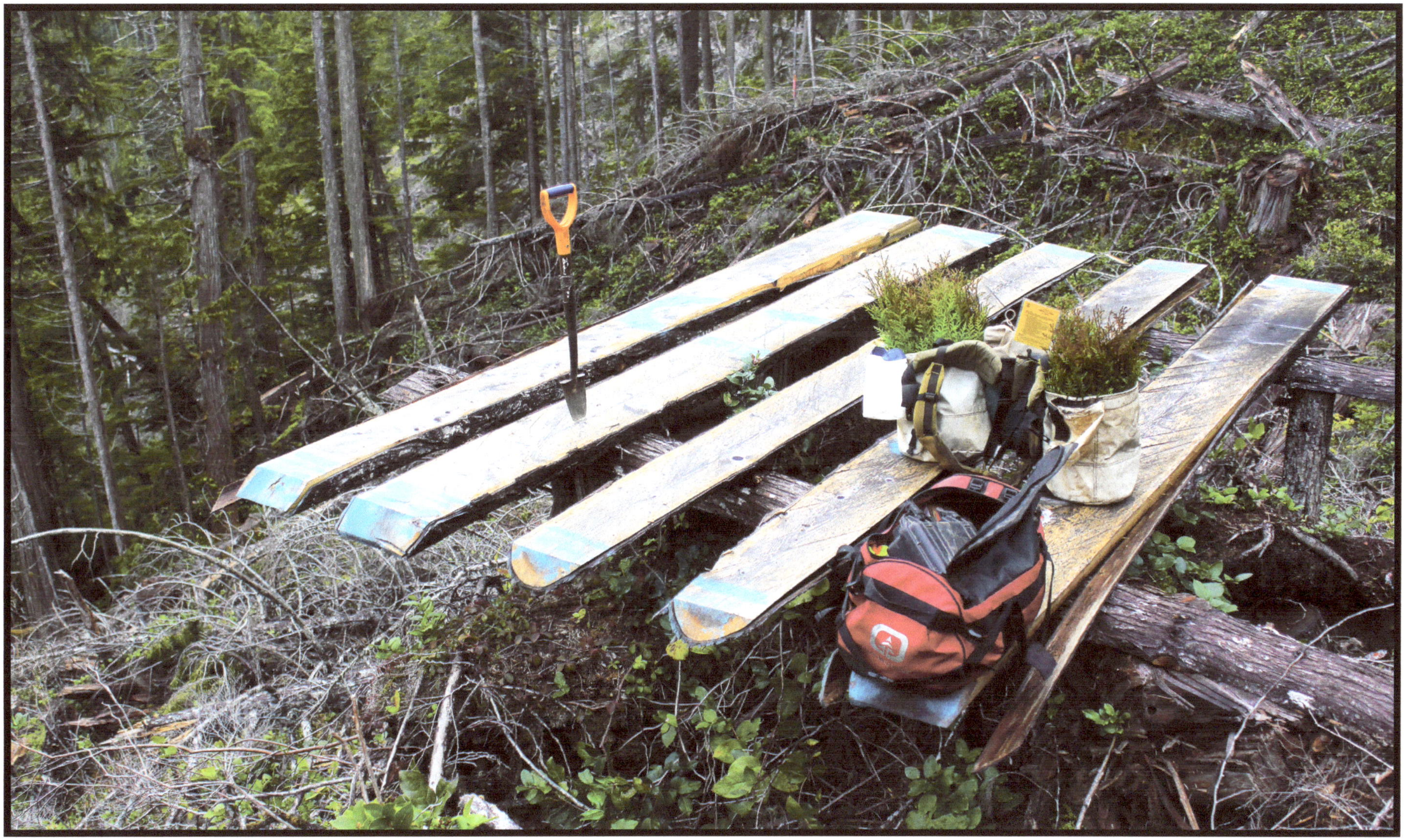

This photo was taken at the start of the day, after I had just been flown out to a heli pad. At first glance, I thought that the block didn't look too bad.

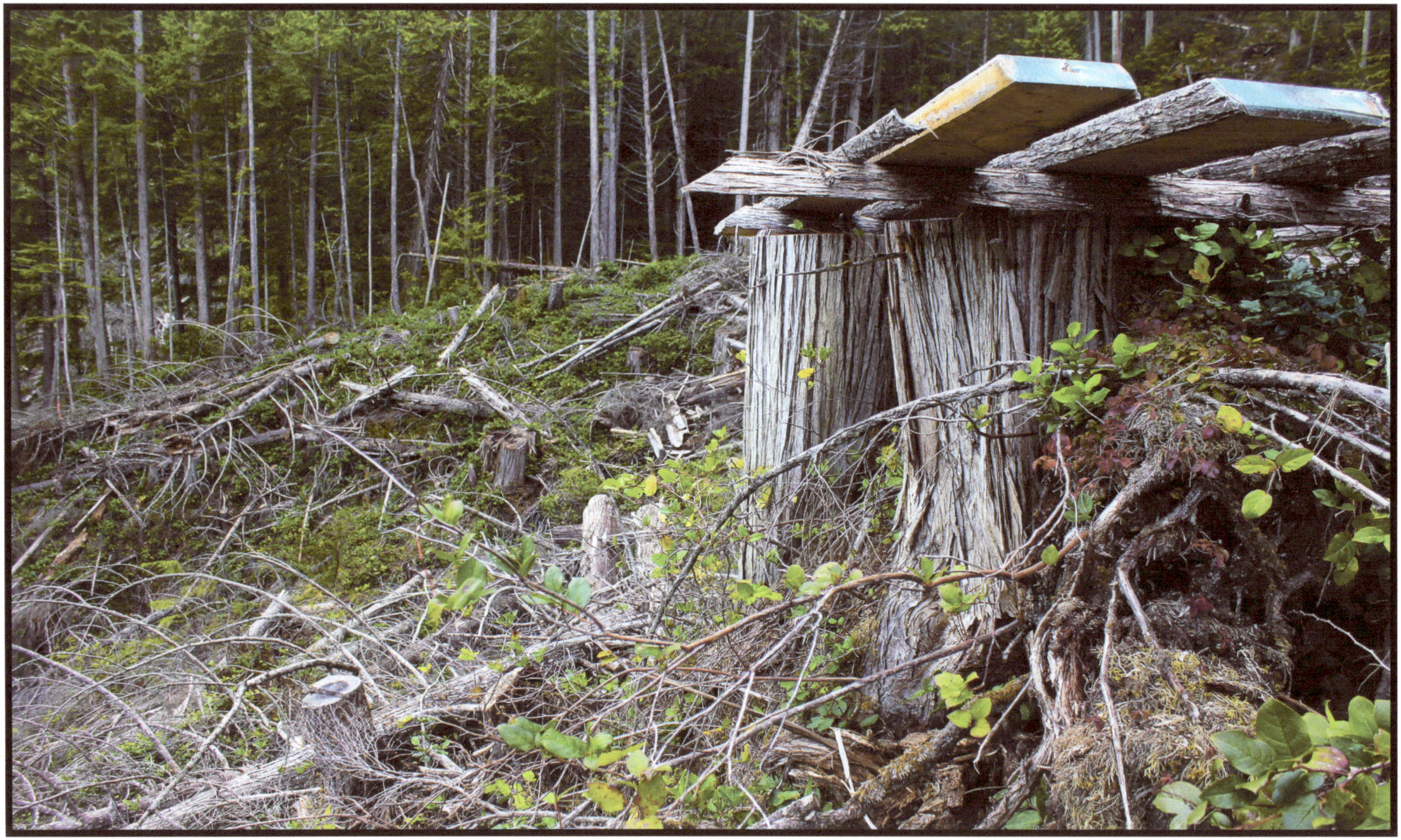

As soon as I stepped below the heli pad, I realized that the block was going to be a bit tougher than I had initially assumed. It was a typically challenging day.

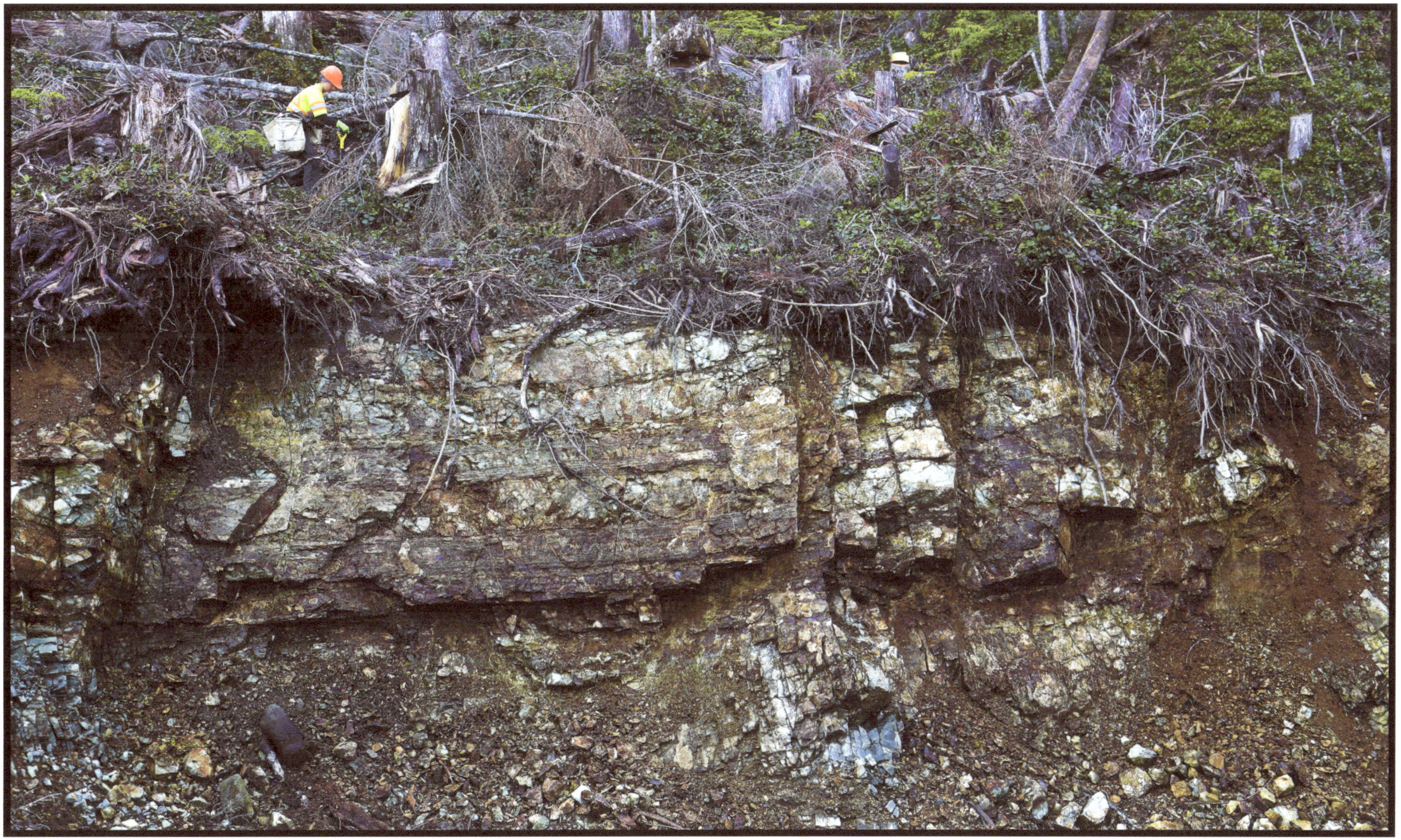

If you're nervous about heights, you may not like some of the blocks. Even if you're not nervous, it pays to watch your step. There's no point in planting along the very edge of a cliff or bluff. That's an unnecessary risk, especially considering how easy it is to catch your foot on something and trip.

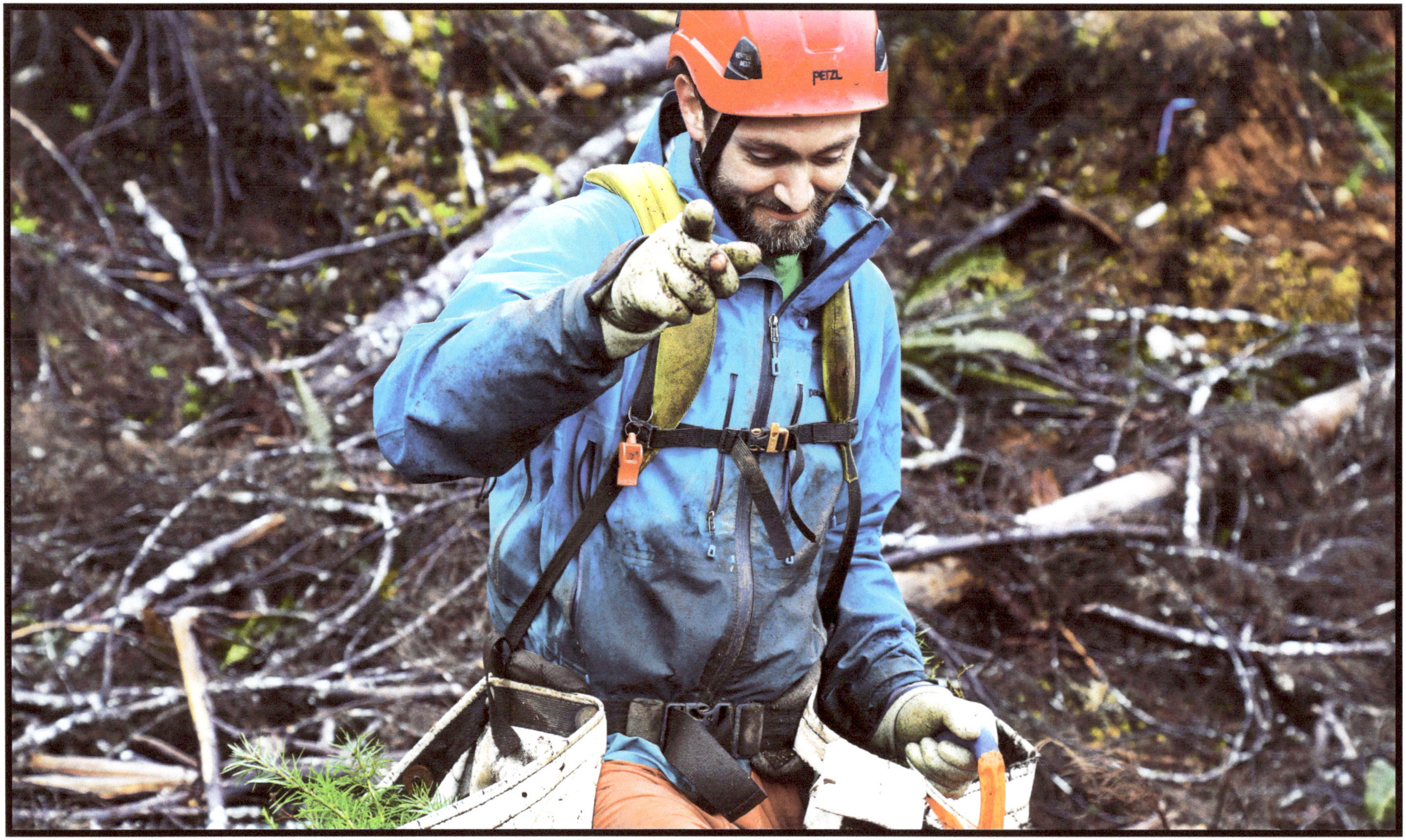

Greg is an incredibly fast planter.

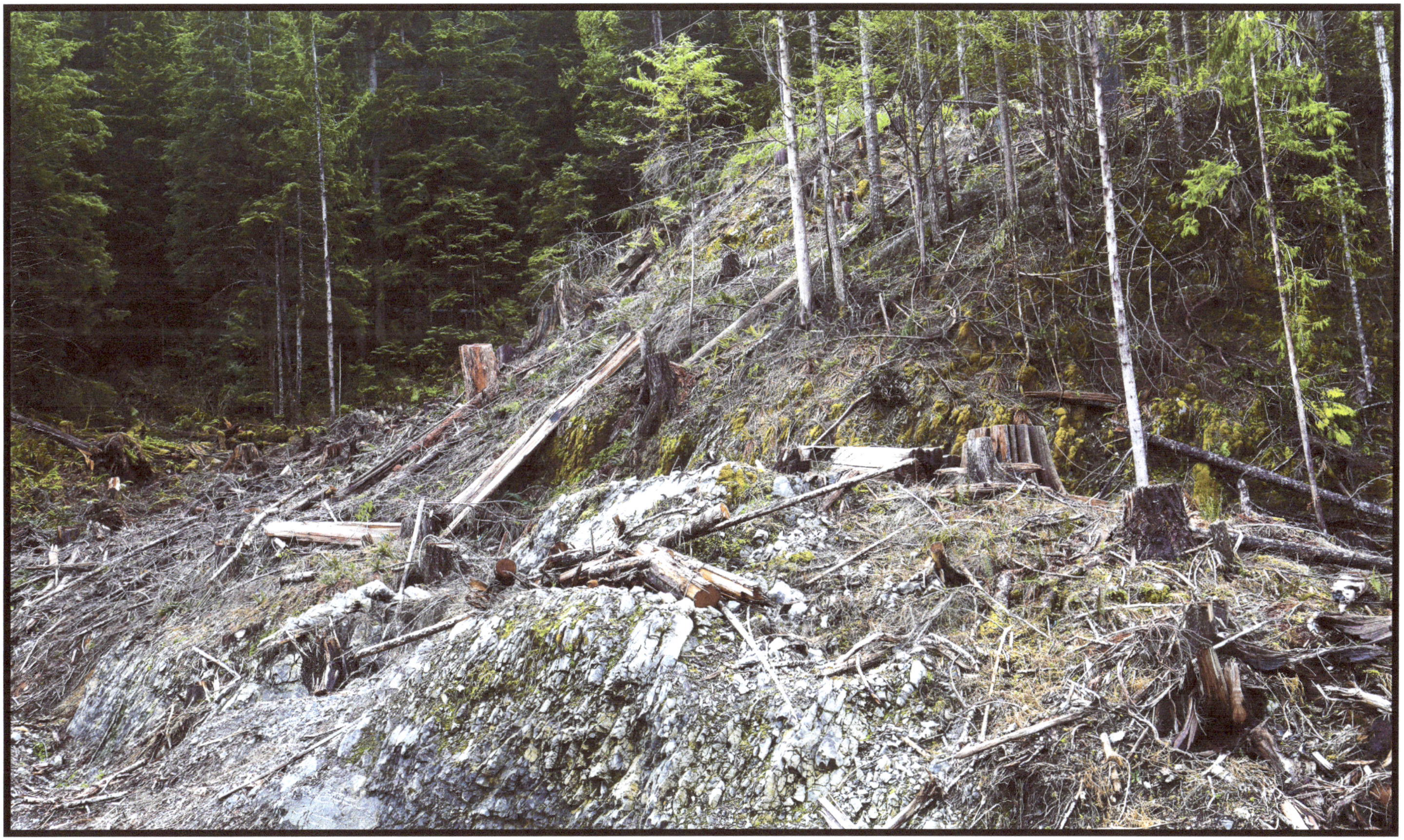

This piece looks uninspiring.

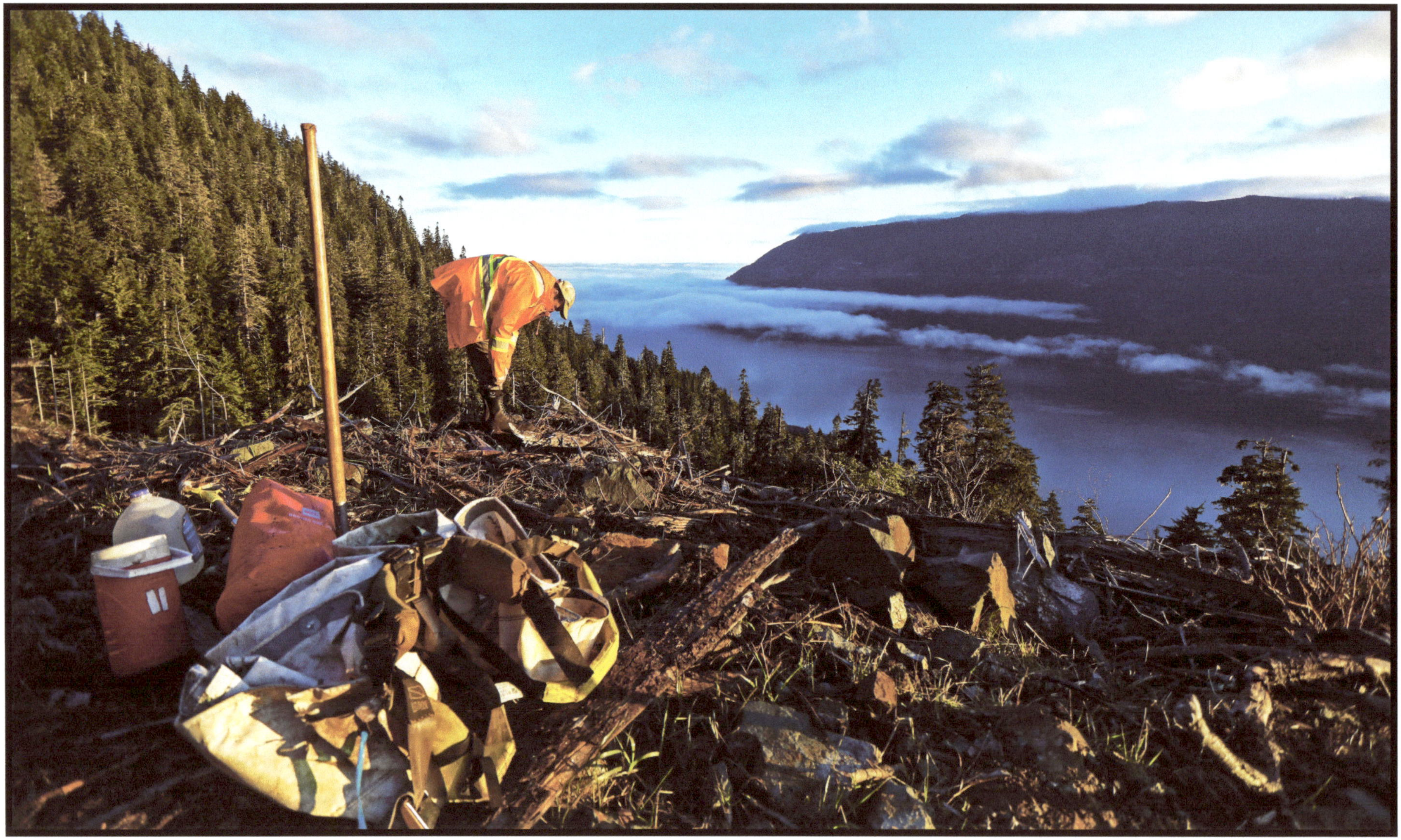

Putting on caulked boots to start the day.

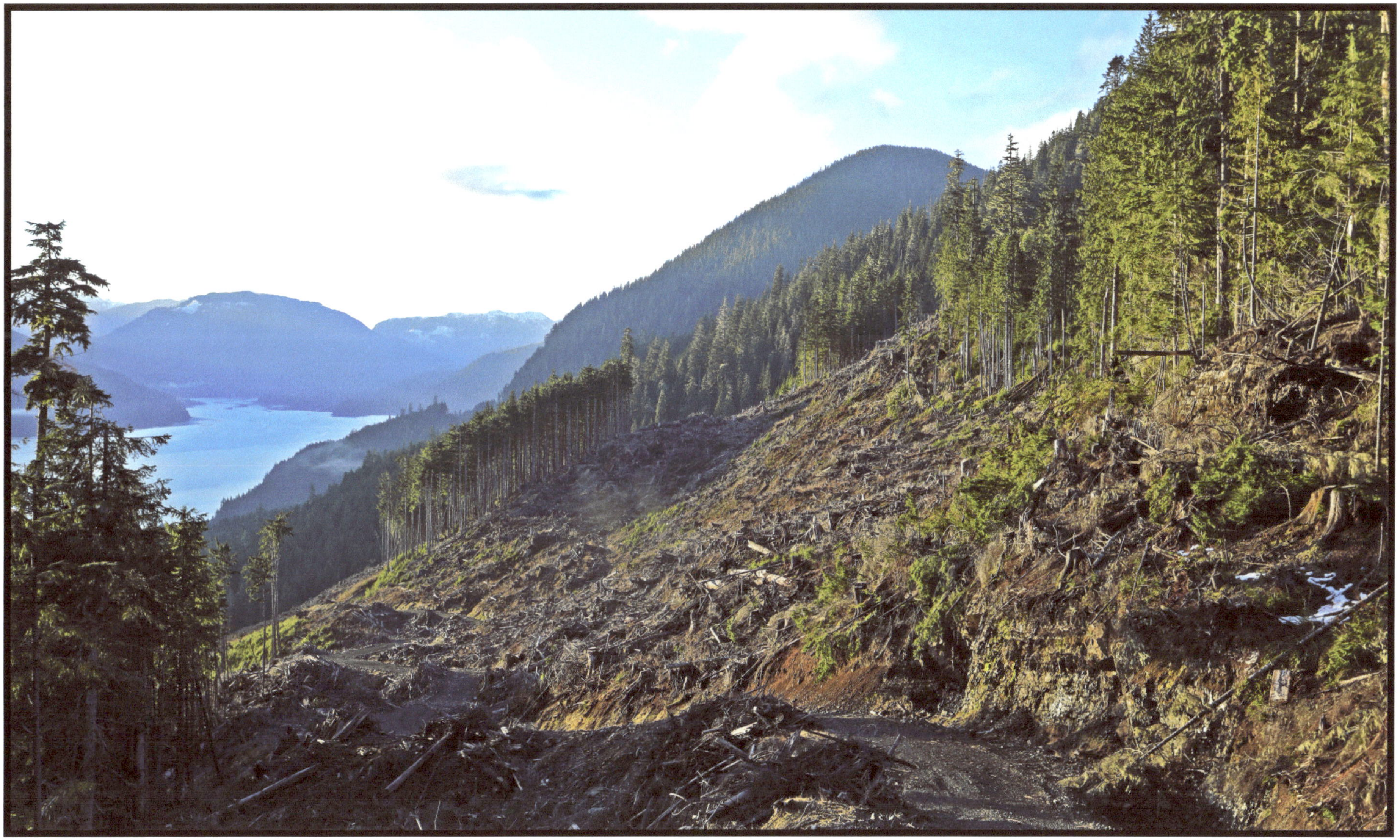

This block had a nice view.

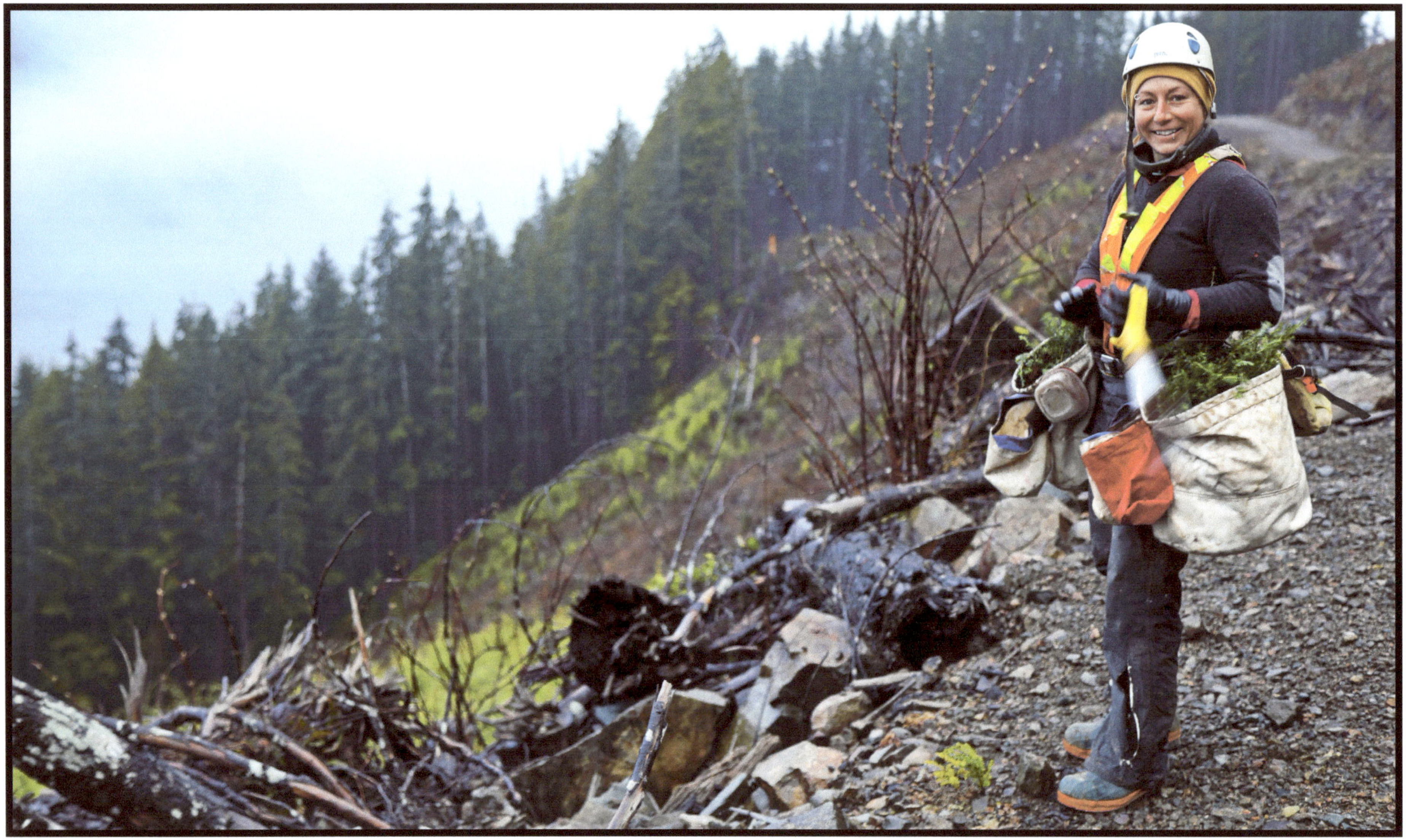

Catherine loves education. She did a program in Technology of Arts & Media (specializing in television) and is an accomplished video editor, graphics designer, and camera-person. She also studied Biology and is a fauna technician, which suits her love of the outdoors. In addition to having planted for more than a decade, she loves any outdoors activity, such as hiking, riding, and climbing.

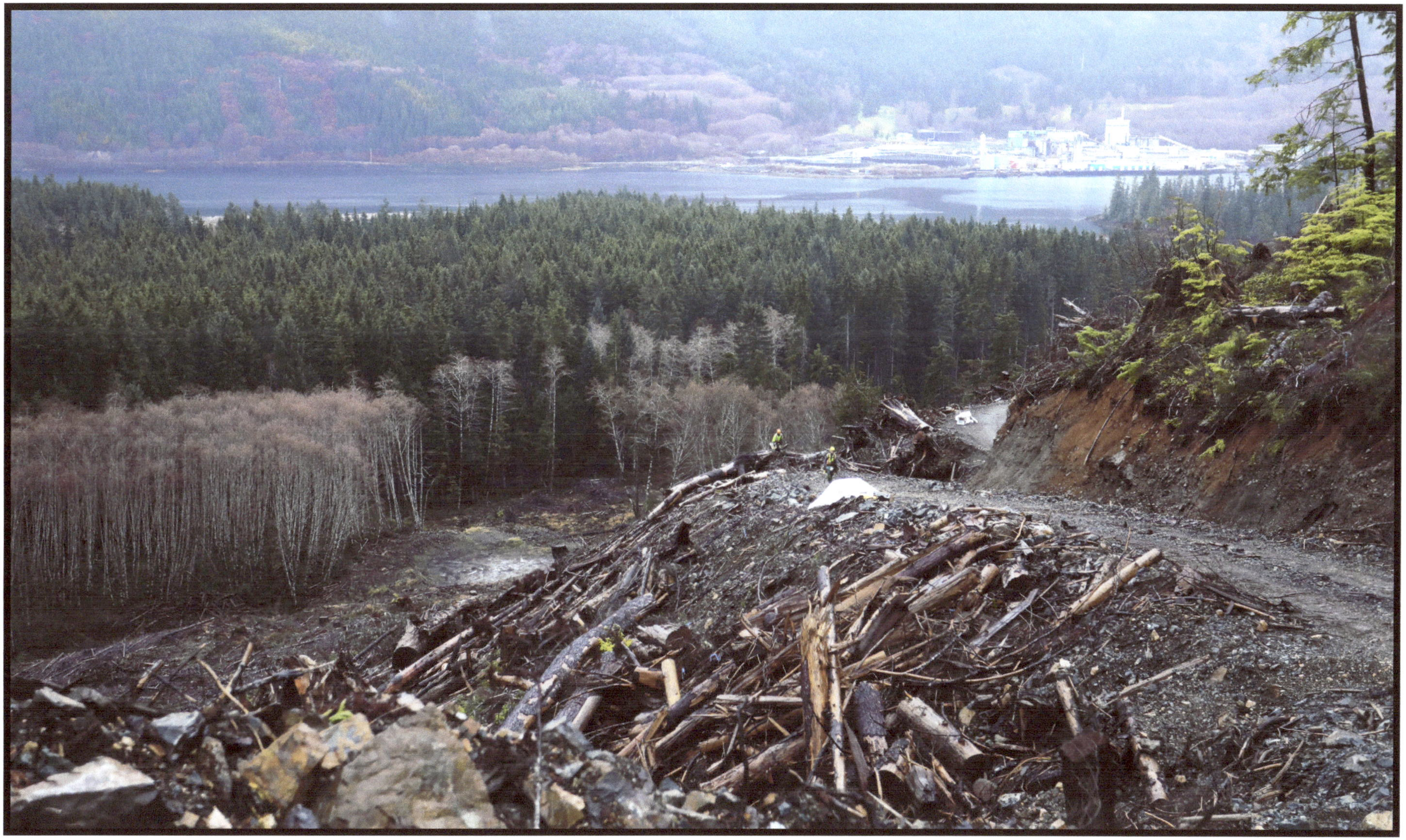

Here's a block across the inlet from the old Neucel Specialty Cellulose factory outside of Port Alice.

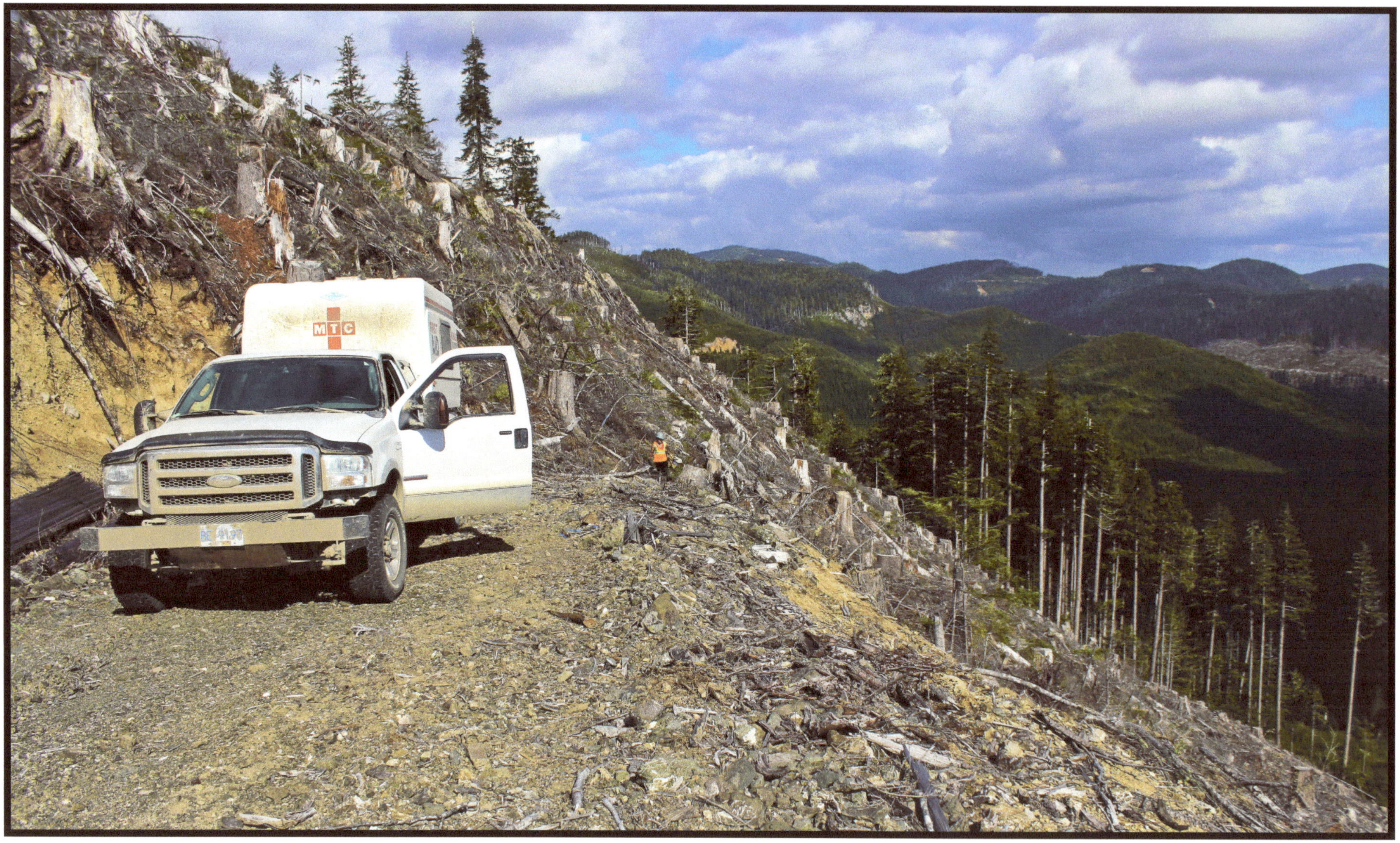

This was a crisp fall day.

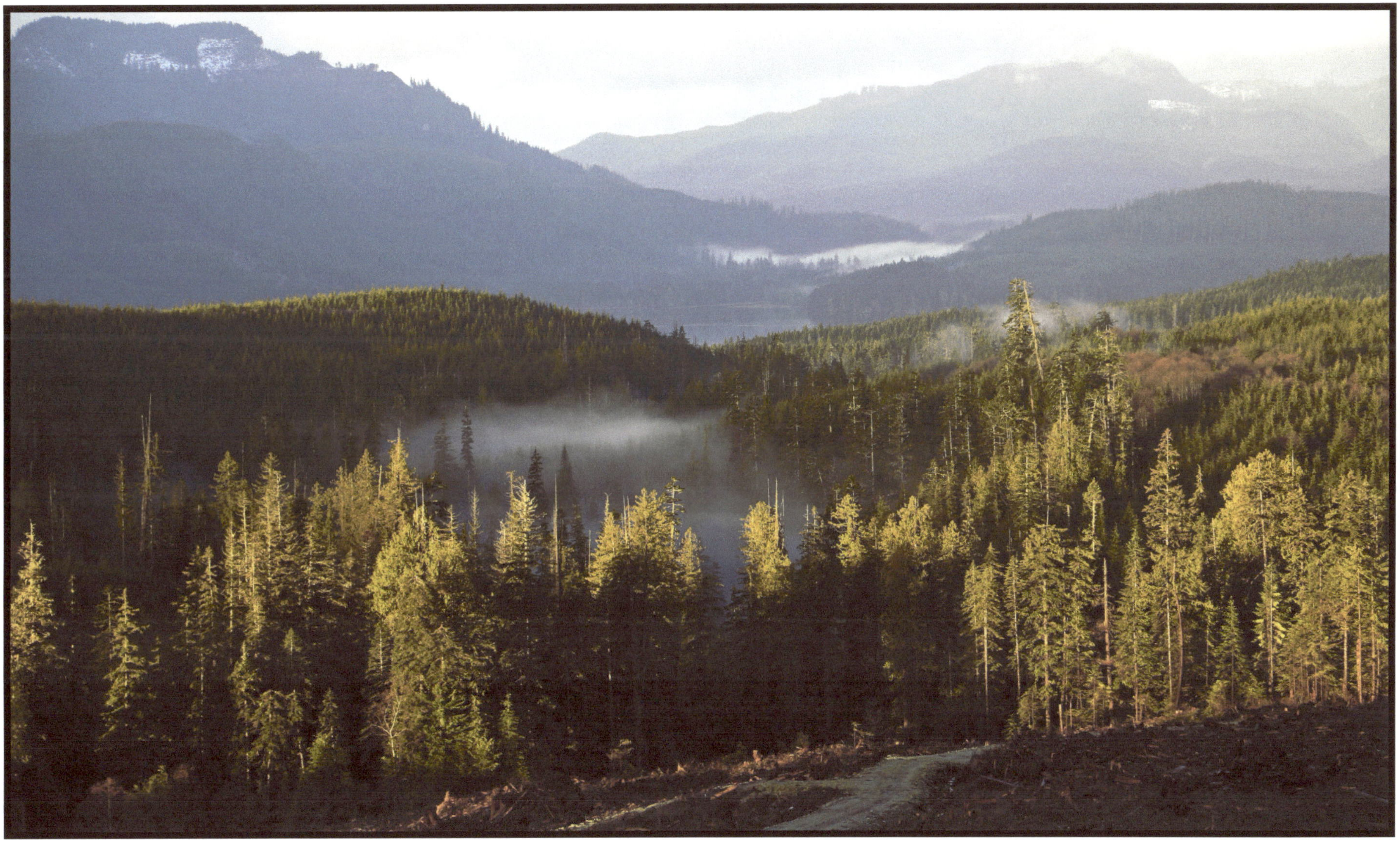

These hints of early morning sunshine are tantalizing.

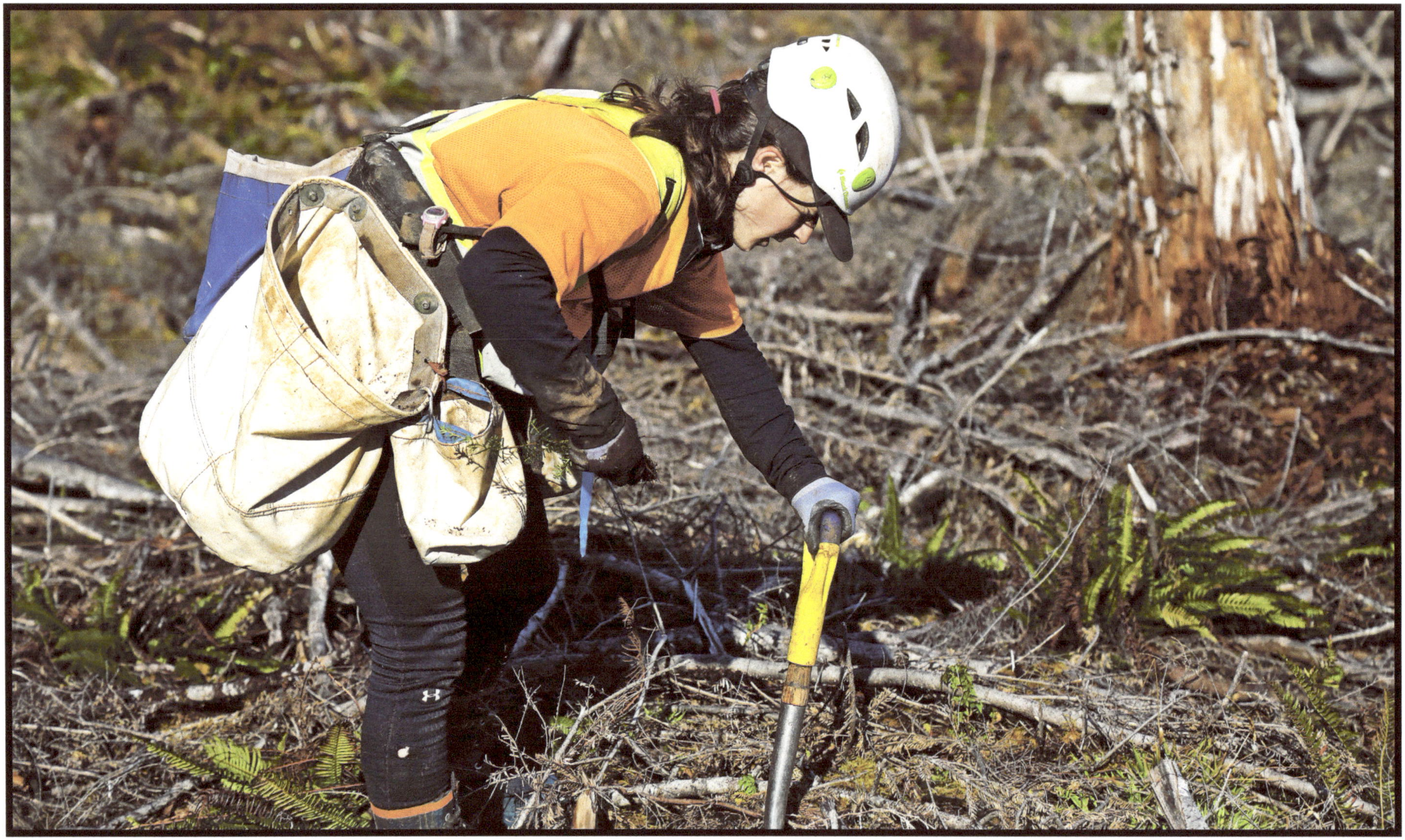

All planters should keep a waterproof watch attached to their bags, so they can keep track of time. Once you get to the block, there's no time to waste. You can sit down and relax at the end of the day.

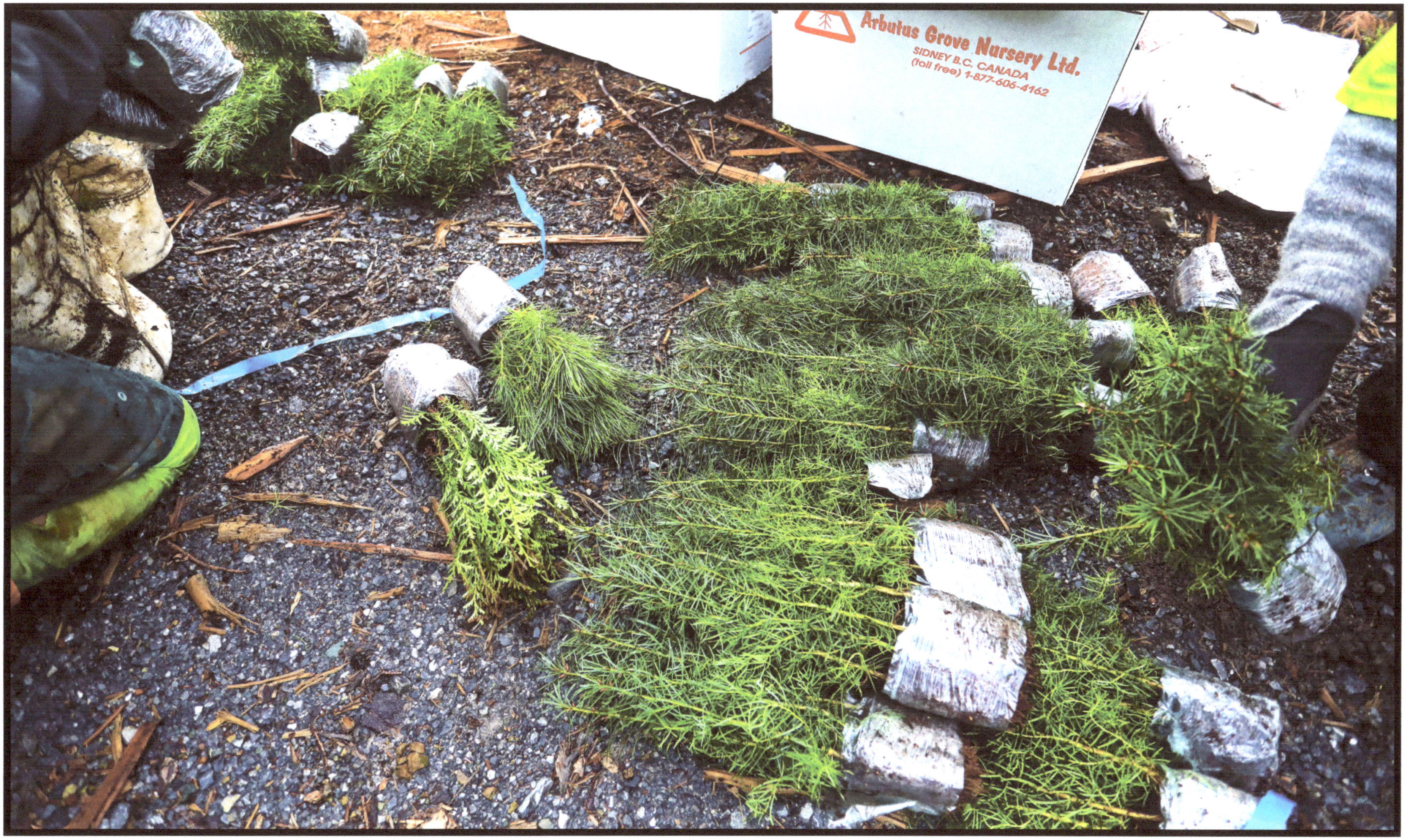

It's common to bag up with multiple species. Biodiversity is always better than planting monocultures. This block looks like it is predominantly prescribed for Douglas Fir, with a small amount of red cedar and white pine.

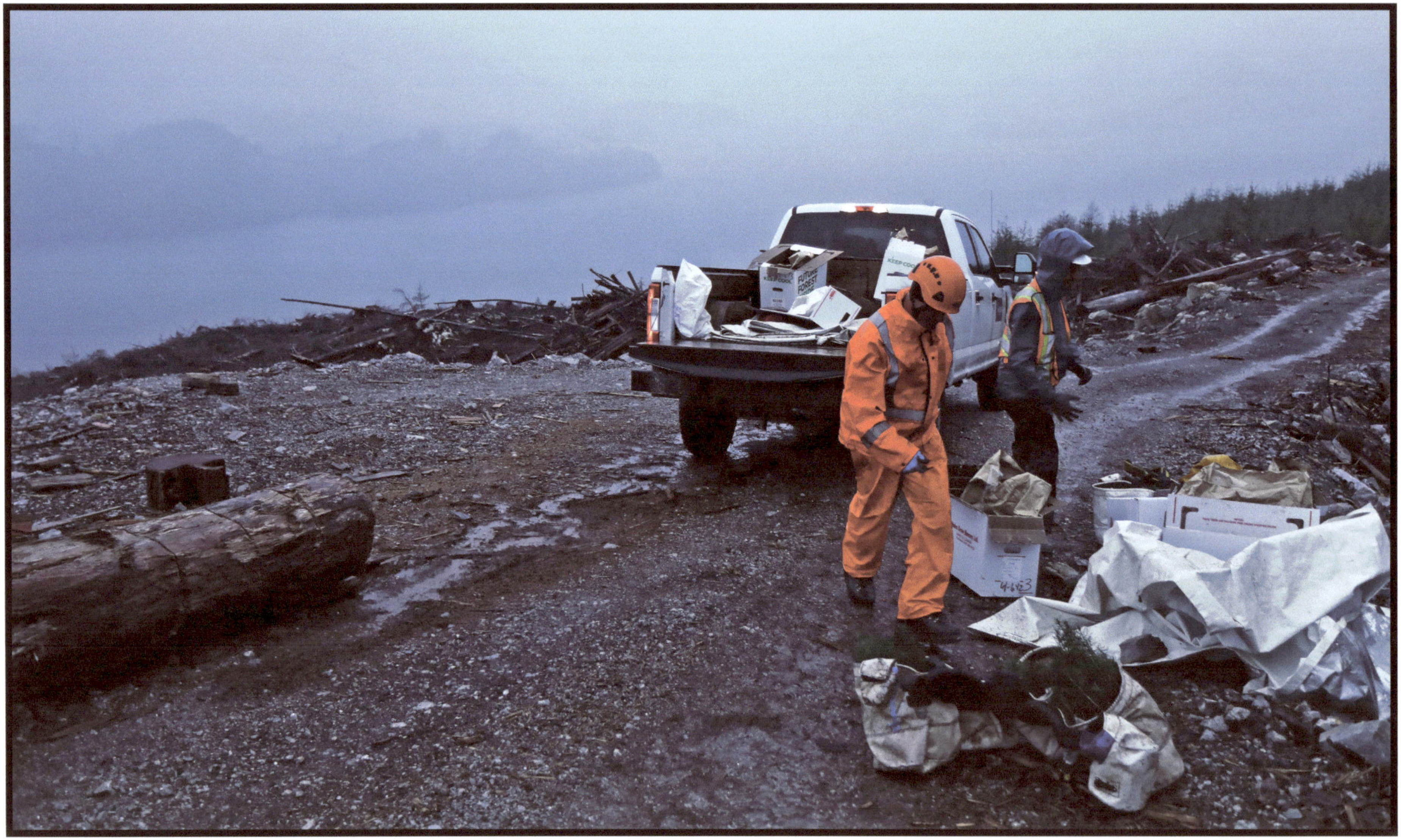

I don't typically take a lot of photos on cold, wet days, because I don't want to damage my camera. However, this is a good representation of a typical spring coastal day: grey, wet, cold, and windy.

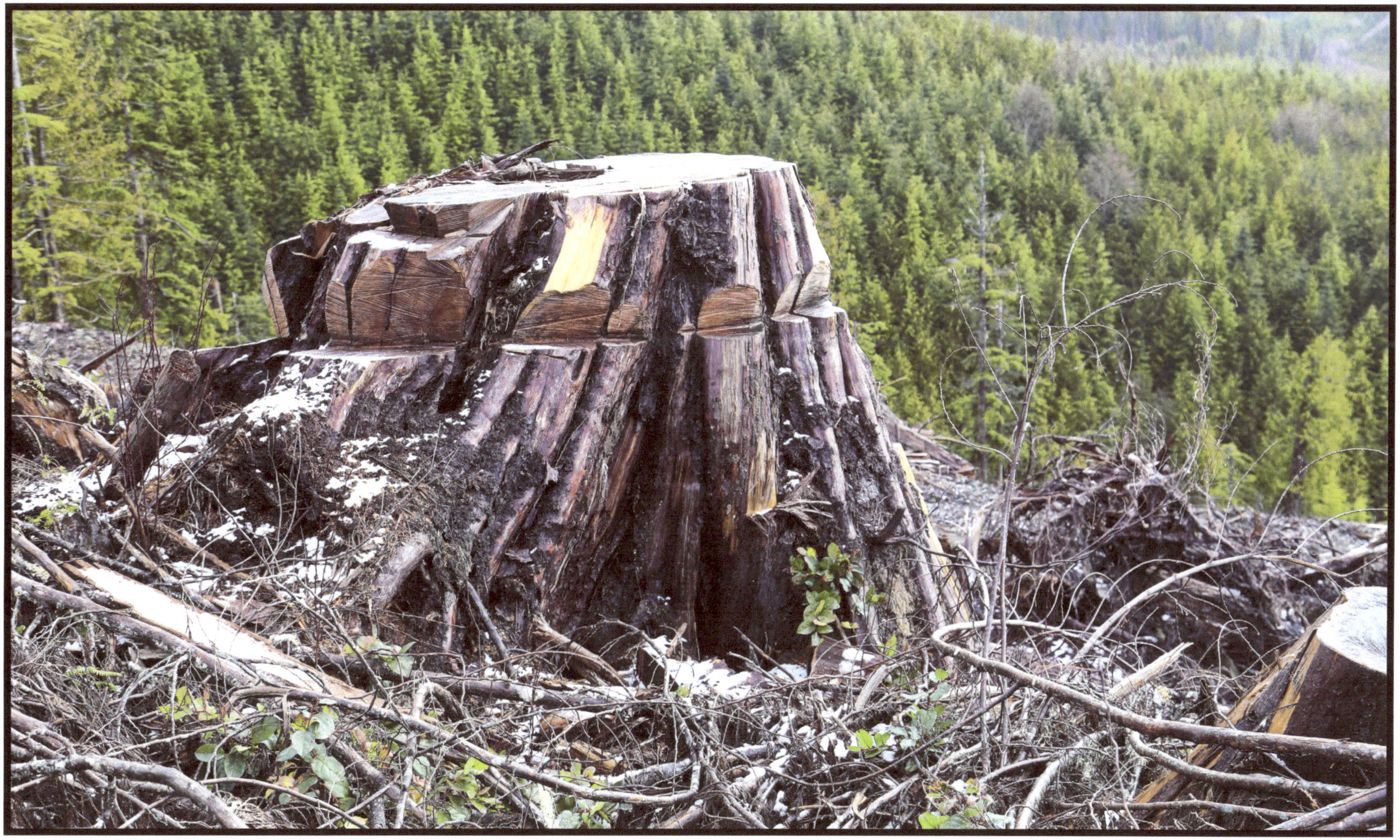

You may wonder about the small notches cut around this stump. A lot of older cedar trees tend to splinter and fragment when they're being felled, which is extremely dangerous. It's a problem that kills quite a few fallers in BC. By notching the outsides of the secondary roots on the outside base of the tree, a faller is able to alleviate some of the tension that increases the likelihood of splintering.

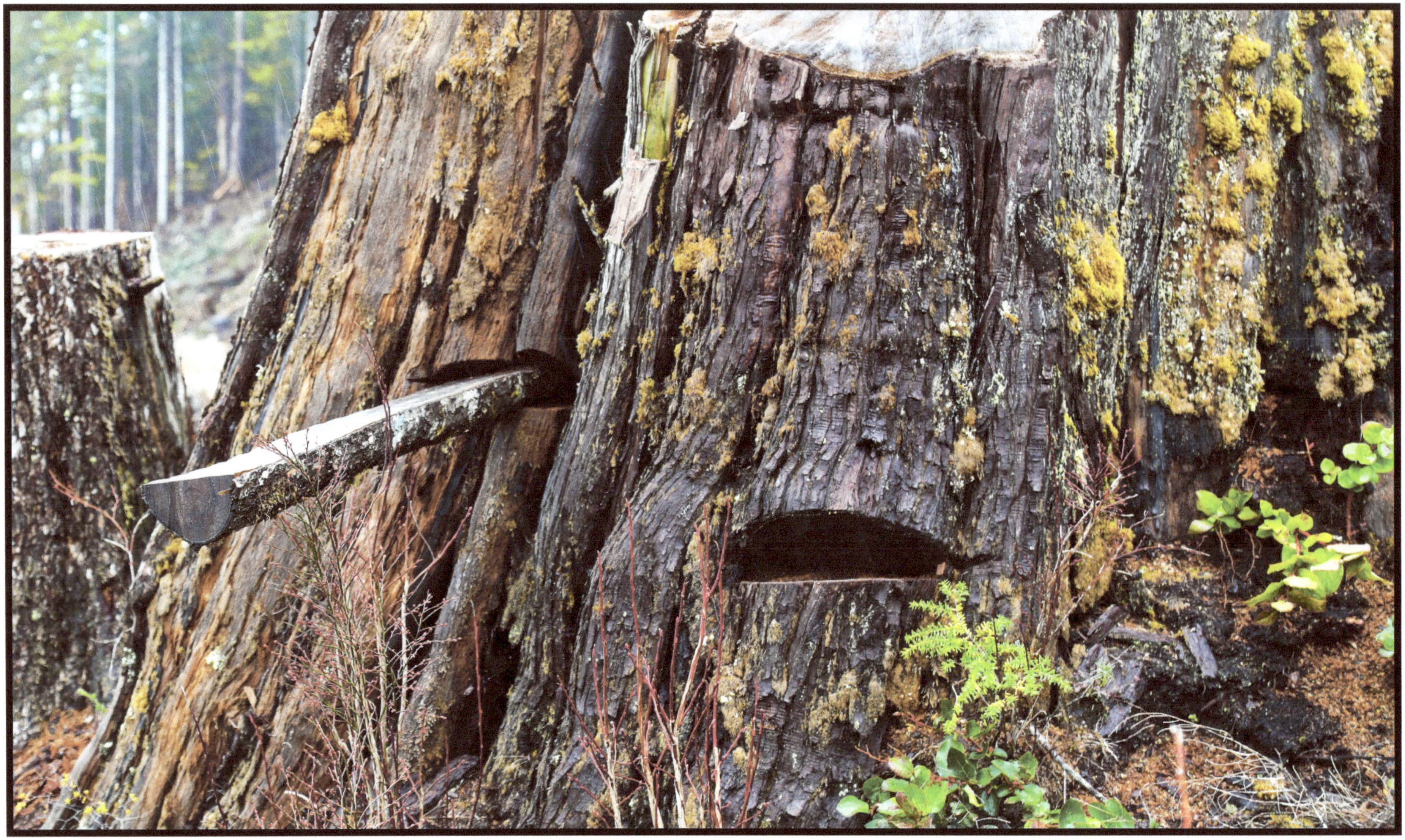

The notches on this stump were cut for an entirely different reason than the ones in the previous photo. If you look closely, you'll see that these notches were designed to allow a plank to be wedged in. The faller is able to stand on this plank while falling a tree, which allows him to cut the tree higher from the ground. This makes it easier to influence the falling trajectory of the tree. A lot of these trees grow on steep slopes, so to make a horizontal "flat" cut through the trunk, the cut on the downhill side of the tree ends up being well above head level.

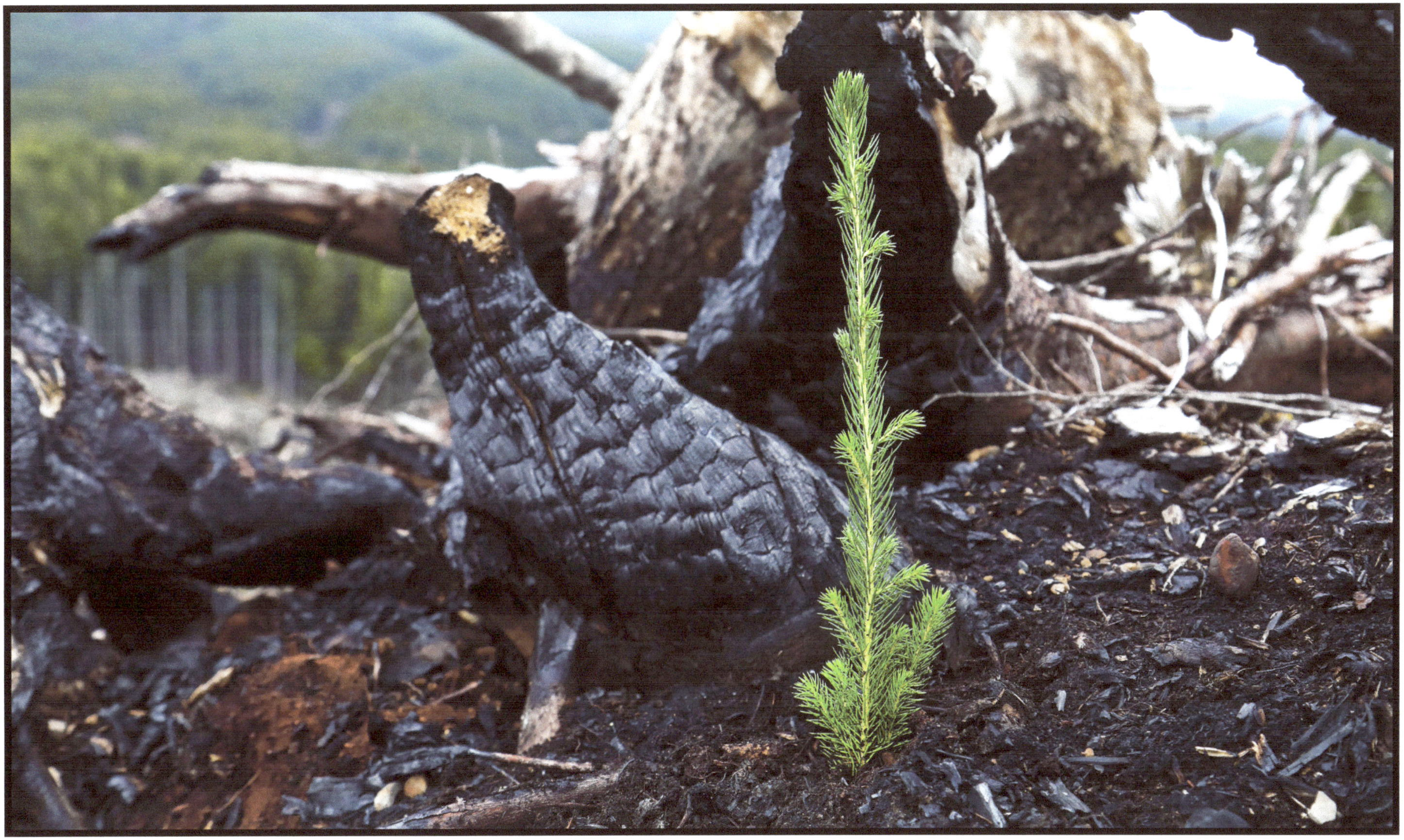

Here's a young Sitka Spruce seedling.

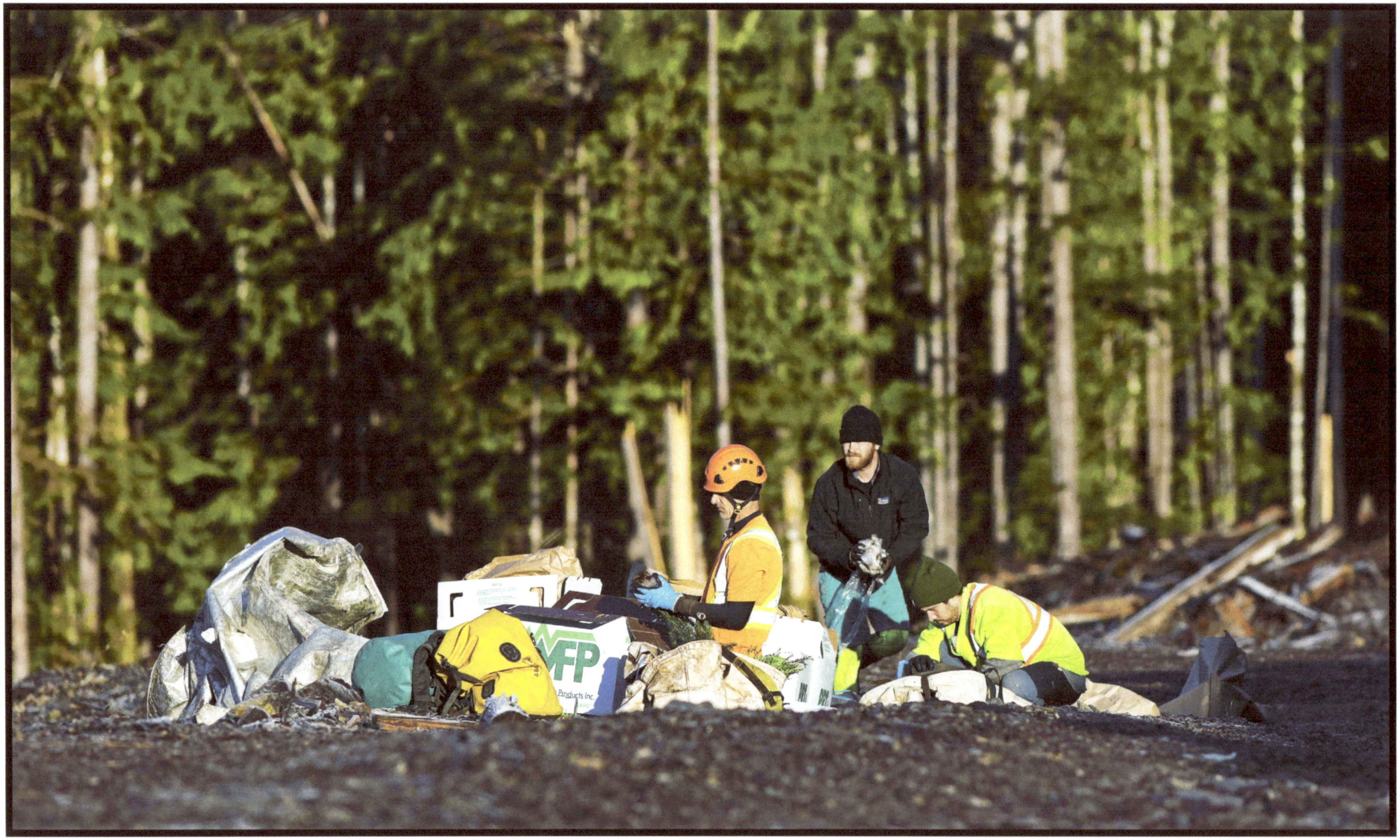

I remember this block quite well. The early morning sunshine and tranquility was deceiving. The block had a large number of holes that were hidden by vegetation. Several of us kept falling unexpectedly into deep holes all day. These weren't karst holes, which are formed over long periods of time when water erodes limestone. Most of these holes were just random deep dips that were four or five feet deep, and hidden with an overgrowth of salal.

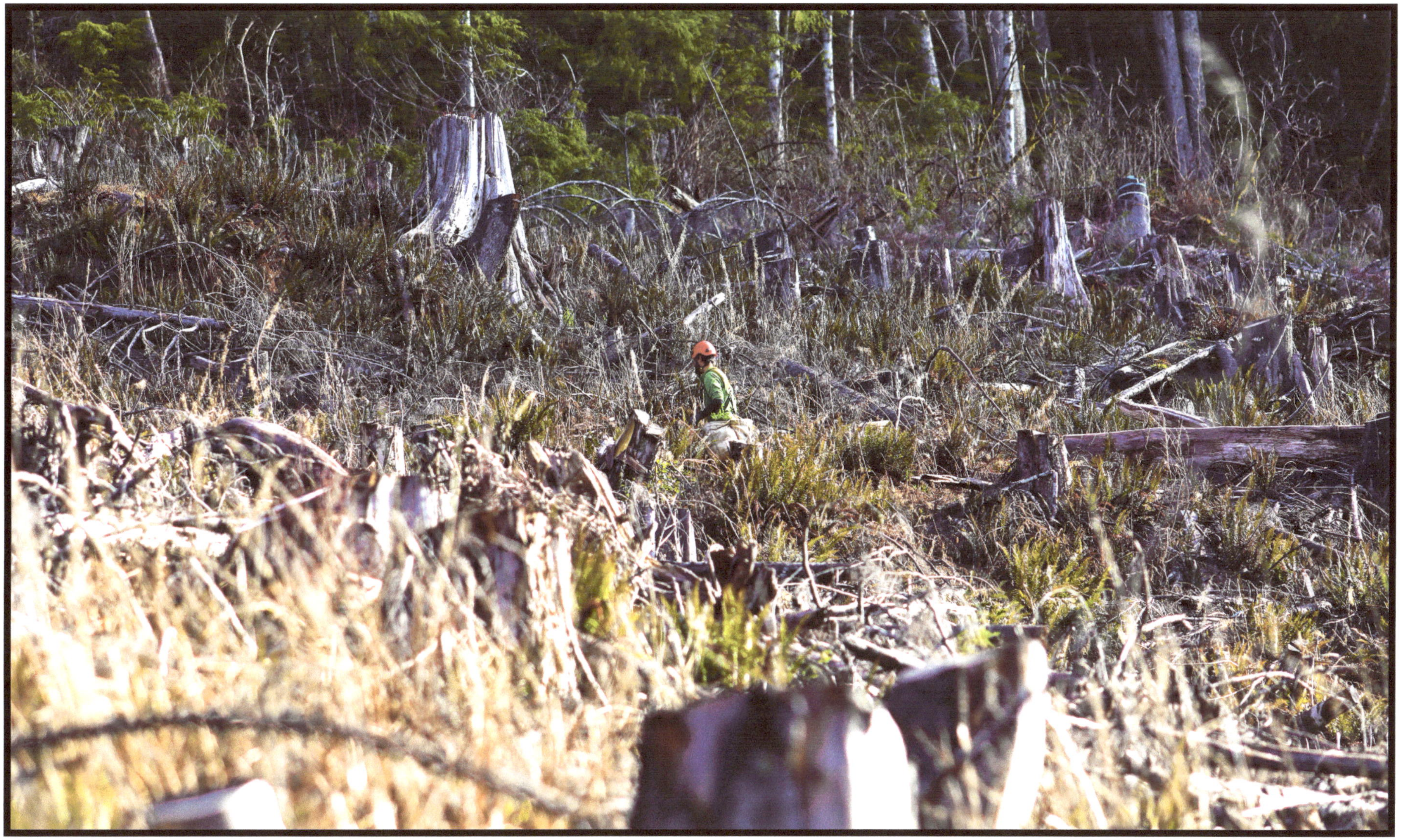

This photo almost does justice to the size of some of the stumps.

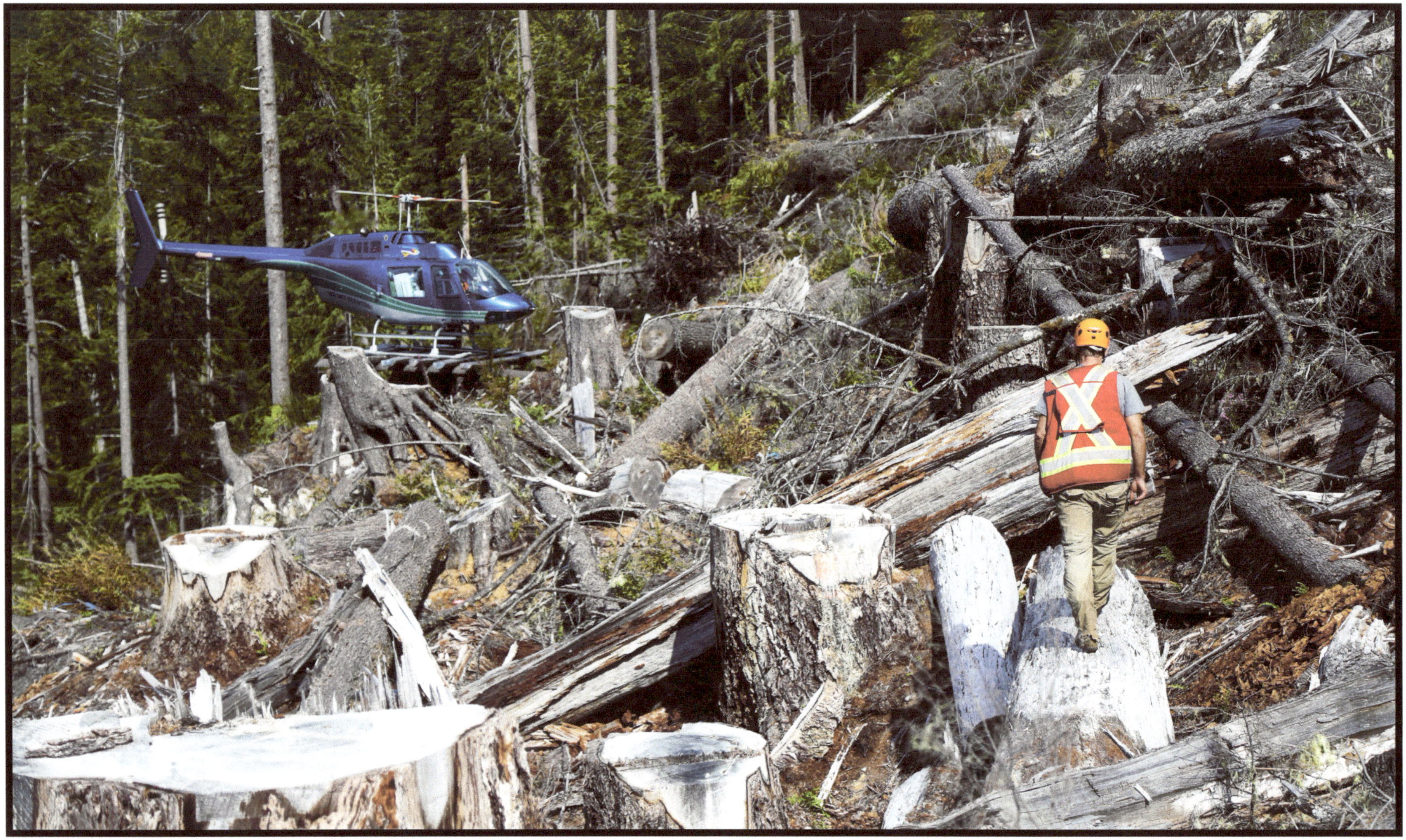

This foreman is moving toward the helicopter at the end of the day. The helicopter has just come in to pick up some of the crew. Heli-pads are generally not to be trusted. Pilots tend to keep the power on while they touch down lightly, rather than trusting the pad as being strong enough to shut down.

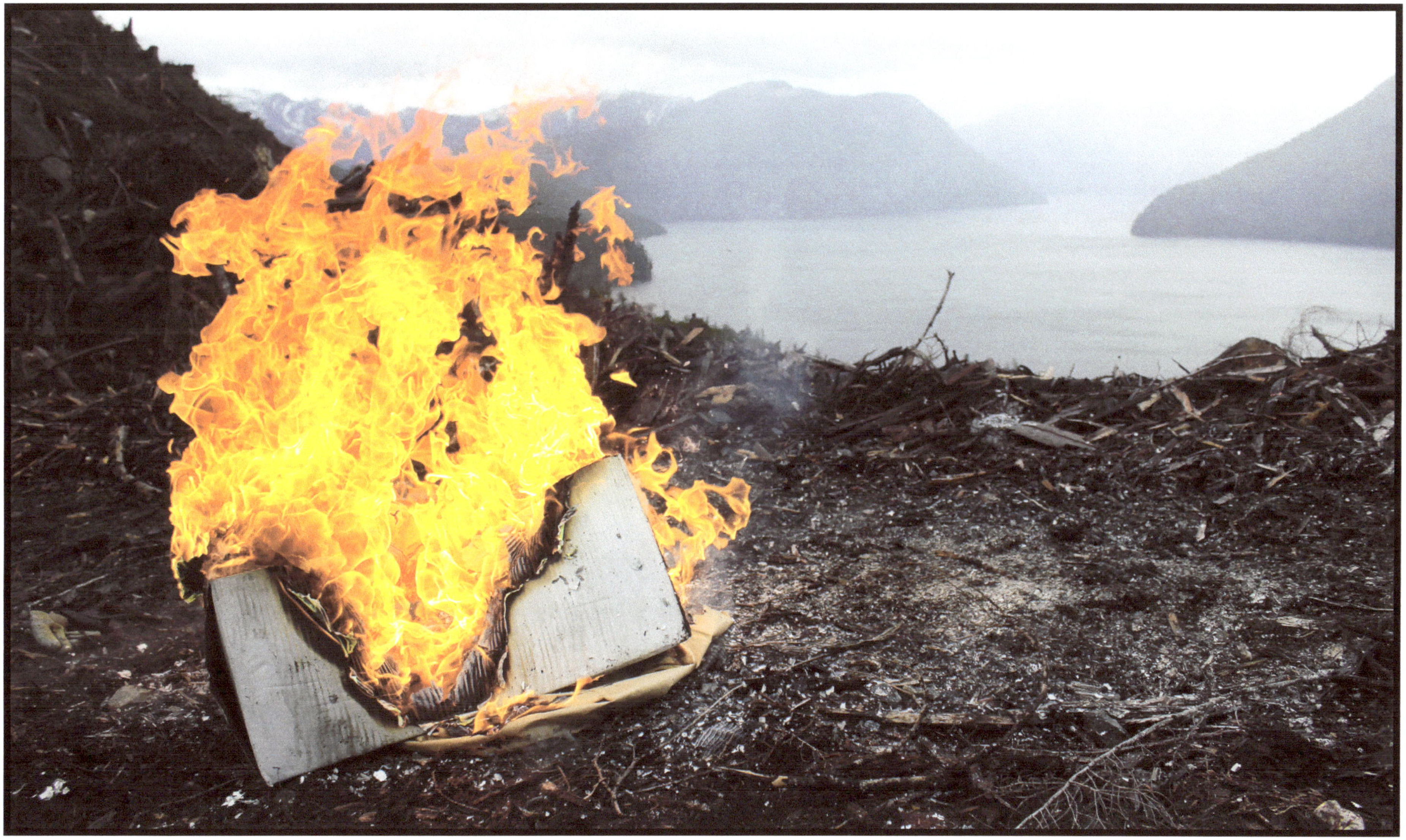

The best way to get rid of an empty tree box on a wet day is to burn it. Unfortunately, the wax content of tree boxes means that they usually can't be recycled.

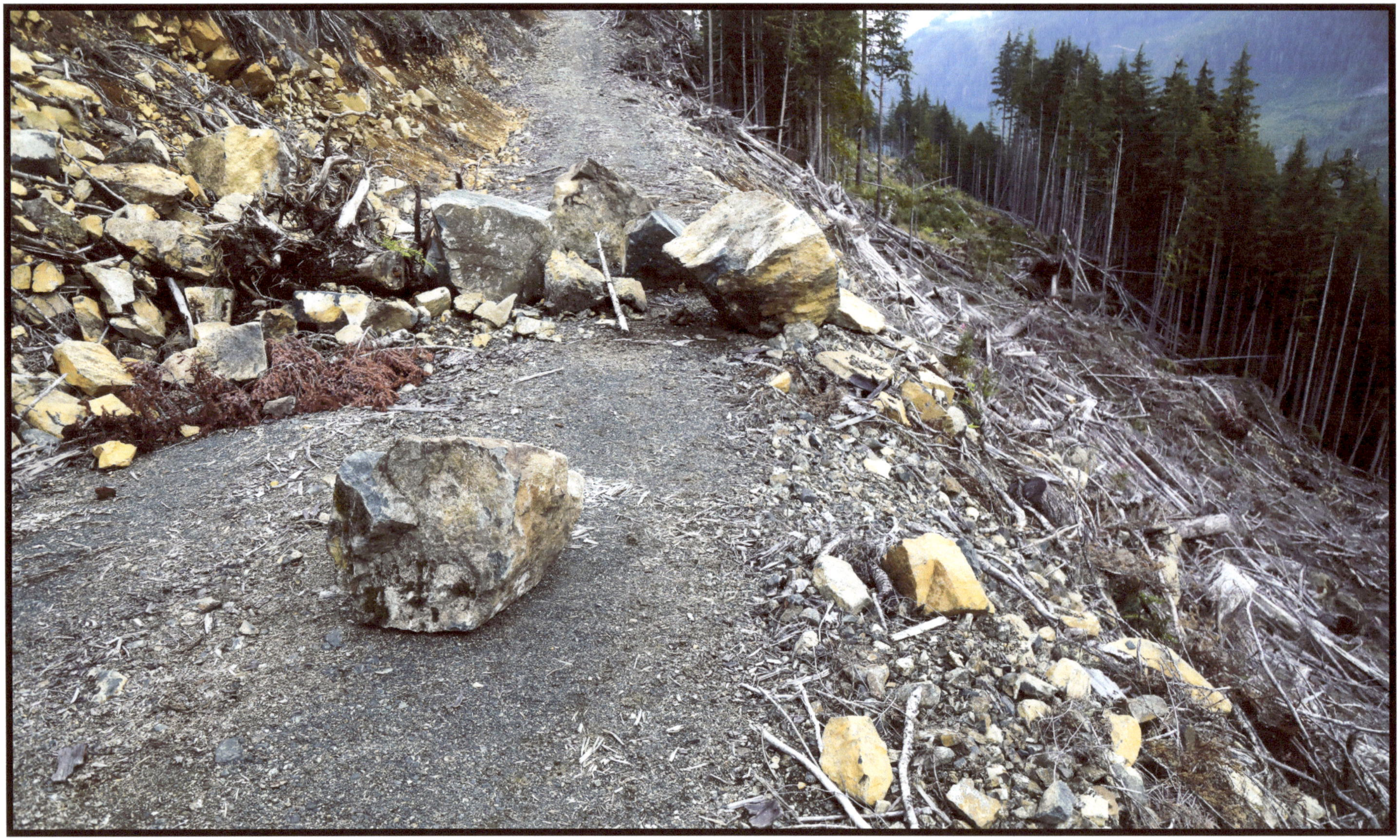

Rock slides are common. This is a very small slide, just barely enough to make the in-block road impassable. Mind you, those rocks aren't small. Each of the bigger ones weighs more than a metric tonne.

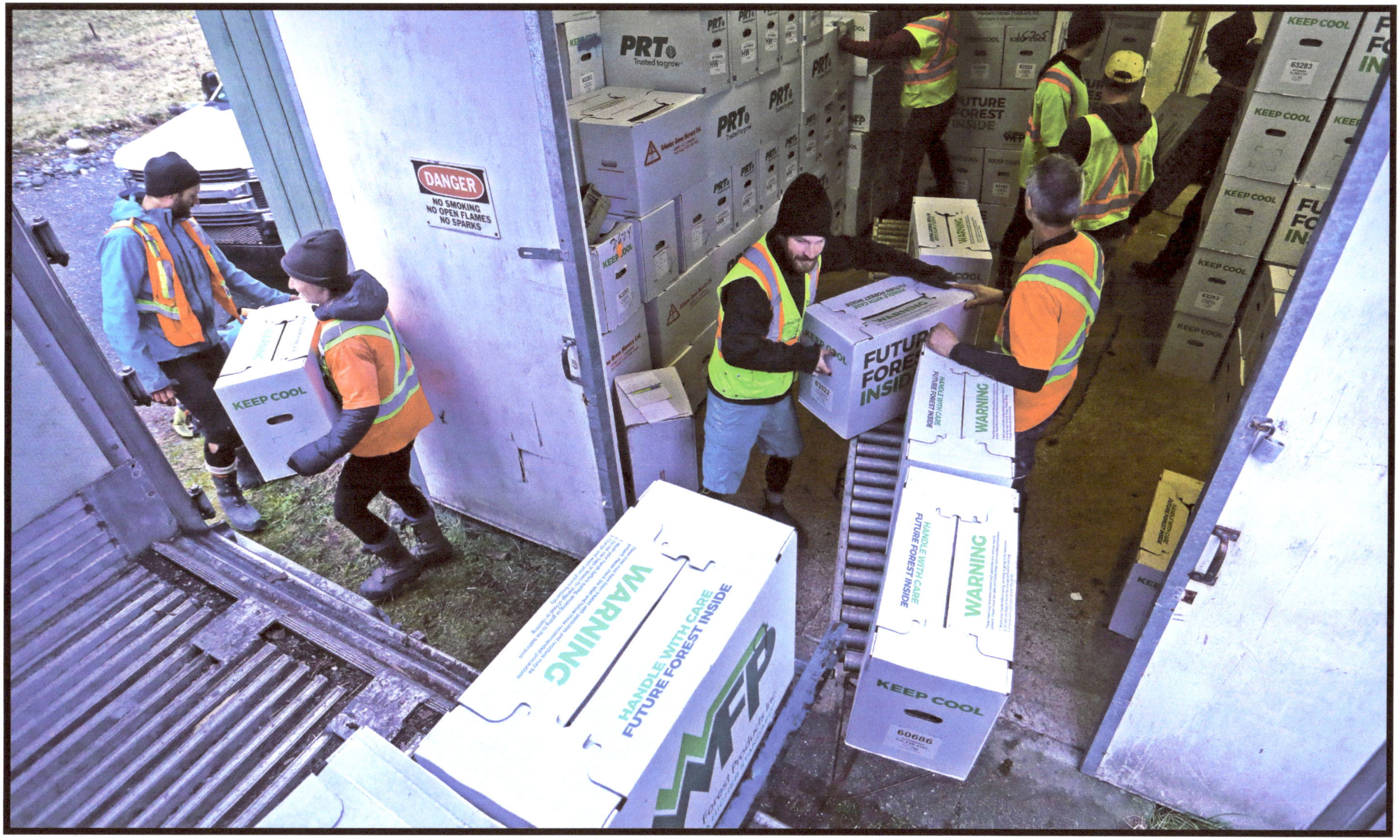

Here we see part of the crew unloading a tree delivery.

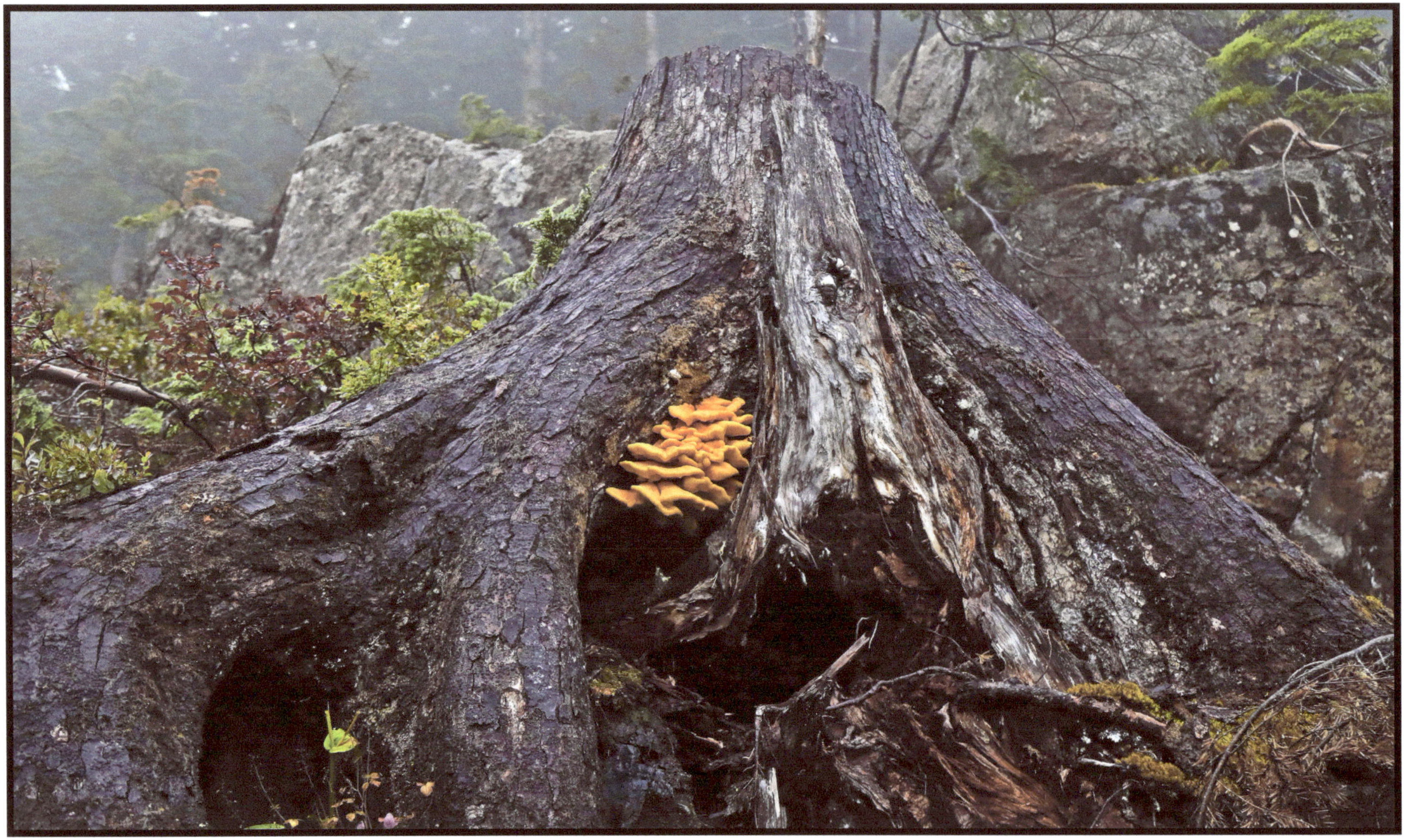

Most of the things that you see on coastal blocks are dark and dreary, but occasionally you'll see some bright fungi.

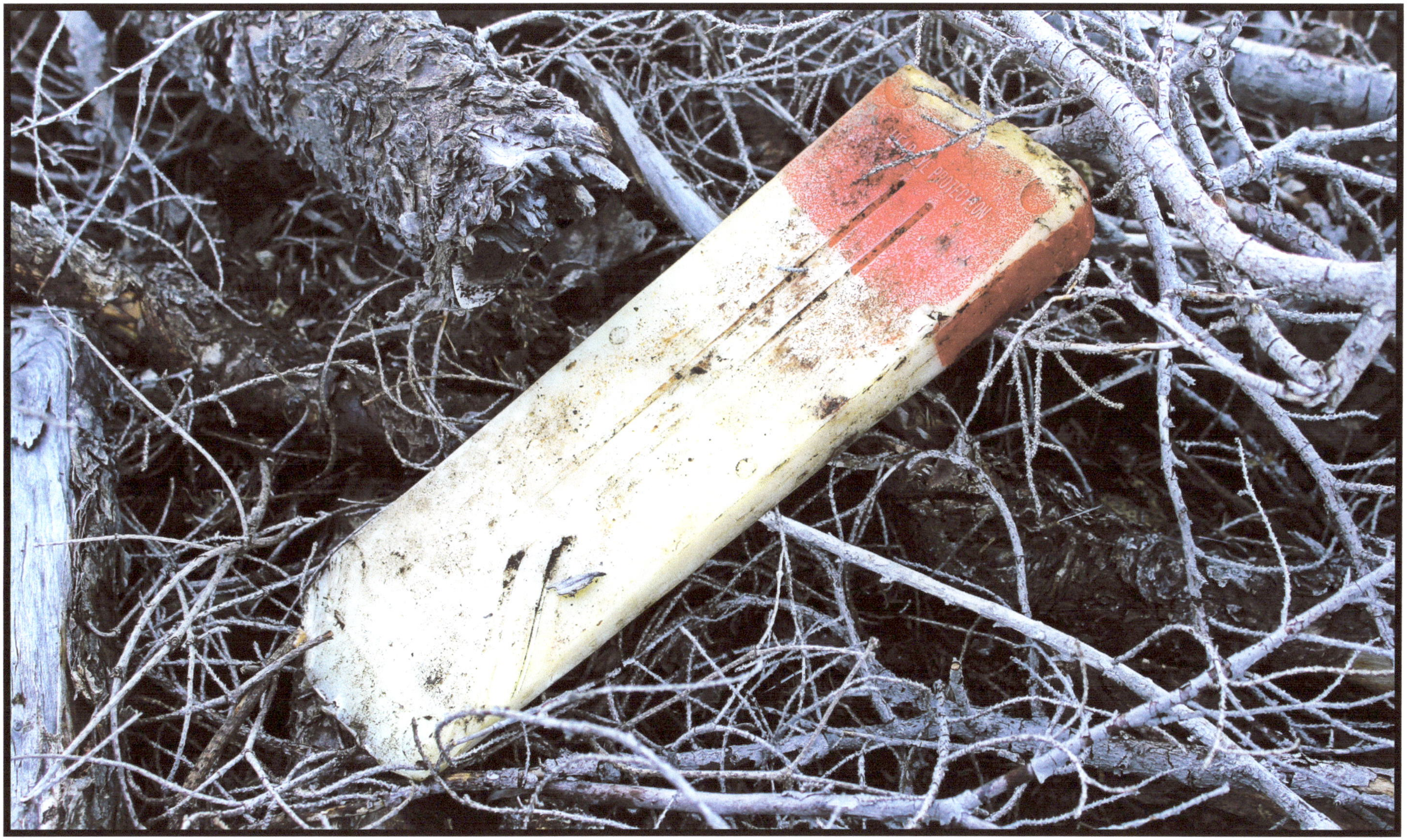

It's common to find fallers' wedges on the blocks. The fallers use these to help control the lean when they're falling a tree. Sometimes, fallers accidentally leave equipment behind. Occasionally, we even find chain saws that were abandoned on the block.

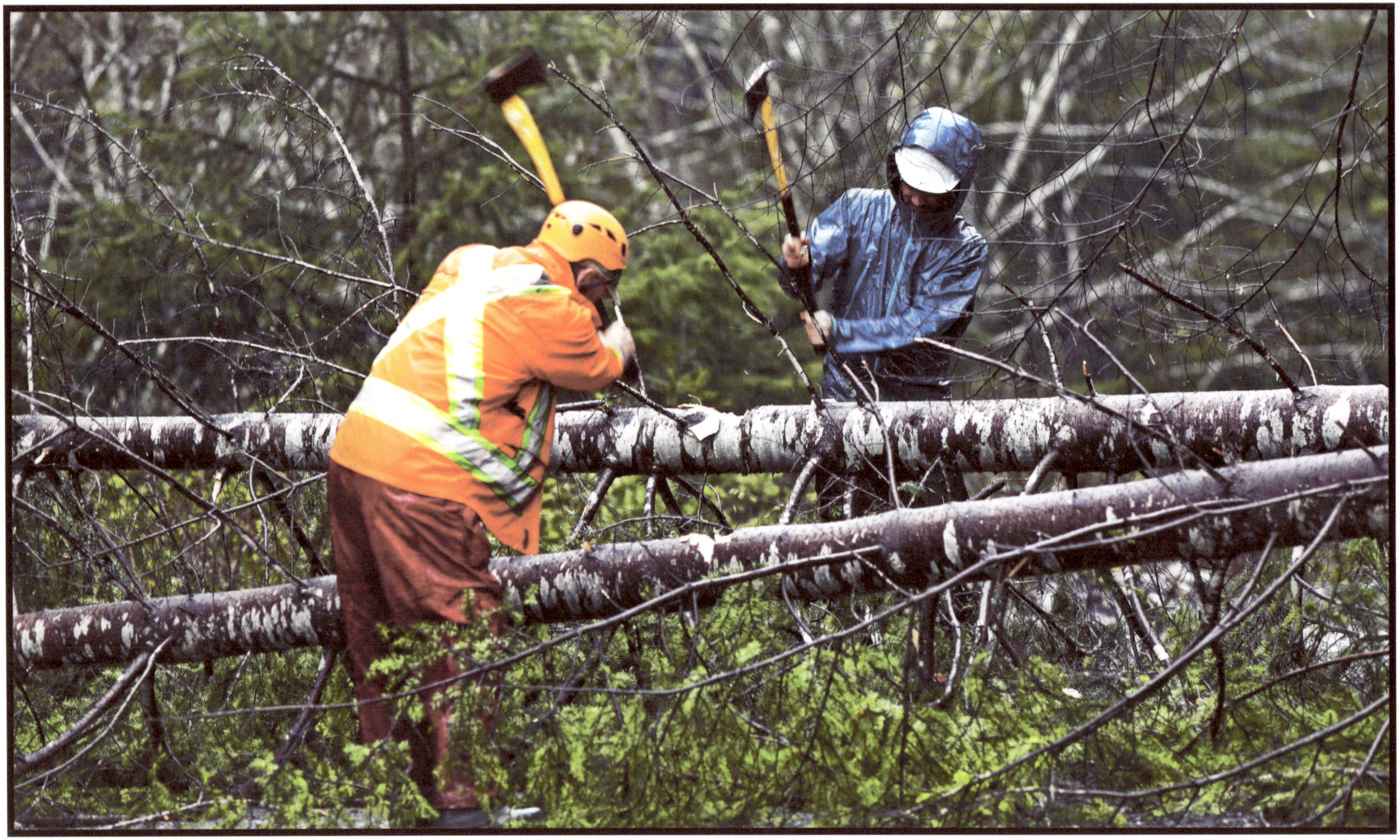

We discovered that these trees had blown down across the road while we were on the way home from work one windy day. It was the only day in a decade of working on the coast that we've had a mid-day shutdown due to strong winds. Considering that I had mature trees being blown down into the piece that I was working in, I wasn't too surprised by the shutdown. At one point, I was startled by something that hit the ground on the slope above me. It turned out to be an empty box that had blown out of the back of my truck, down a hundred meters of roadway, and then tumbled another hundred meters into the piece where I was working.

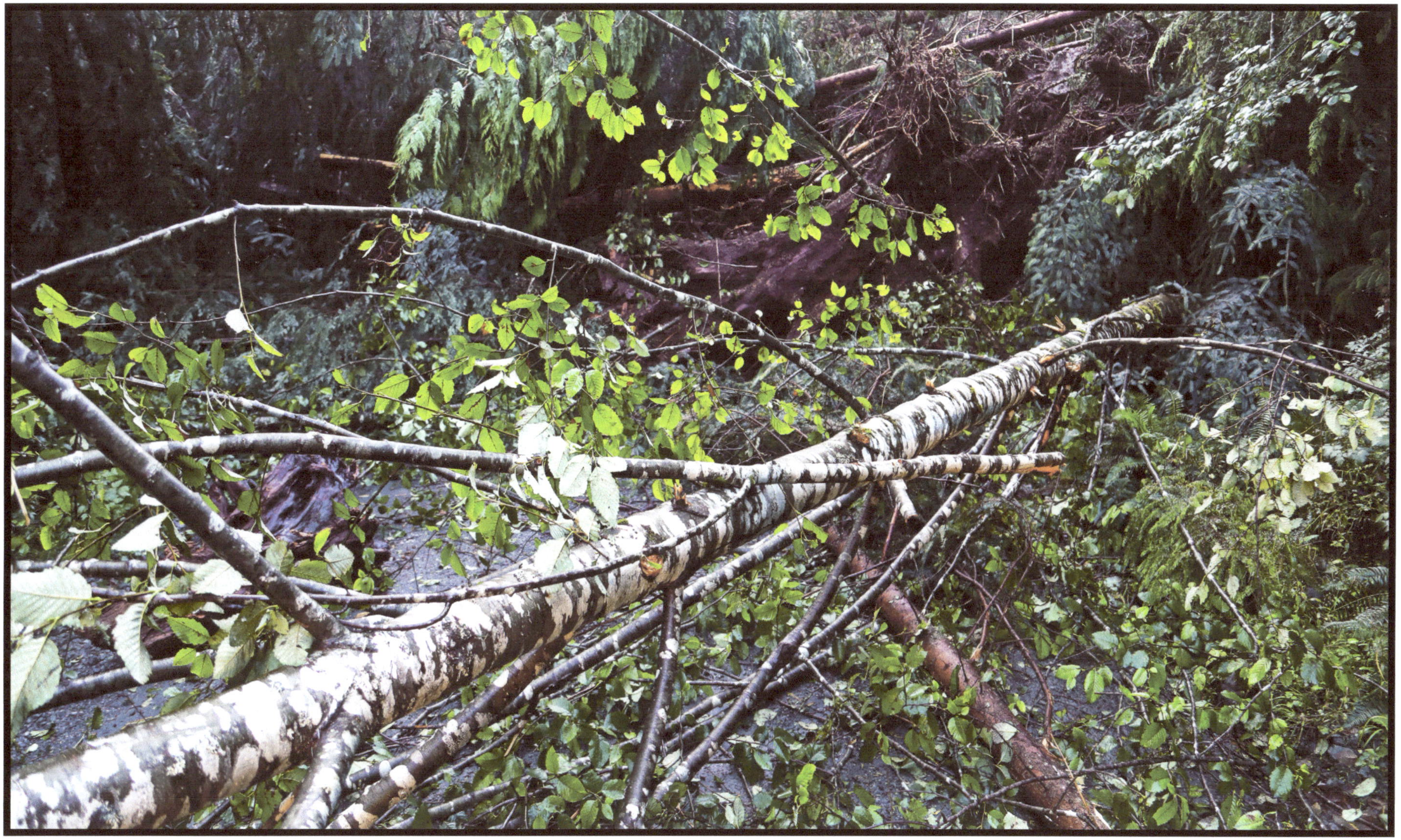

The day after the windstorm, we ran into this mudslide on the way to the block. The entire side of the mountain had collapsed. Needless to say, we had to divert to a different block that day. Mudslides are common on the north Island. It's typical that crews shut down for a day if the region gets more than 120 millimeters of rain in 24 hours. If you're wondering how much rain that is, I believe the phrase is "Biblical floods." Working in 50-100 millimeters of daily rain is completely normal. Remember that Vancouver Island is part of a temperate rain forest.

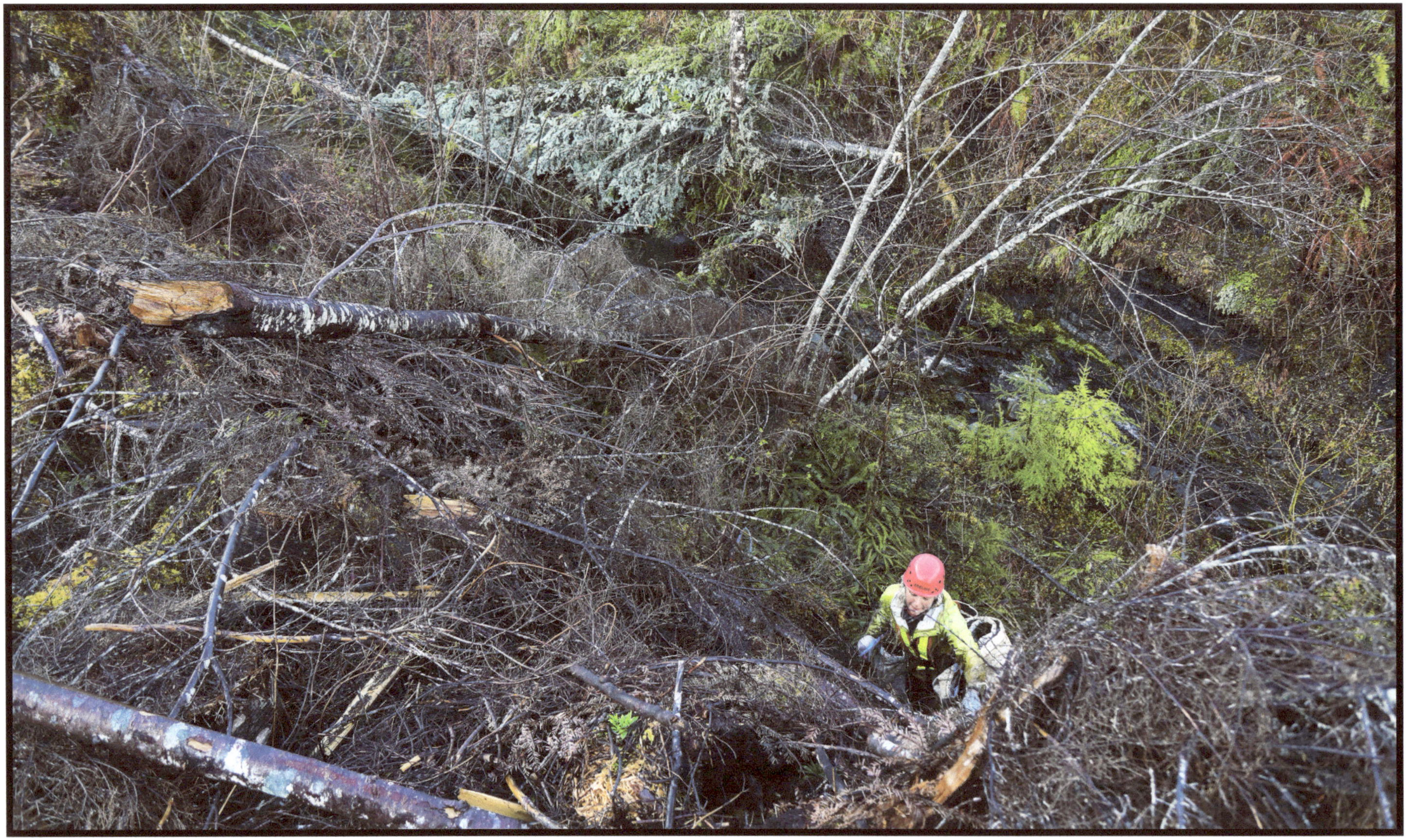

This was not part of the mudslide. This was just a normal block.

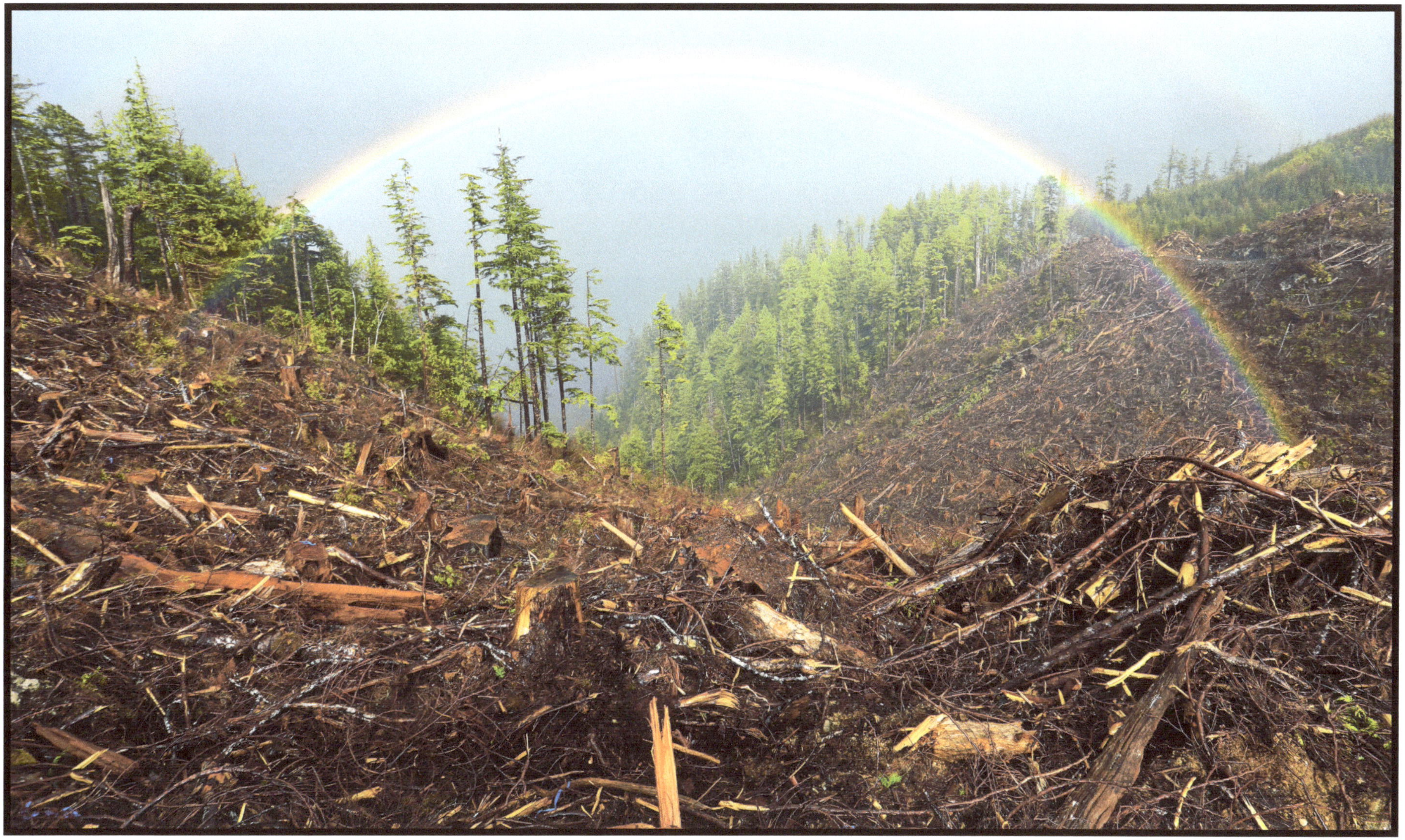

Another rainbow. If you look closely, there are actually two rainbows.

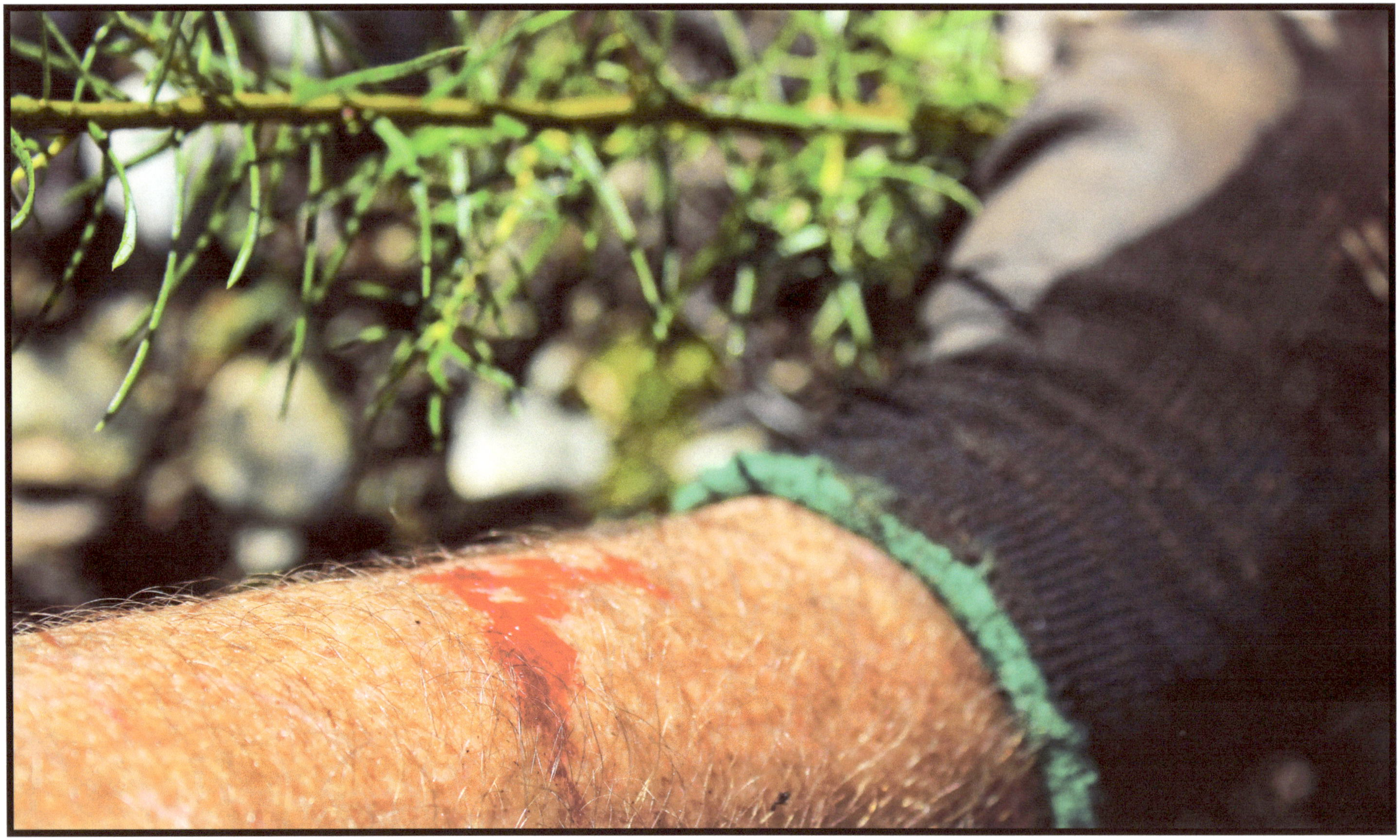

If you don't bleed twice a day while you're planting, you're doing better than I am.

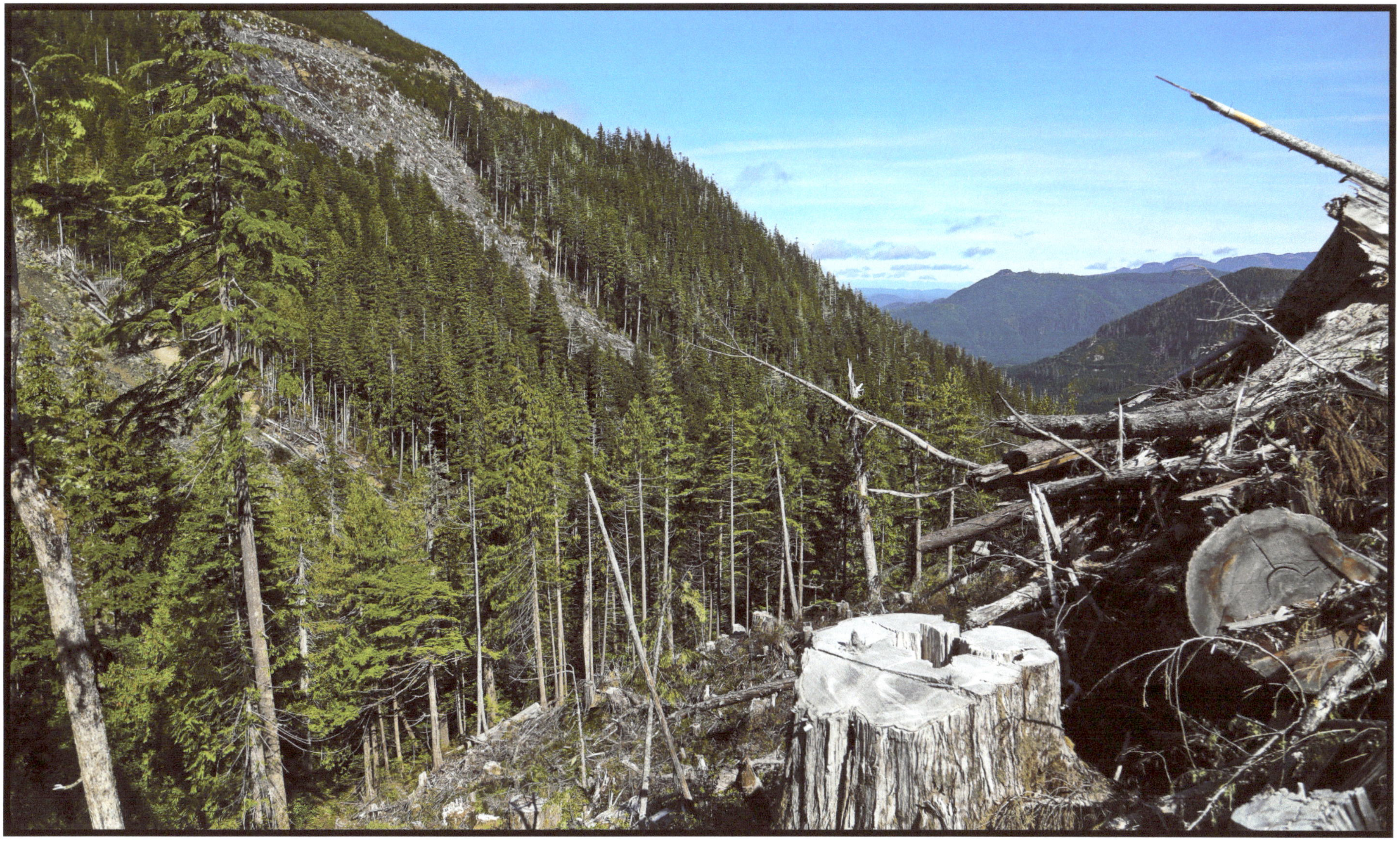

The block across the valley appears to be steeper than the one that I was working on.

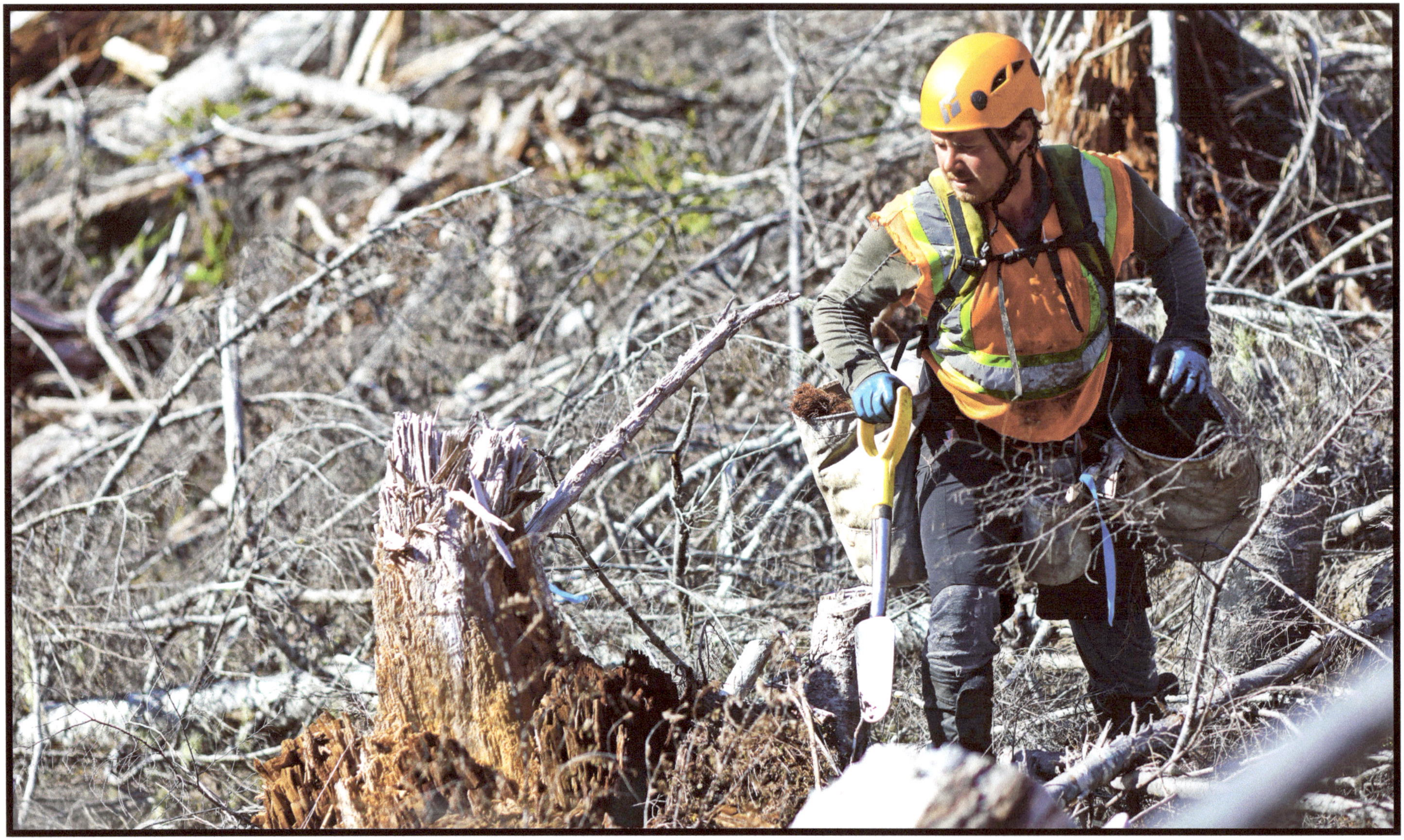

Alex is from the Yukon, but took a Political Science degree from UBC. He's an example of the highly educated workforce found on many coastal projects. Alex loves being in the bush. He plants for a significant portion of the year, then spends the winter skiing.

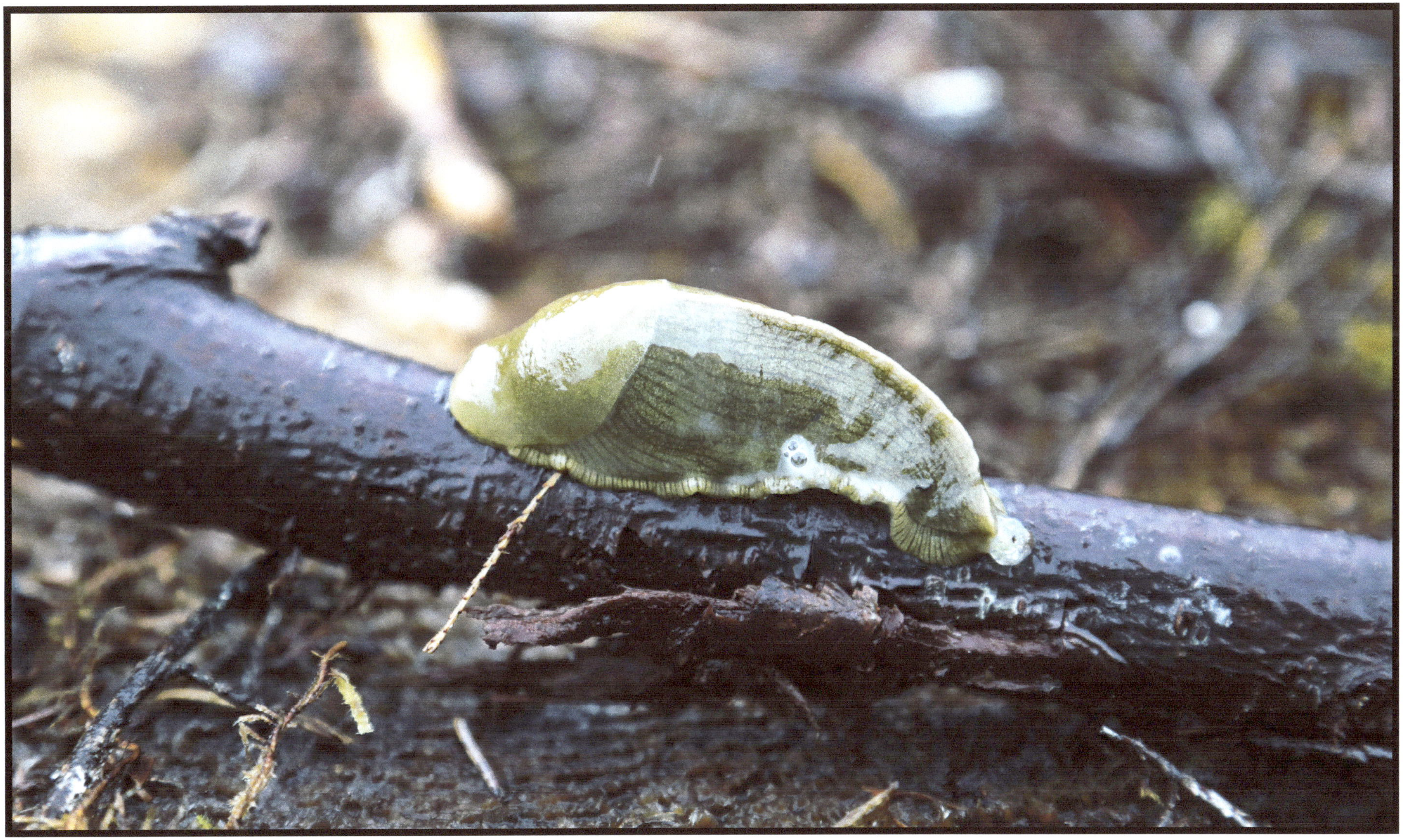

This is a small banana slug. Some banana slugs grow up to ten inches in length, and can live for as long as seven years. Some slugs are covered with mottled brown spots, making them look like an overripe banana. Banana slugs have a single lung, with an opening on the right side of their head that allows them to breathe. Banana slugs are also found outside of BC, down the west coast of the US. In fact, they're the school mascot of the University of California at Santa Cruz, as any fan of Pulp Fiction is well aware.

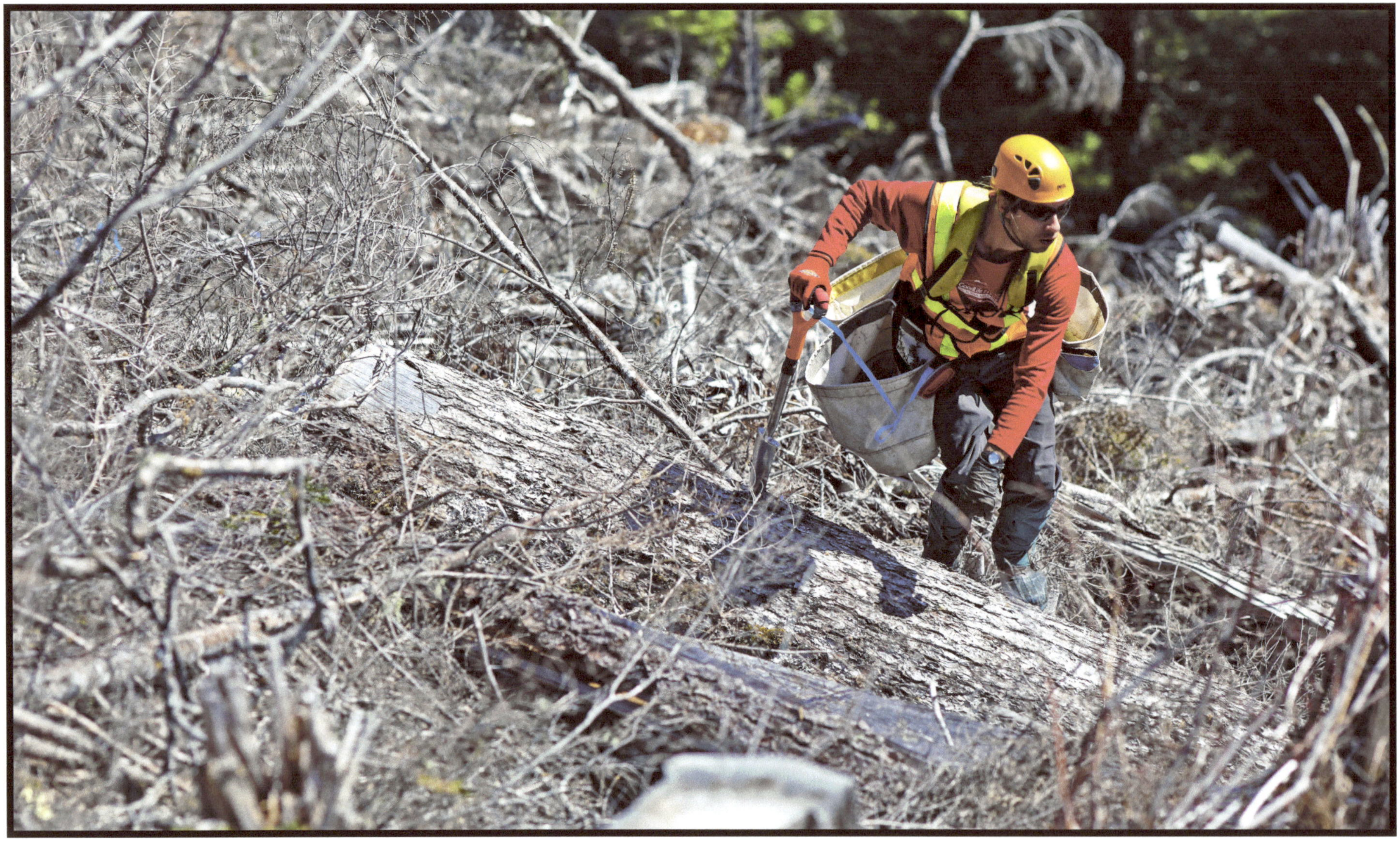

Marty has a Masters degree in Economics, and is currently completing a civil engineering program. This slash doesn't bother Marty.

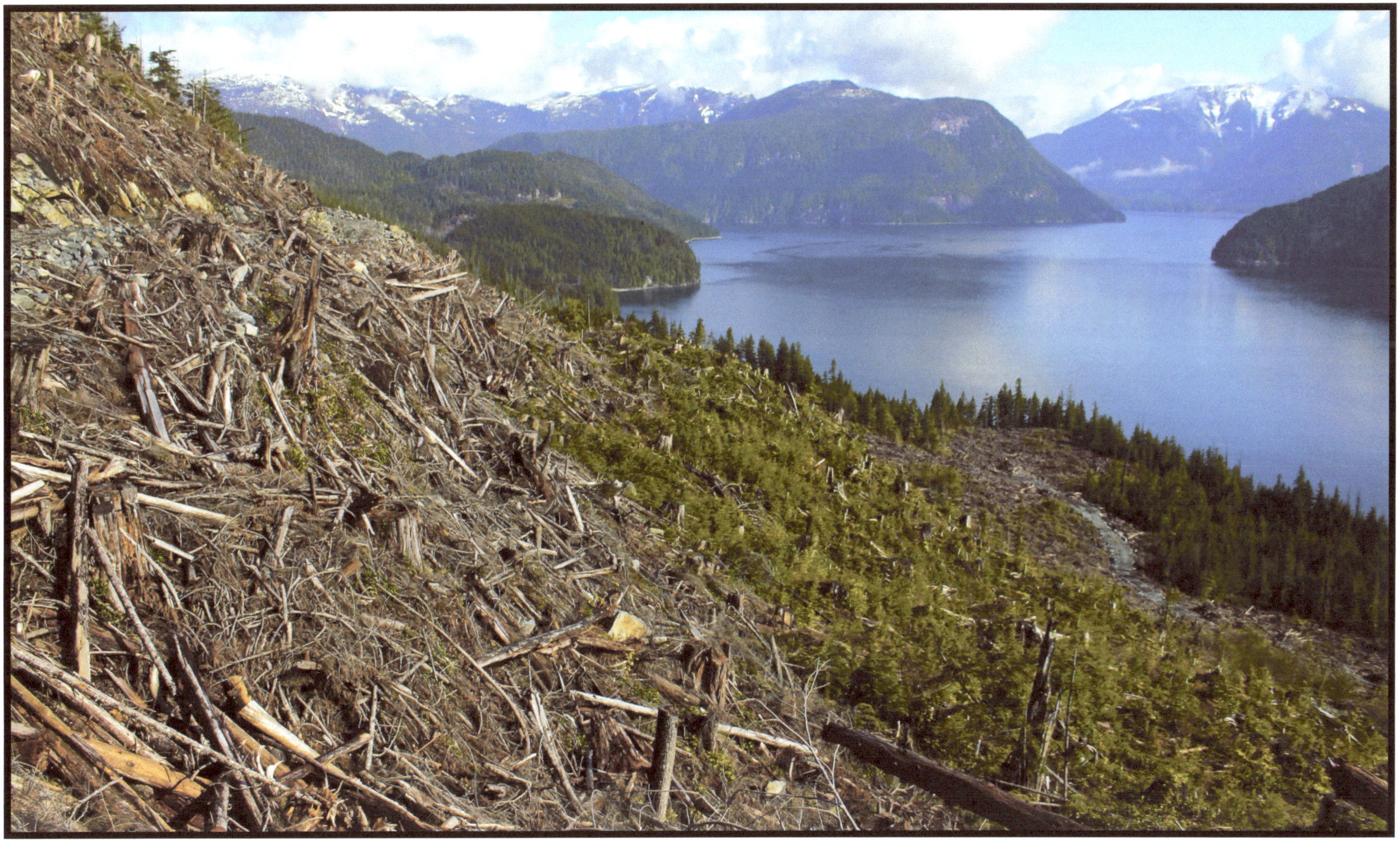

A scenic view from one of our blocks.

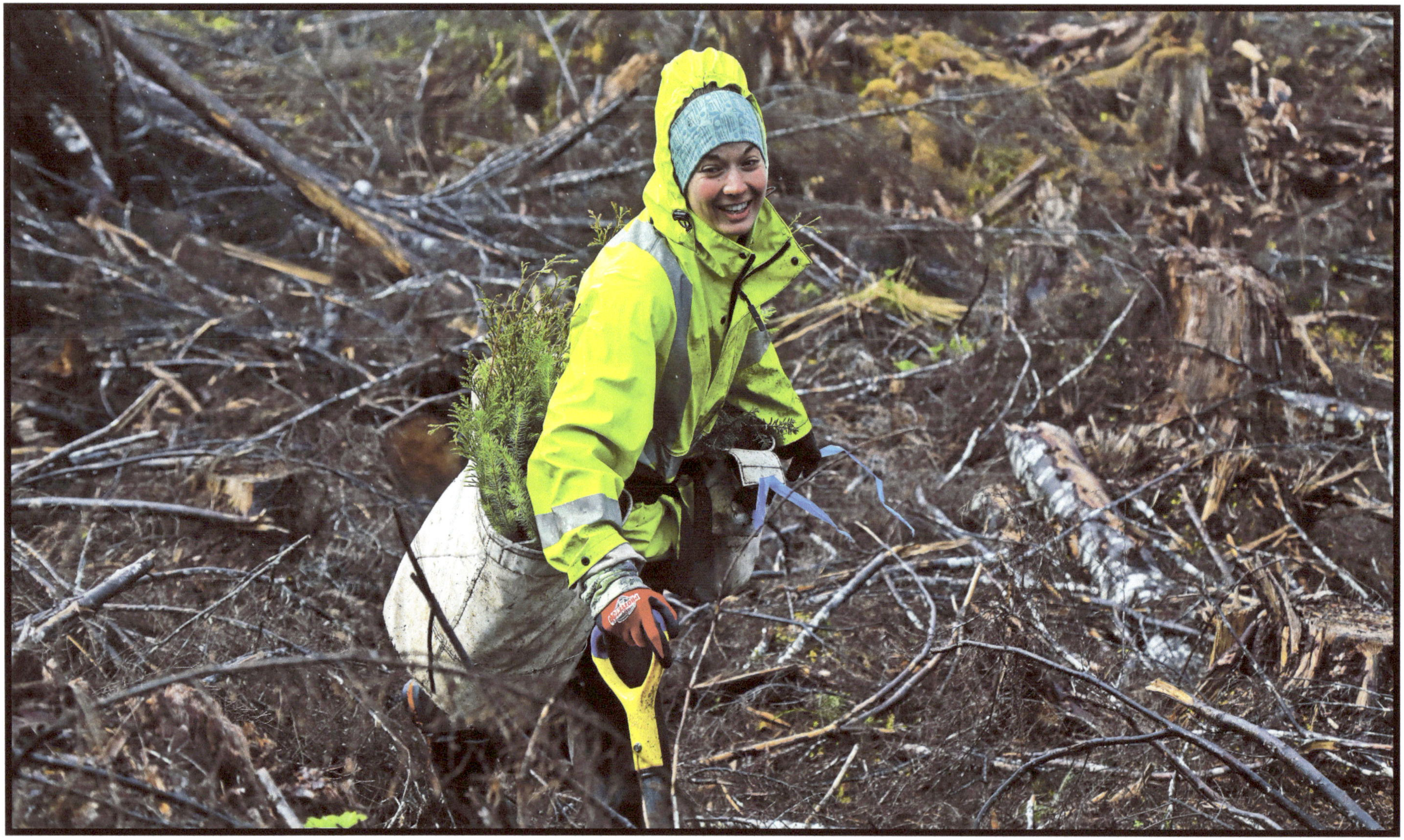

Somebody is spending more time laughing than planting.

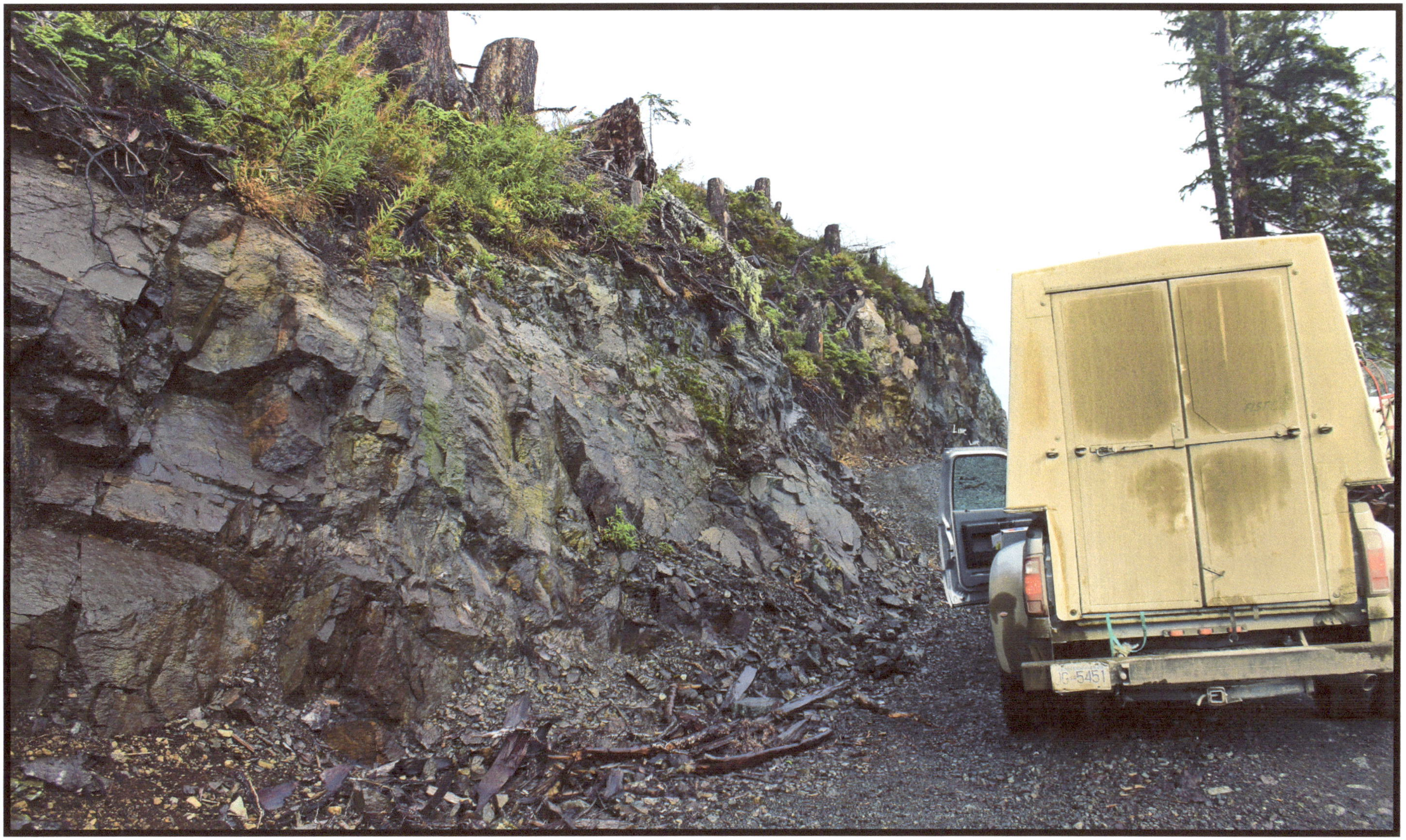

On some blocks, the hardest part is getting up into the piece.

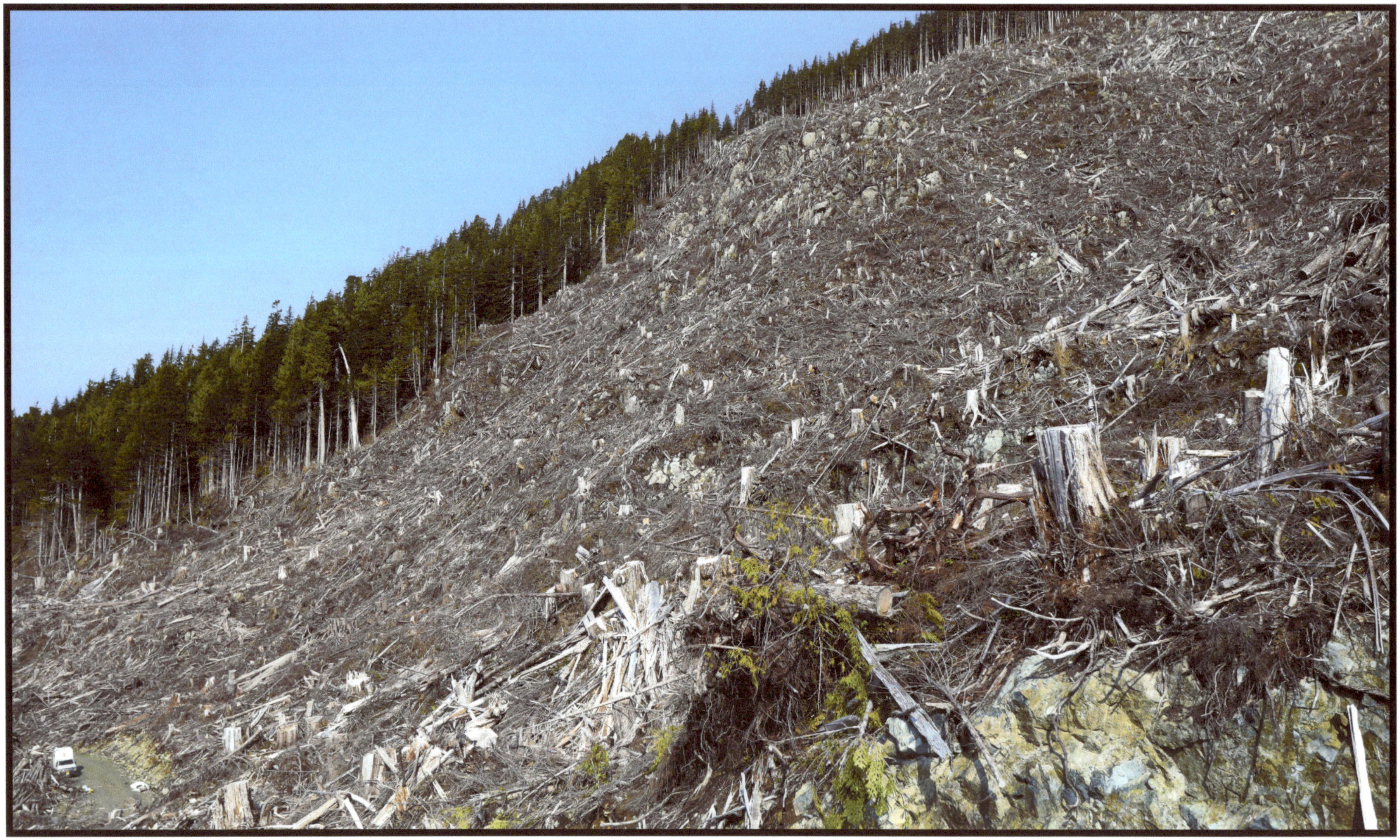

Find the planter. This is a tough one.

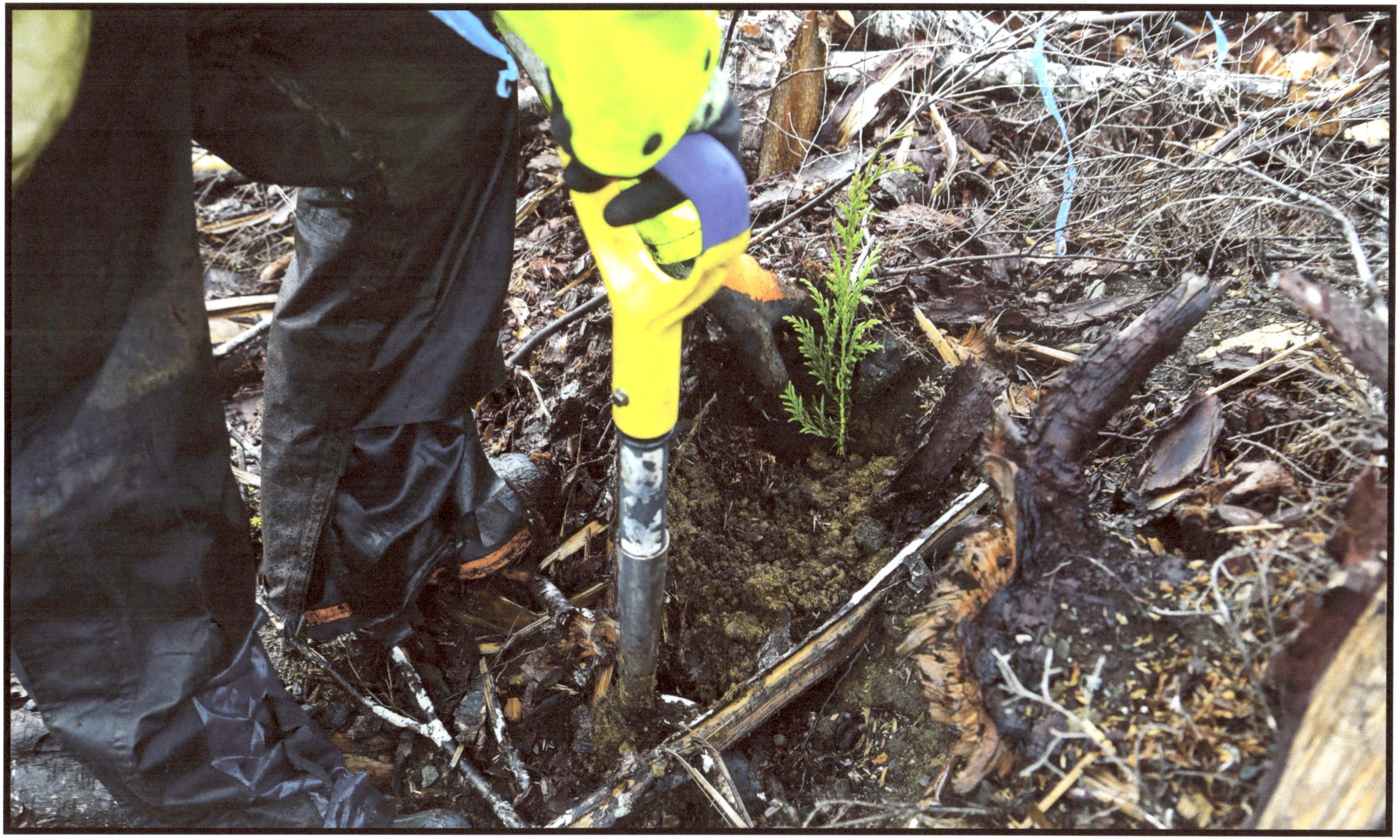

Planting a red cedar seedling.

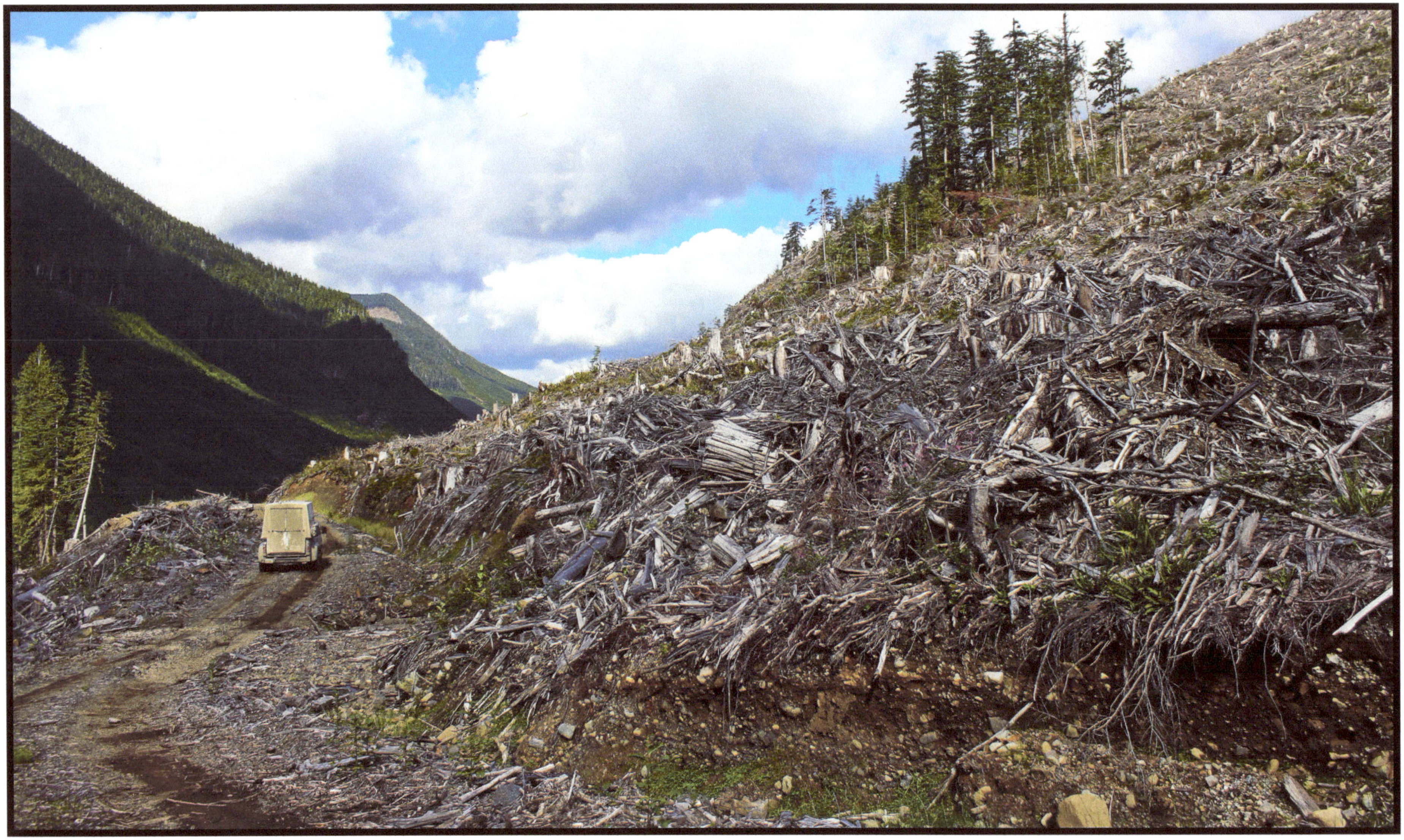

Heading home at the end of the day.

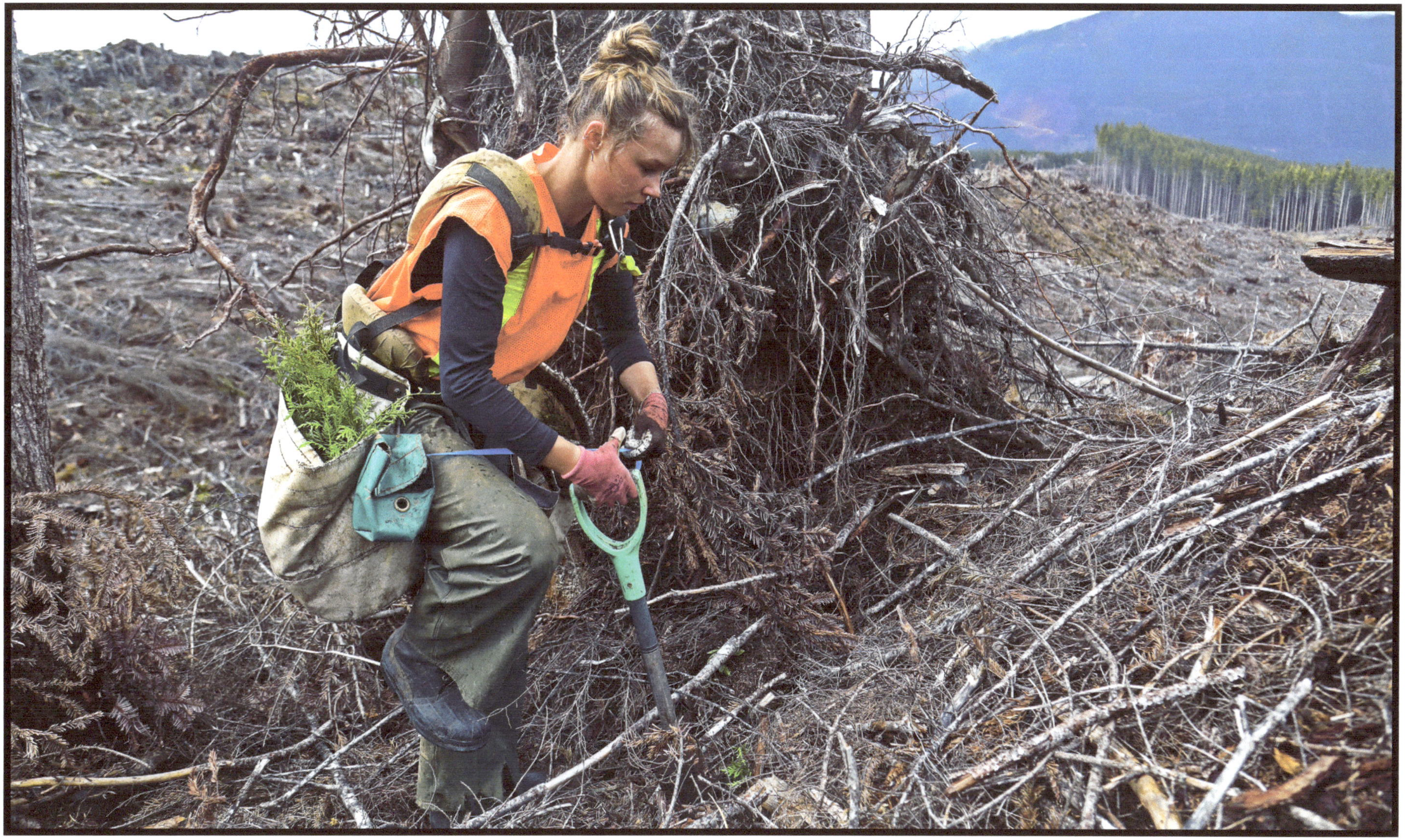

There's hardly any big slash on this block, just a few sticks.

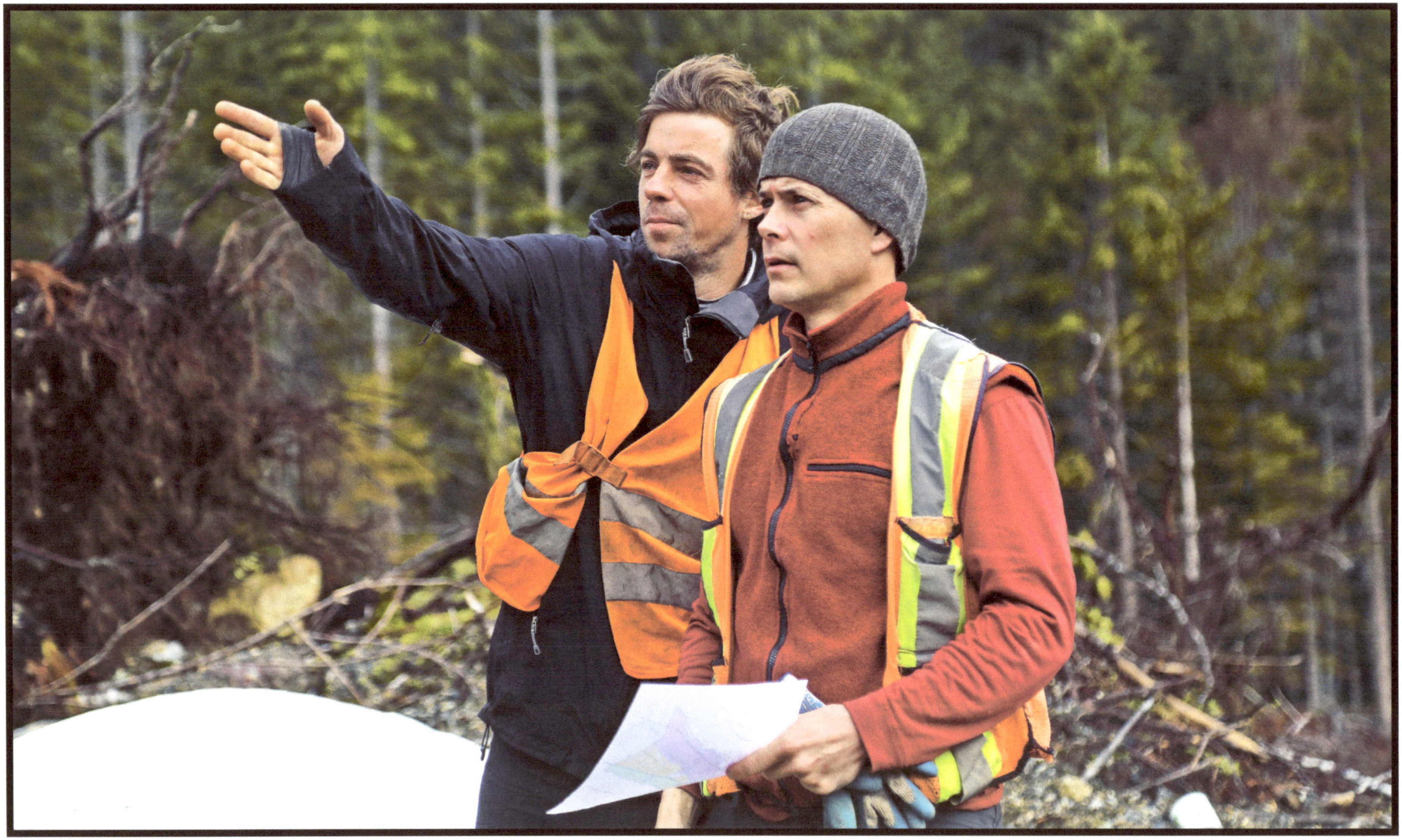

Lukas is confirming the boundaries of his piece with Rene.

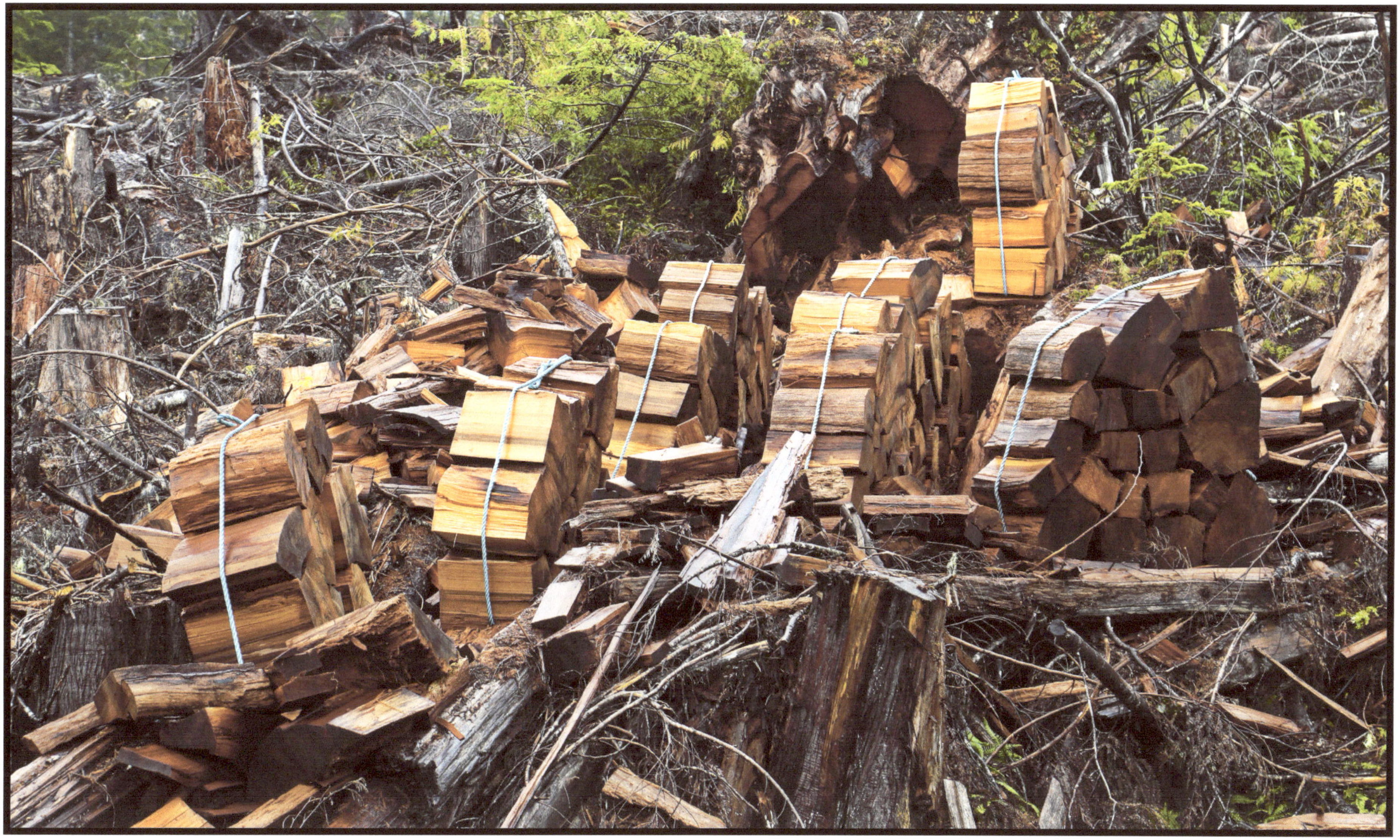

These are pieces of cedar, cut and prepared by the shake makers. Shake makers are people who go through the blocks to look for cedar pieces that they can salvage. They cut scrap wood into pieces that are then tied up into bundles. Eventually, they'll bring a helicopter into the block to pick up the bundles and move them out onto a road.

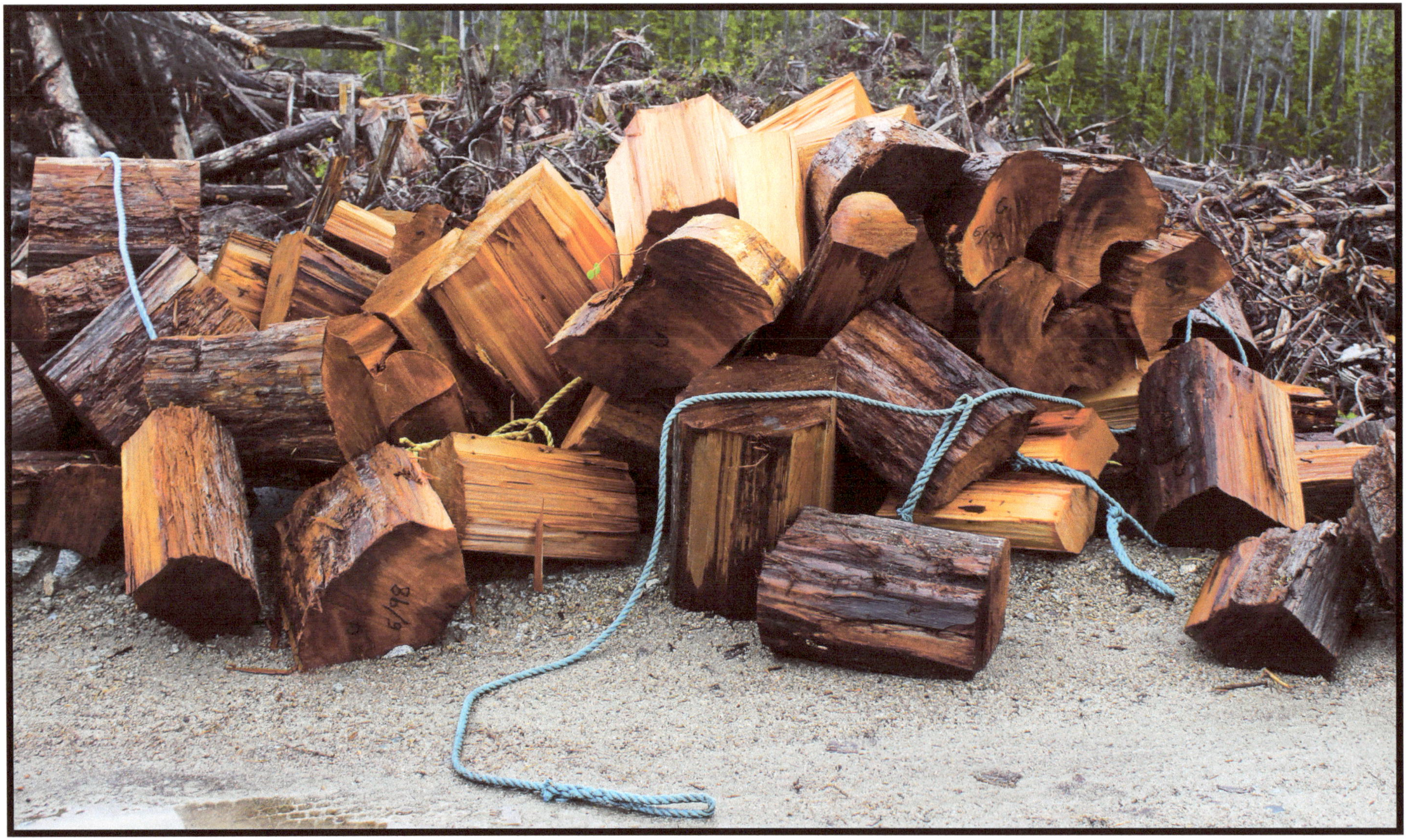

This pile of cedar pieces has been moved to the side of a road, and is ready to be picked up by a truck.

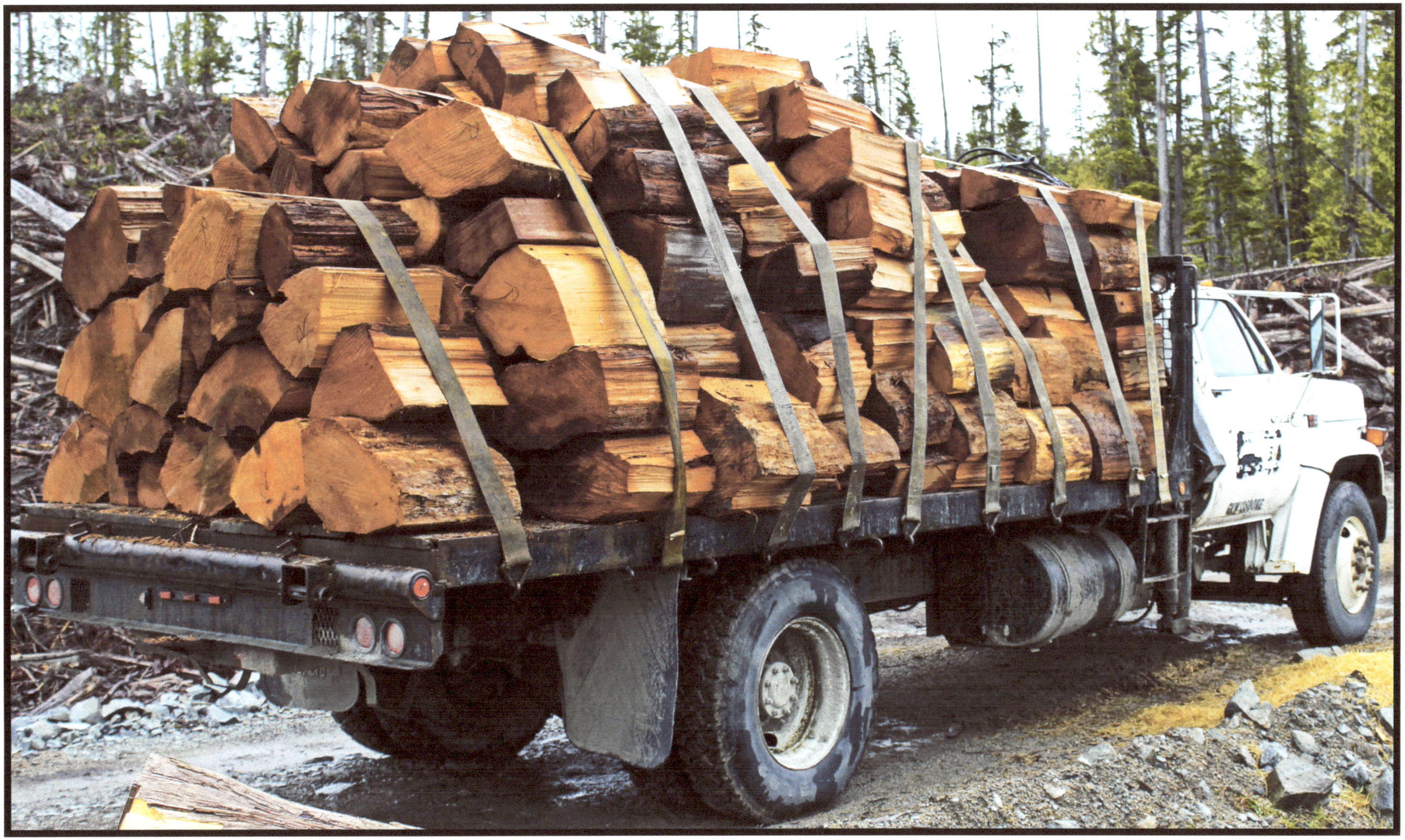

This truck will move the cedar pieces into town, where they'll be processed carefully and turned into shakes and shingles that are used on the exterior of houses.

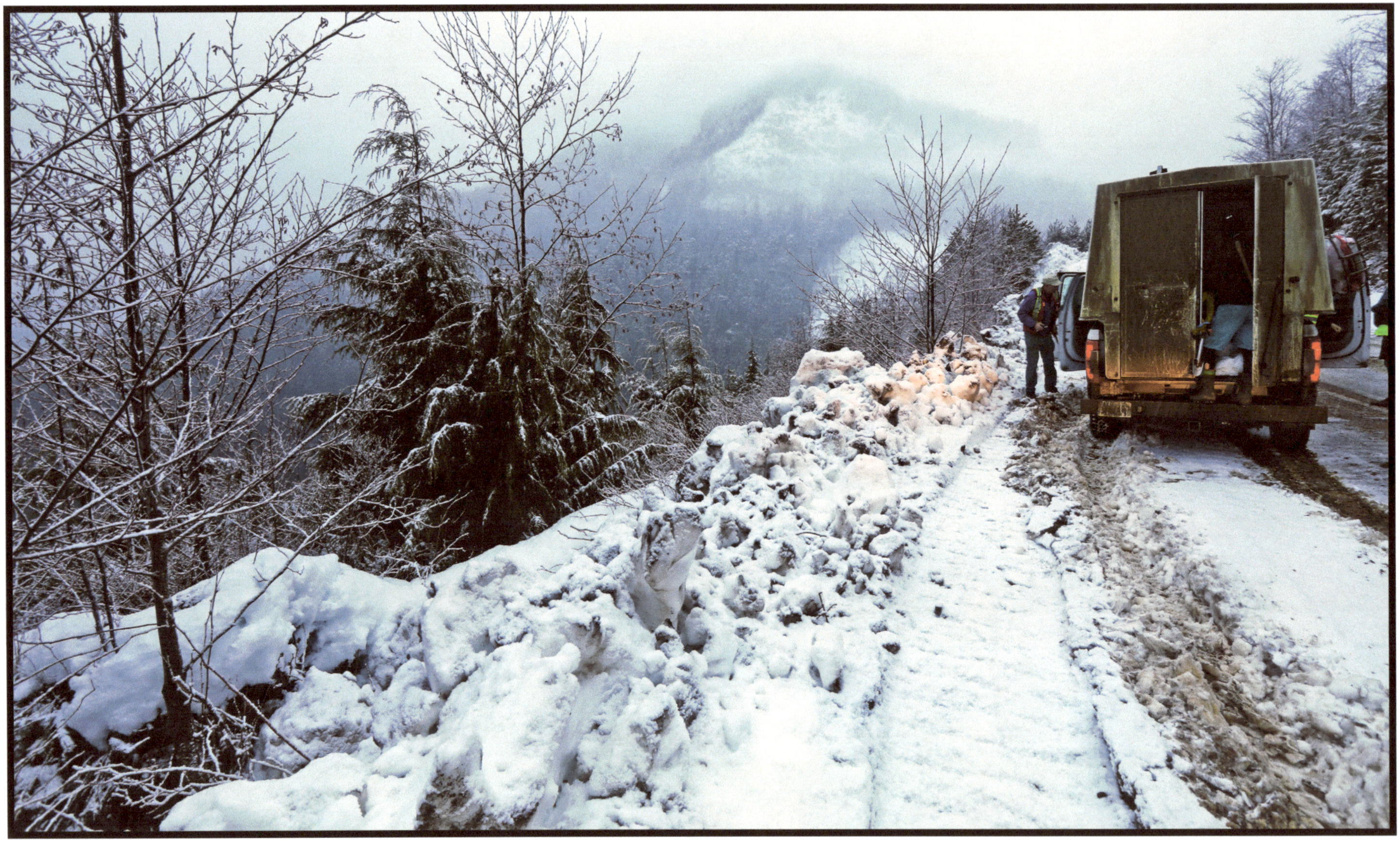

Snow on the Mahata Hump. Luckily, the blocks were at a lower elevation, and mostly free of snow.

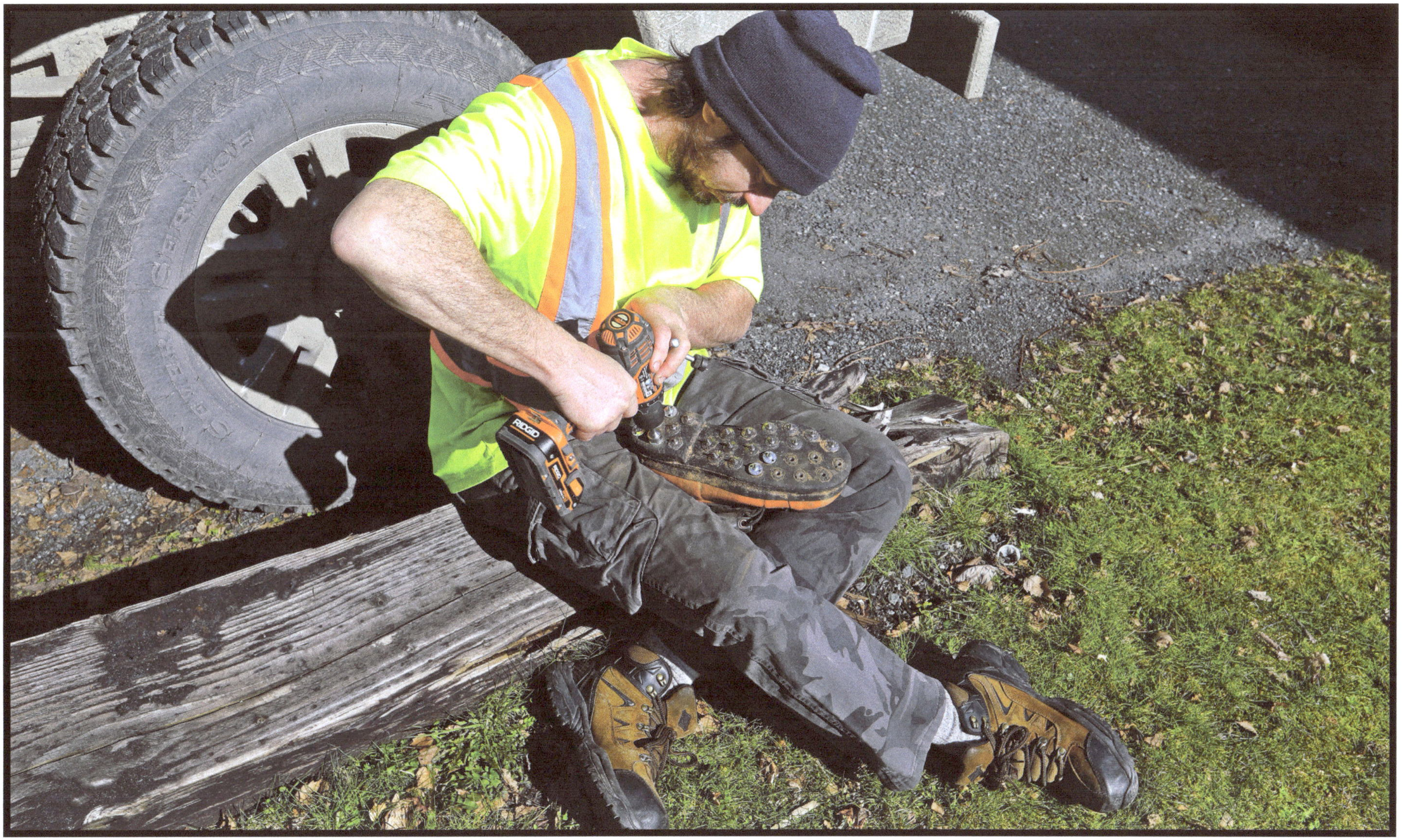

Changing out some of the caulks on my work boots.

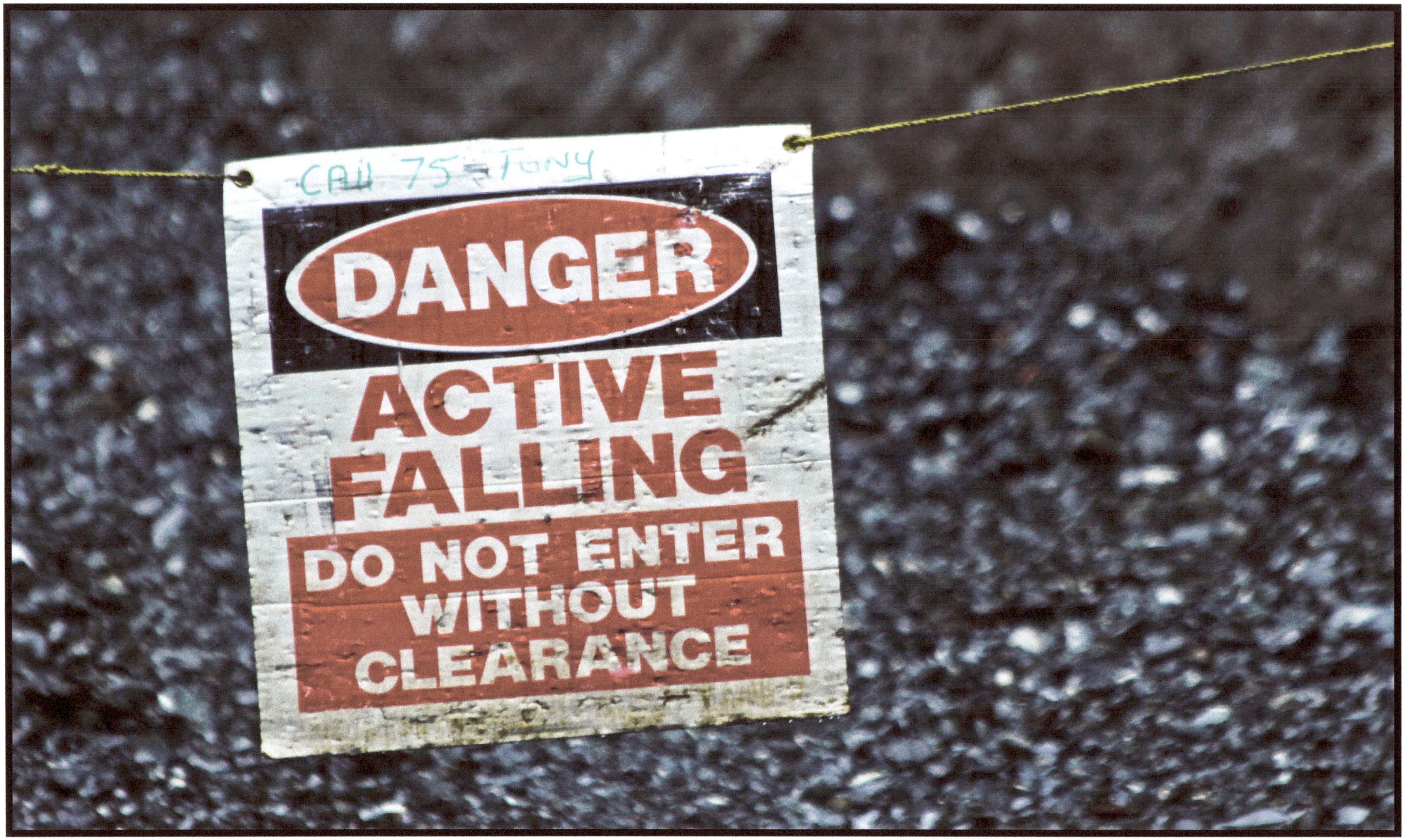

Coastal hand falling is such a dangerous activity that nobody else is allowed in an area where the hand fallers are working, to prevent distractions.

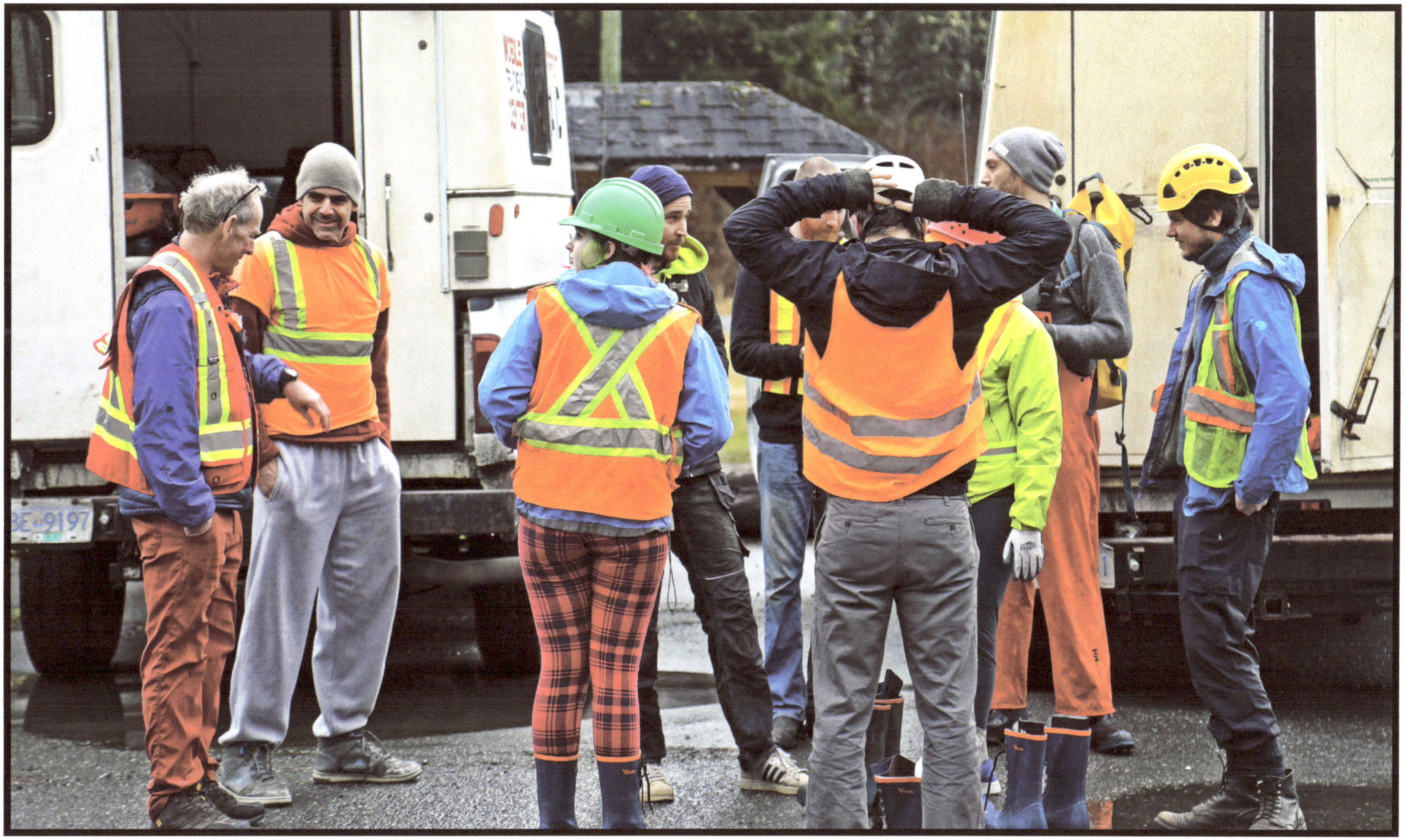

Part of the crew, waiting to head to work in the morning.

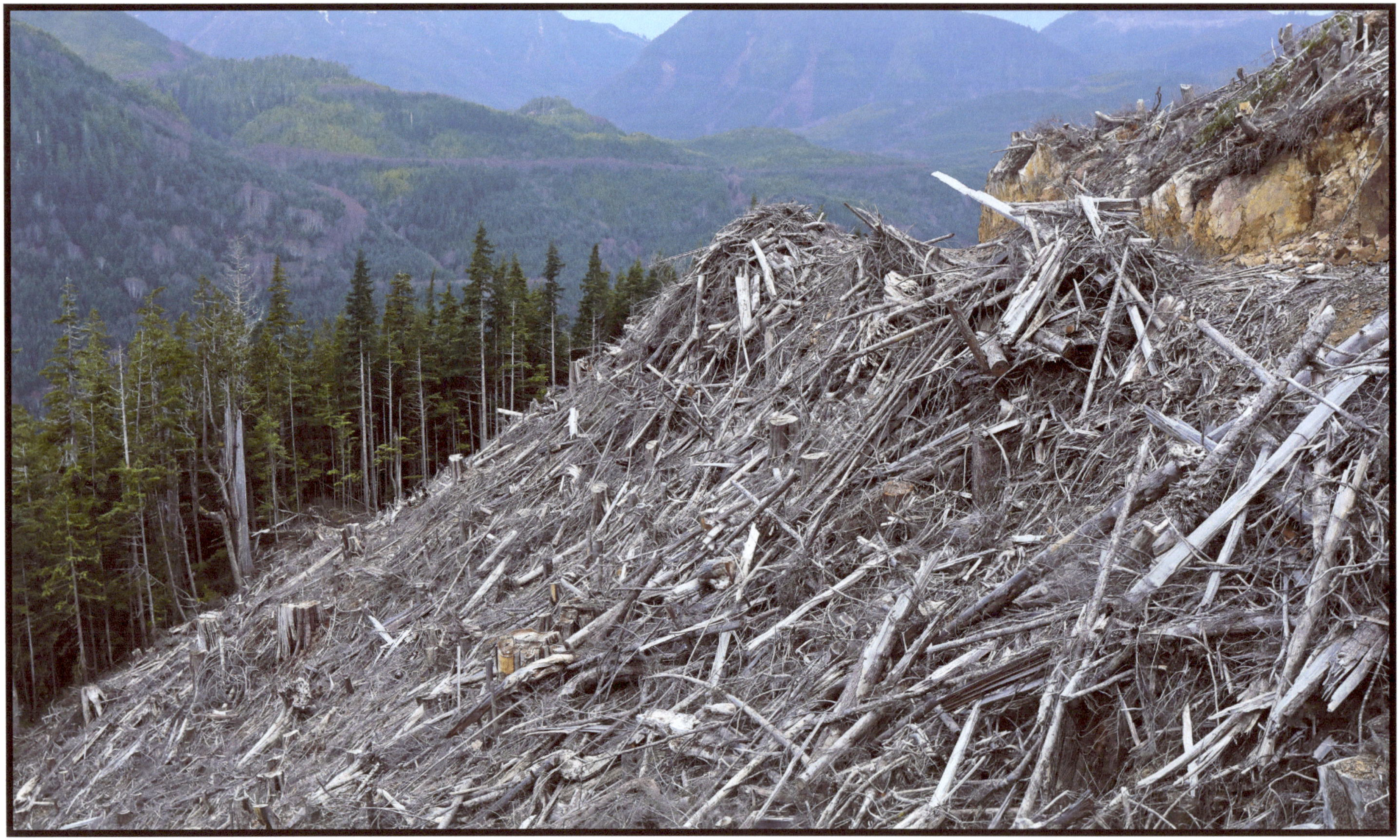

Another typical block.

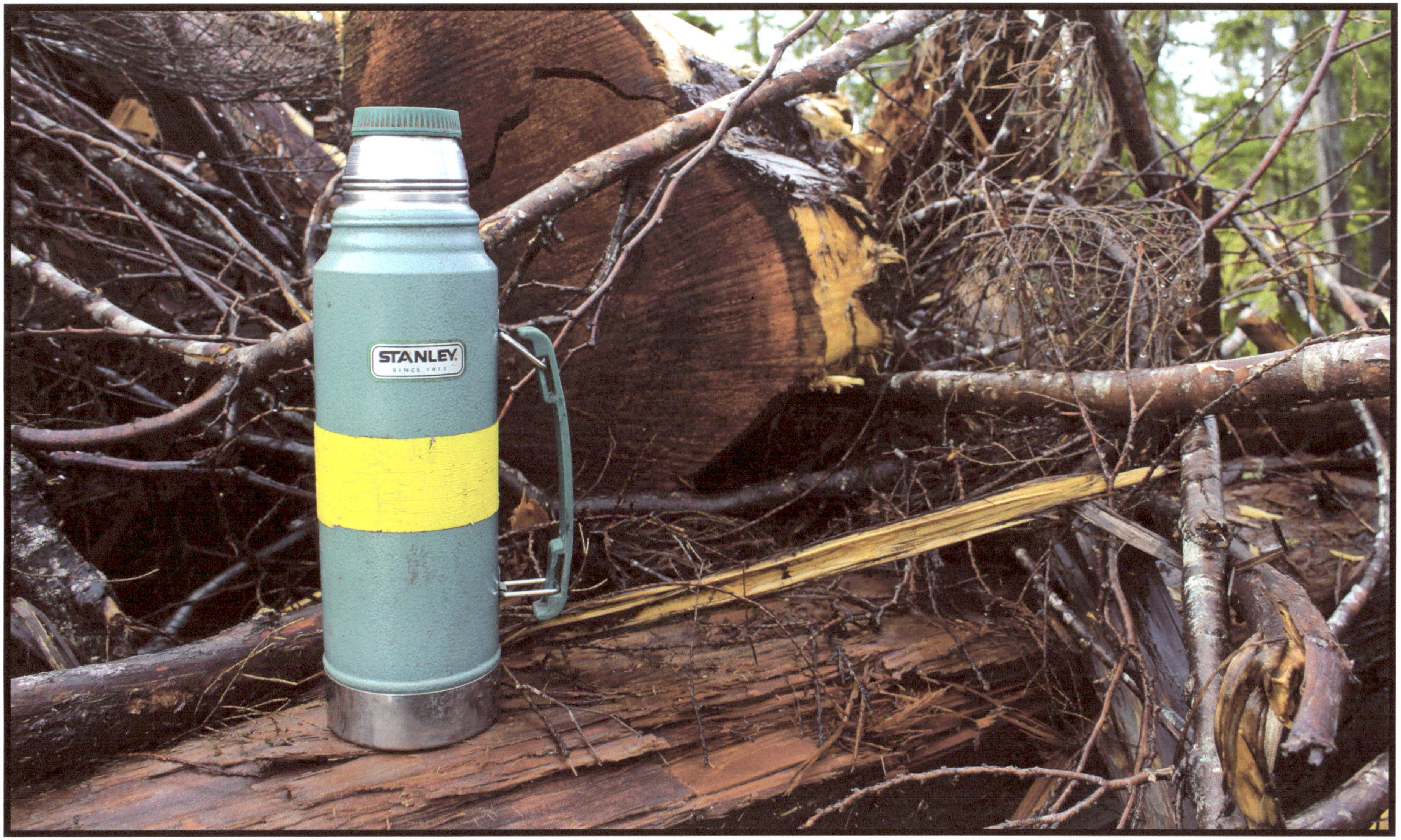

On a cold spring morning, nothing beats a thermos of hot coffee with chocolate.

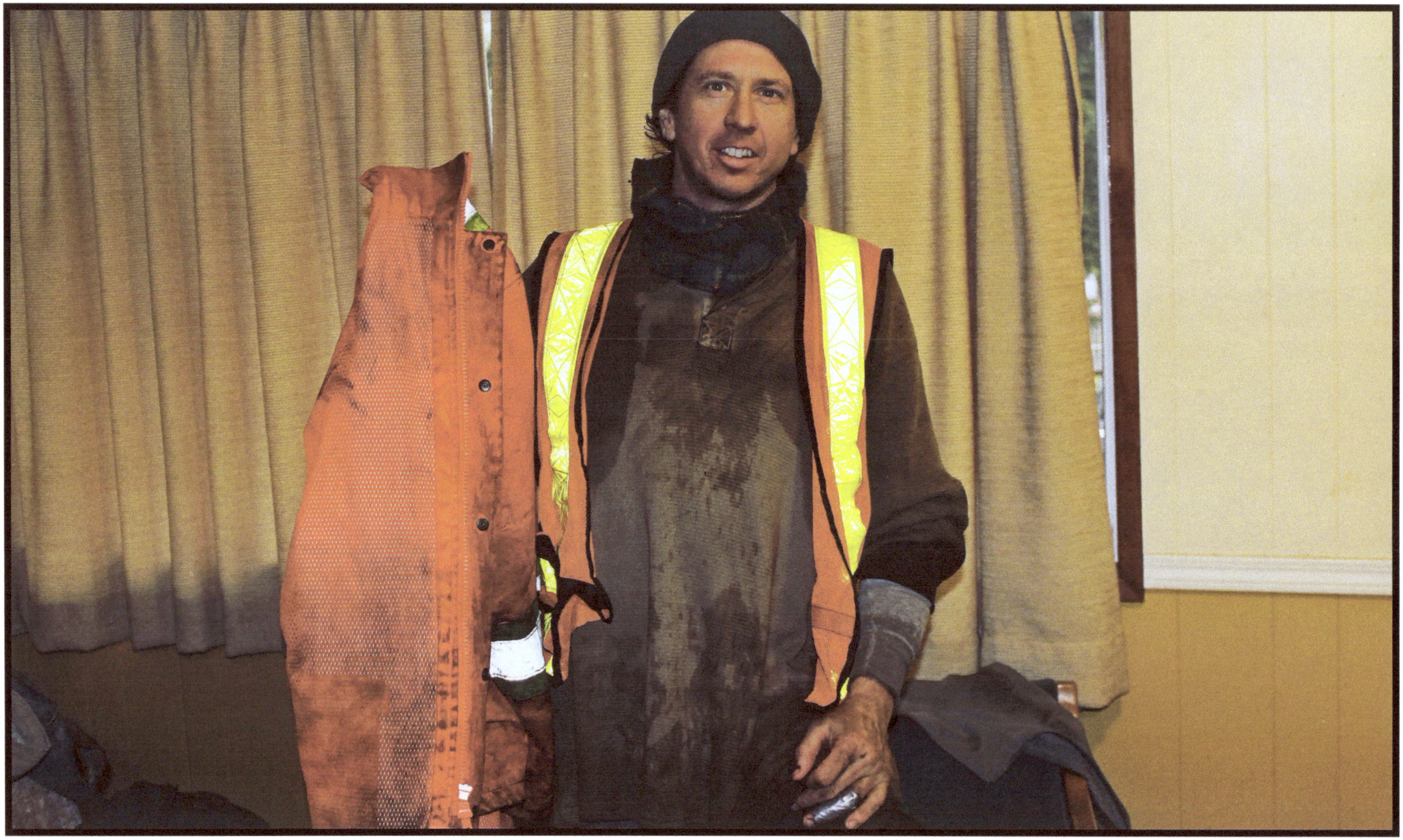

Although most planters don't use scarves, I find that a good wool scarf is extremely helpful for keeping me warm when I'm working in freezing rain.

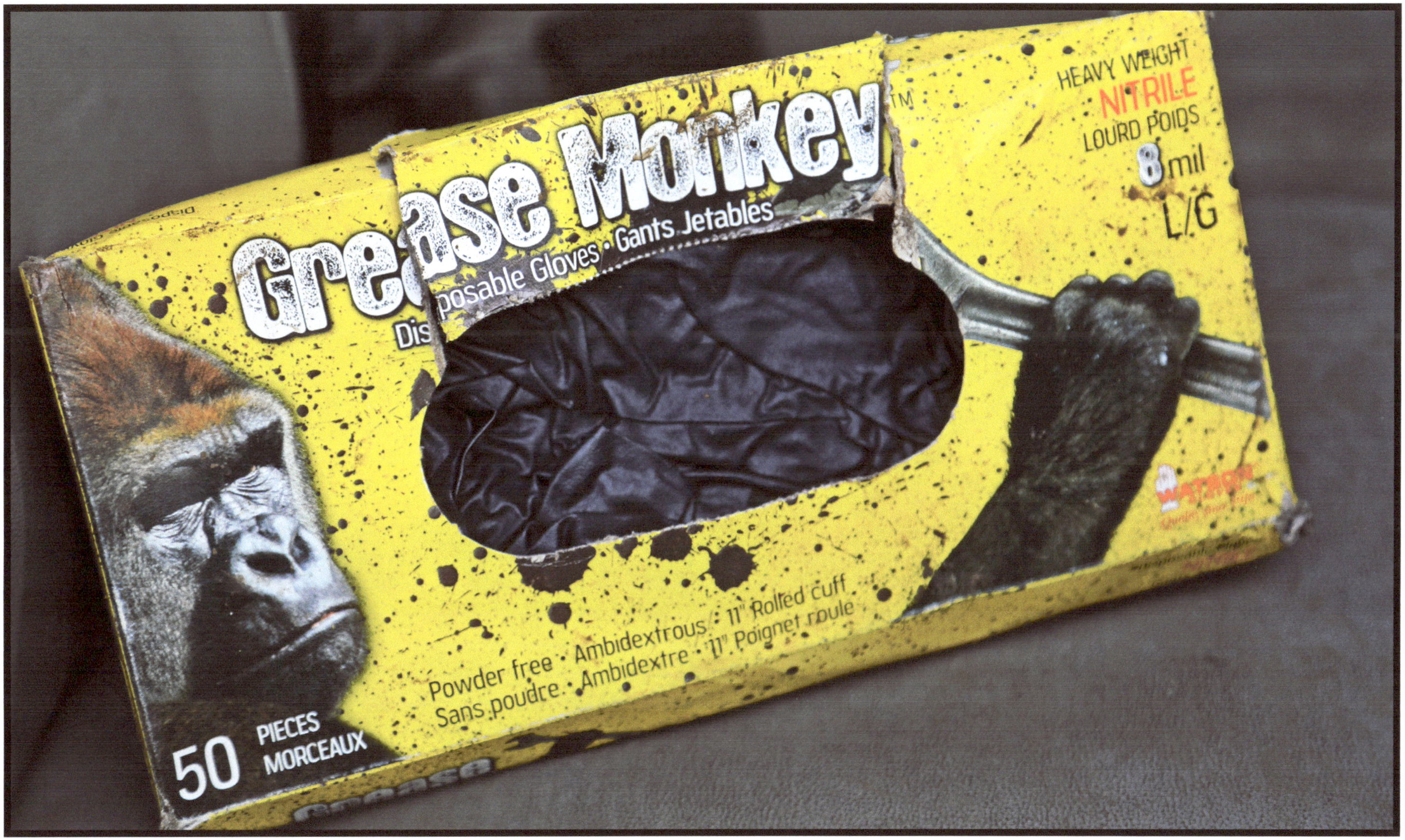

Nitrile gloves are rarely used at most Interior planting companies, but they're extremely common on the coast. When worn under a pair of thin work gloves, the waterproof nitrile gloves keep a planter's fingers much warmer. Wearing a nitrile inner glove under a light outer work glove is the only thing that keeps my fingers nimble enough to keep planting in wet ground that is hovering just slightly above the freezing mark. The best nitrile inner gloves have a thickness of 8mm, and can be found at any auto parts supplier.

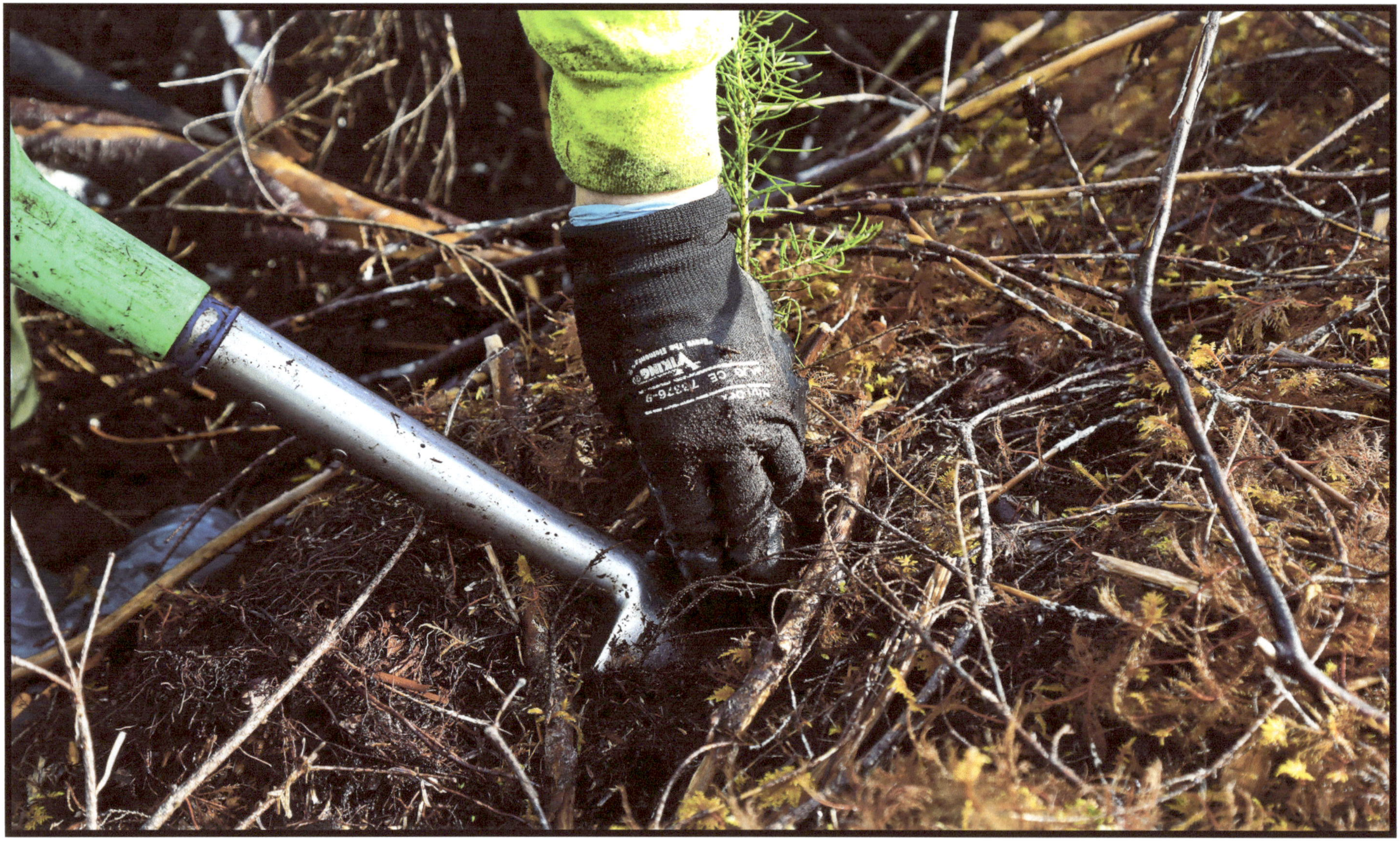

The "Nitridex" glove is a popular outer glove, which protects the nitrile inner glove from ripping.

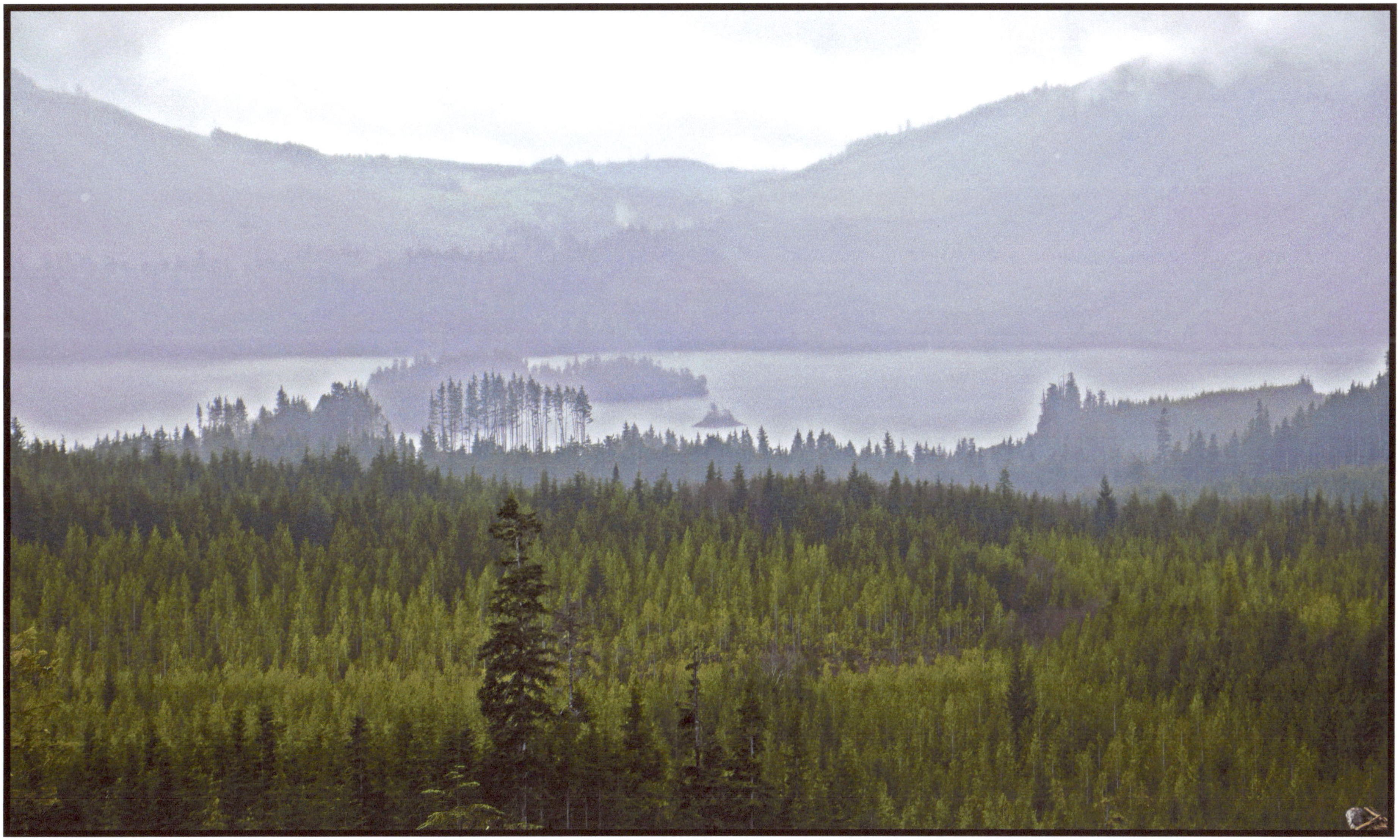

A typical misty morning.

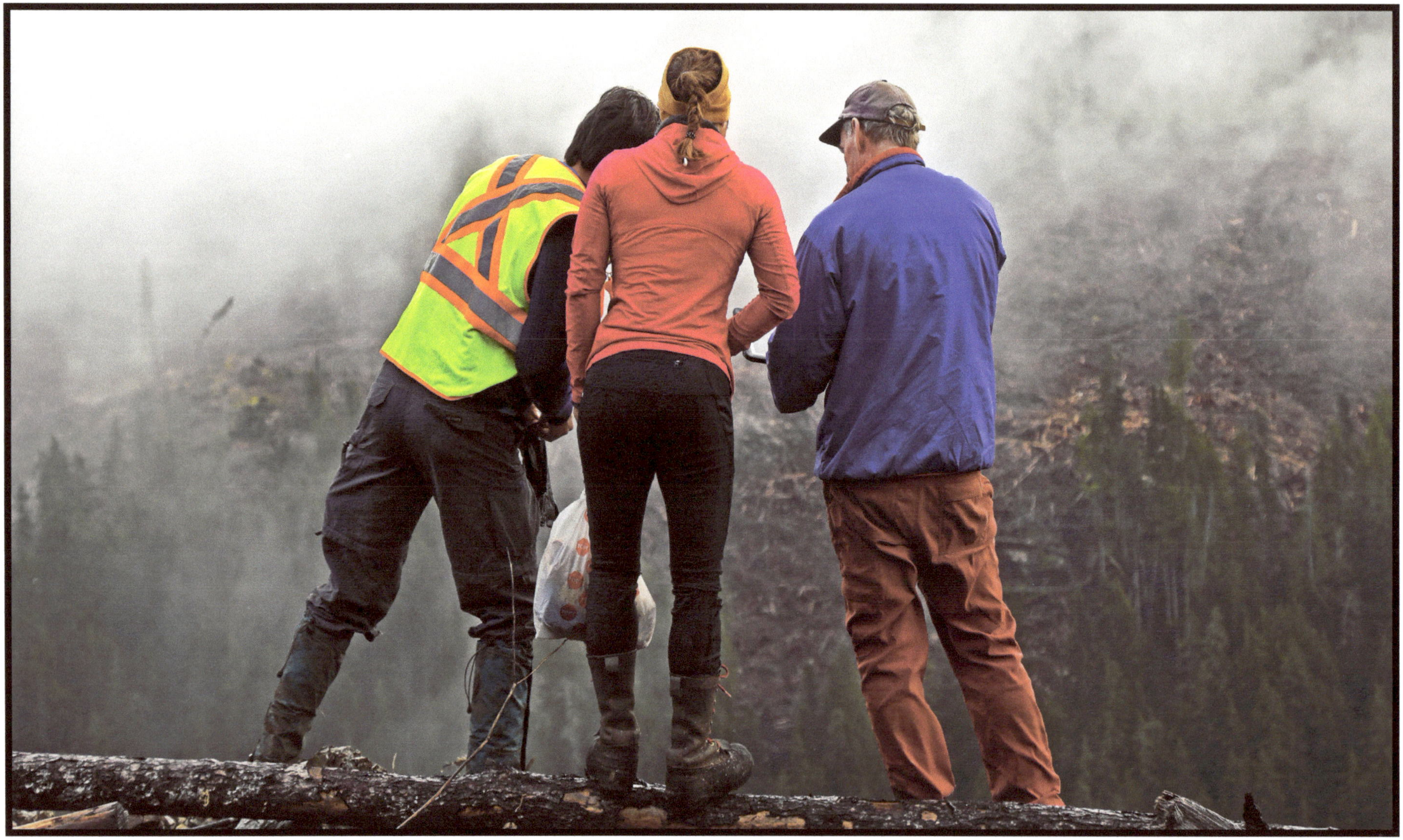

You know that your piece is steep when you stand at the edge and you can't see the bottom.

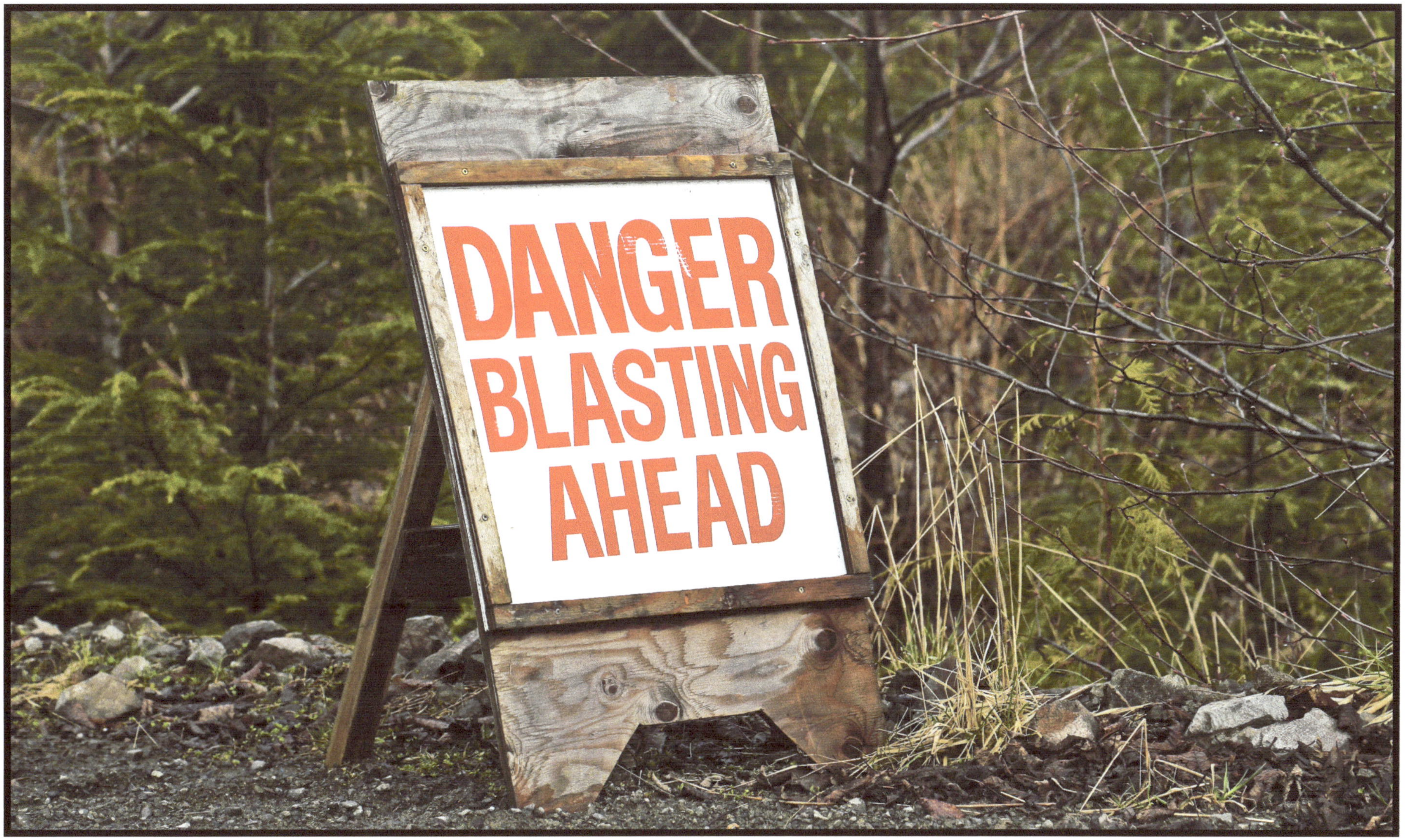

Trying to build roads into coastal blocks is a major engineering challenge. In some areas, logging companies spend hundreds of thousands of dollars per kilometer, using large quantities of dynamite and other explosives. Planters may frequently hear blasting during the day.

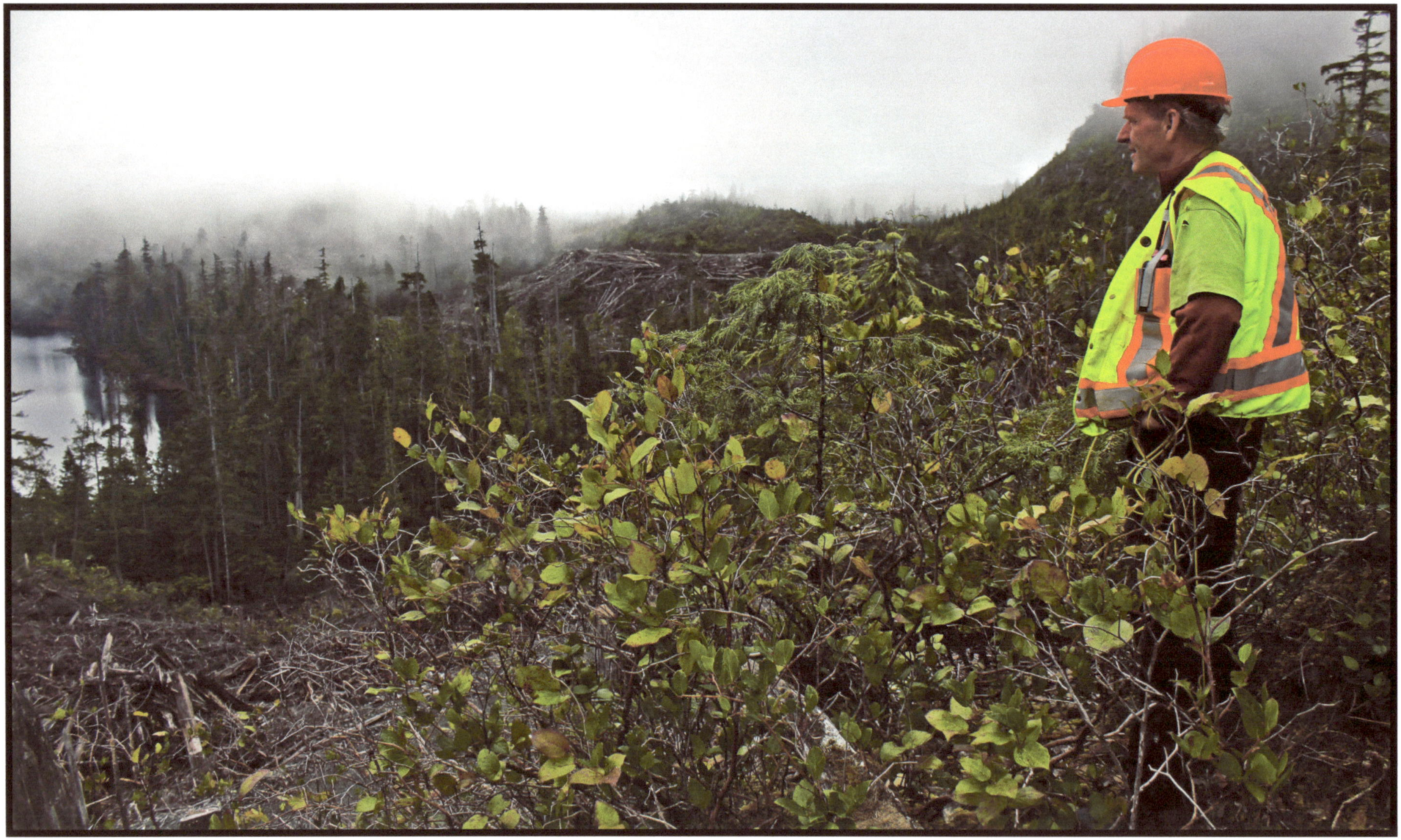

Foreman Mark, looking out over a block.

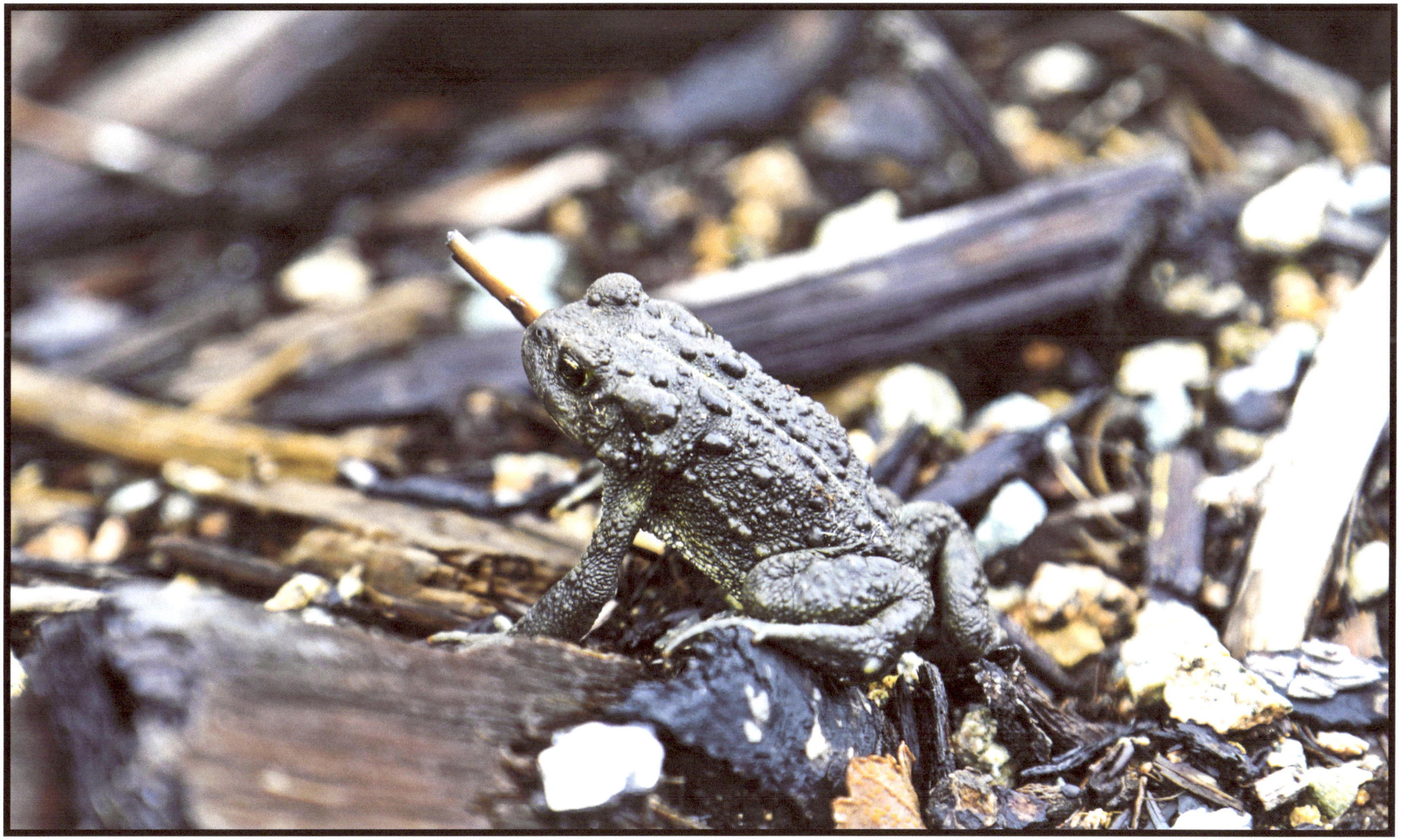

The coastal climate with its frequent precipitation is ideal for many amphibians. Frogs, toads, and salamanders are all common on the blocks.

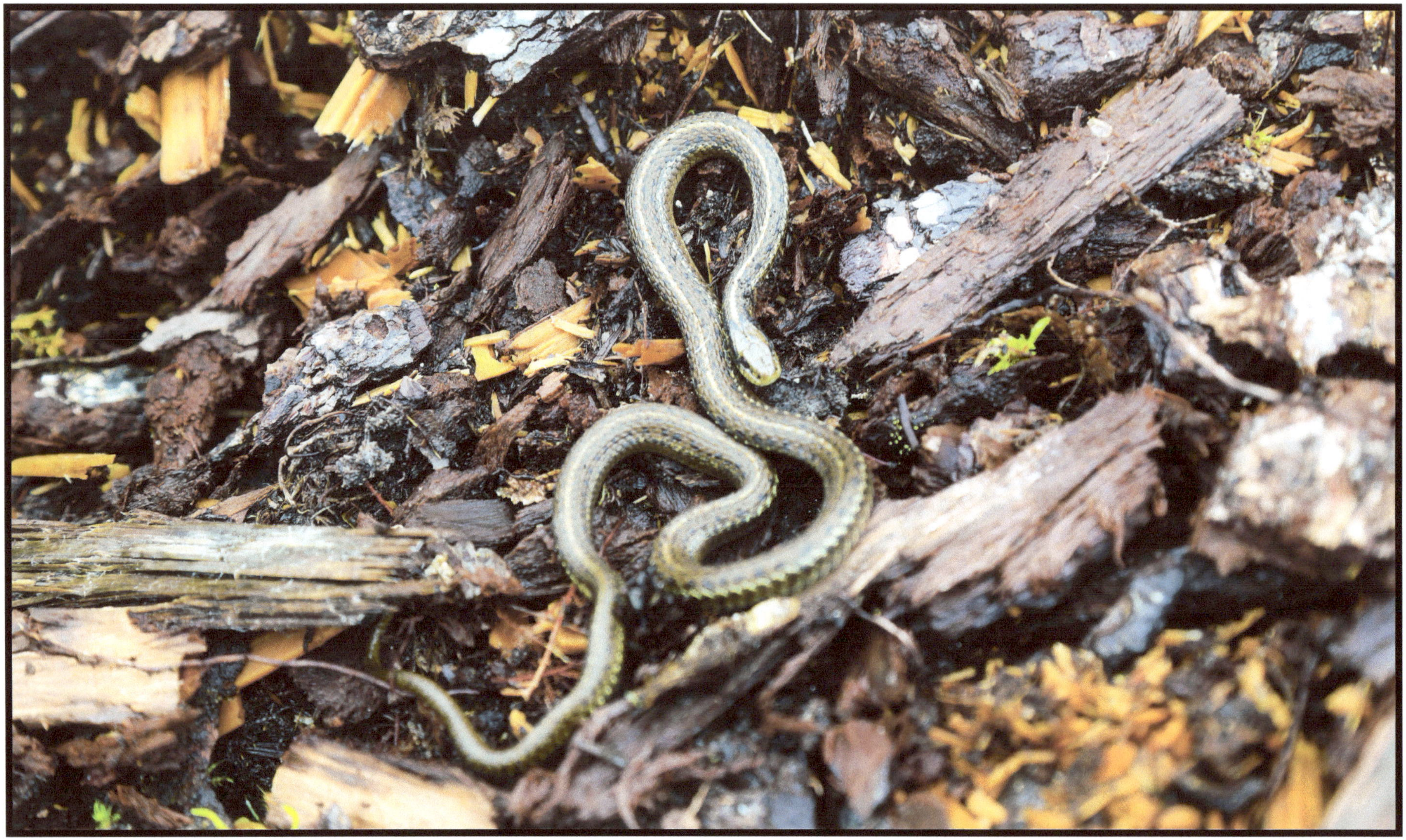

Snakes aren't that common on the northern half of Vancouver Island, but they're a daily sight on the southern Island, especially once the weather warms up in the fall.

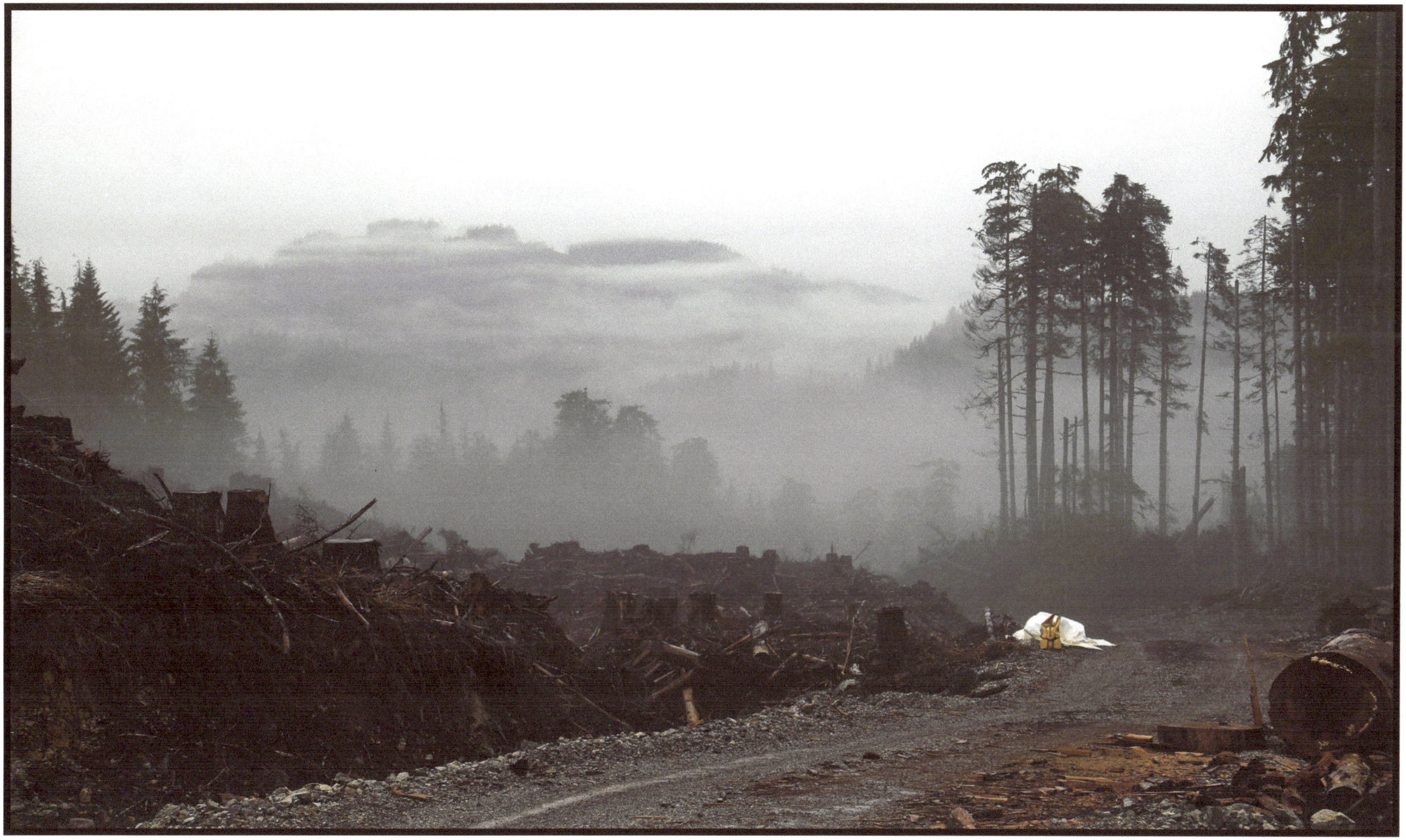

A typical spring day.

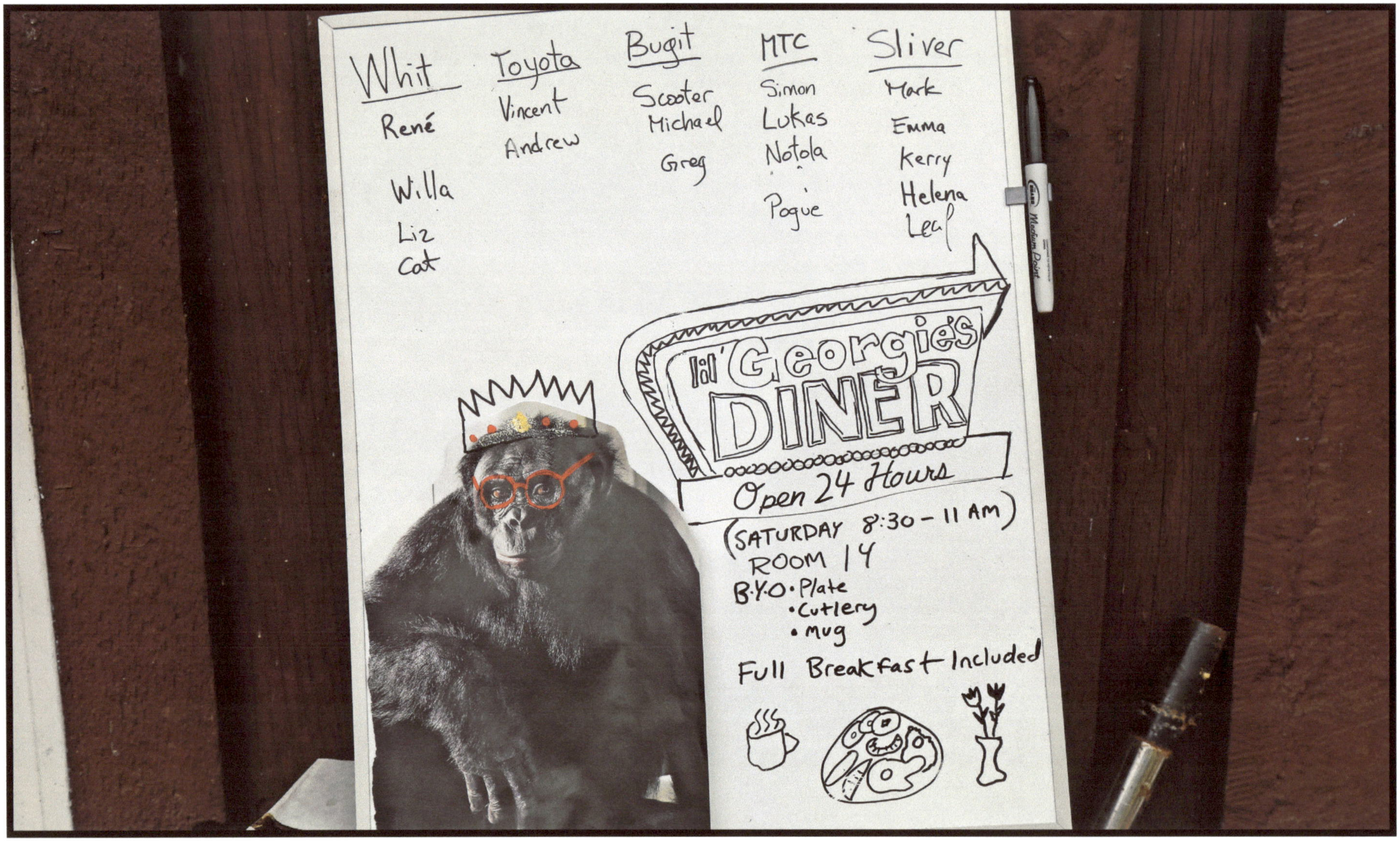

Lil Georgie's Diner will be serving breakfast on the upcoming day off. George is the mascot for King Kong Reforestation. Don't be confused by the fact that George is a monkey and King Kong was an ape.

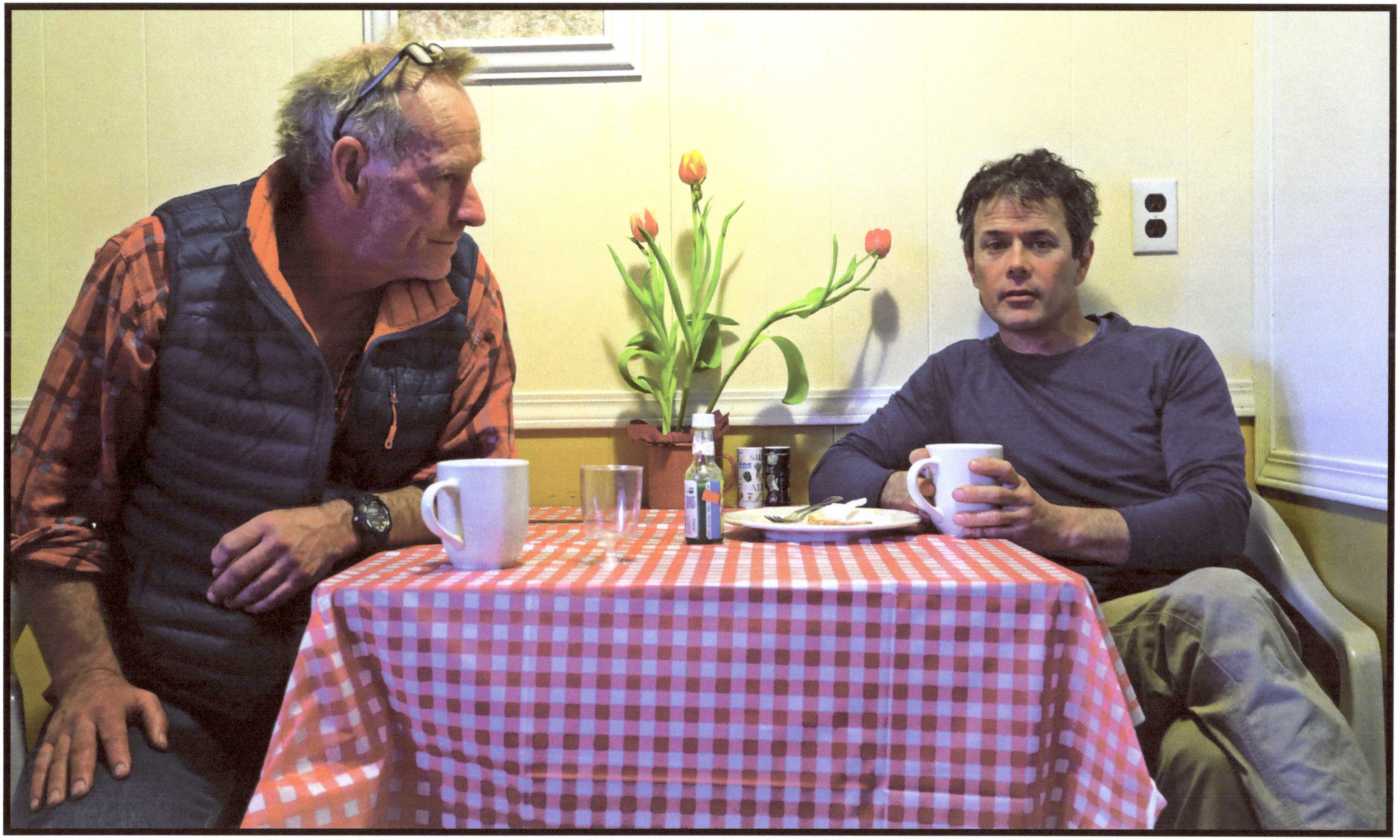

Breakfast at the diner, nothing could be finer.

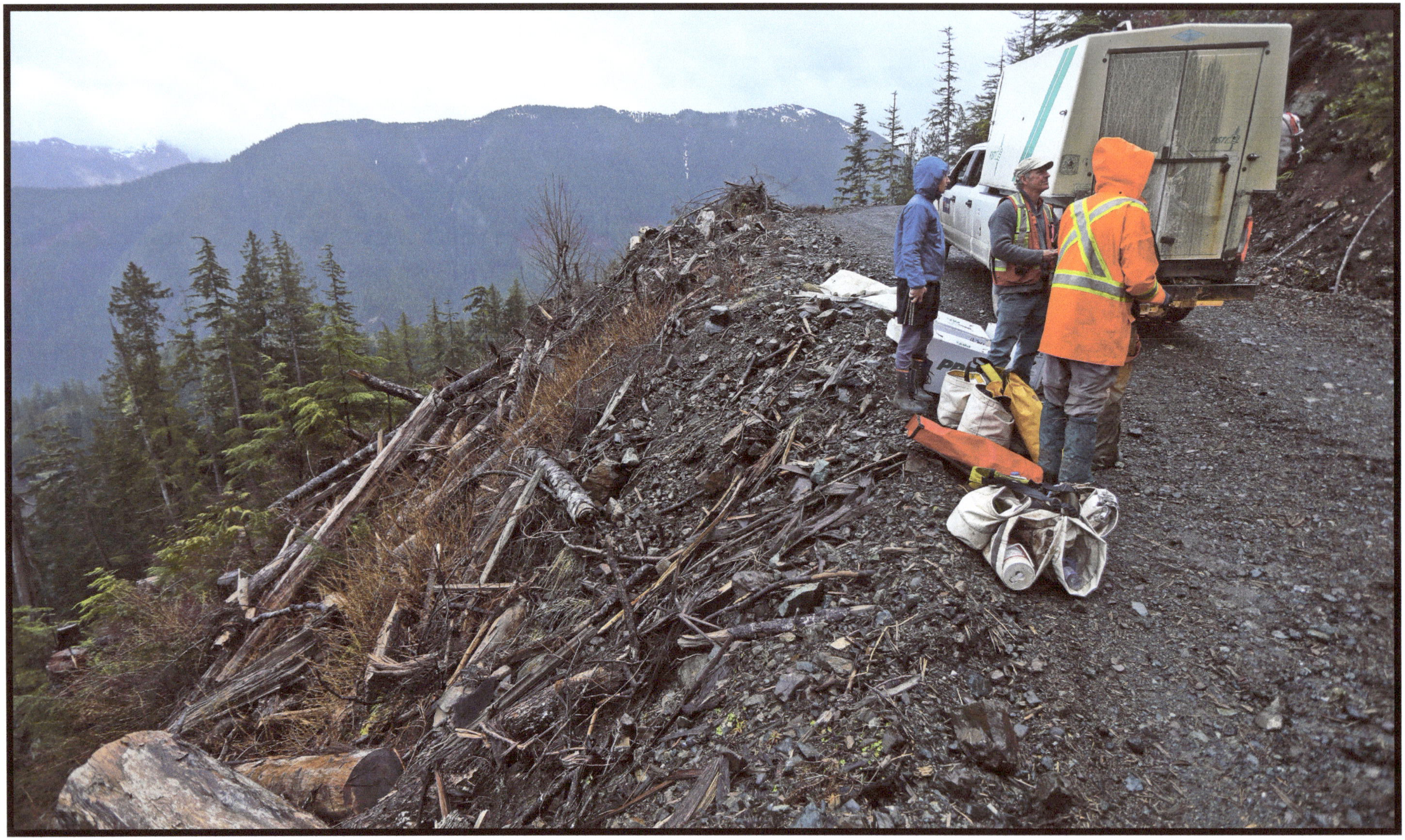

The piece above the road didn't look any better than the piece below the road.

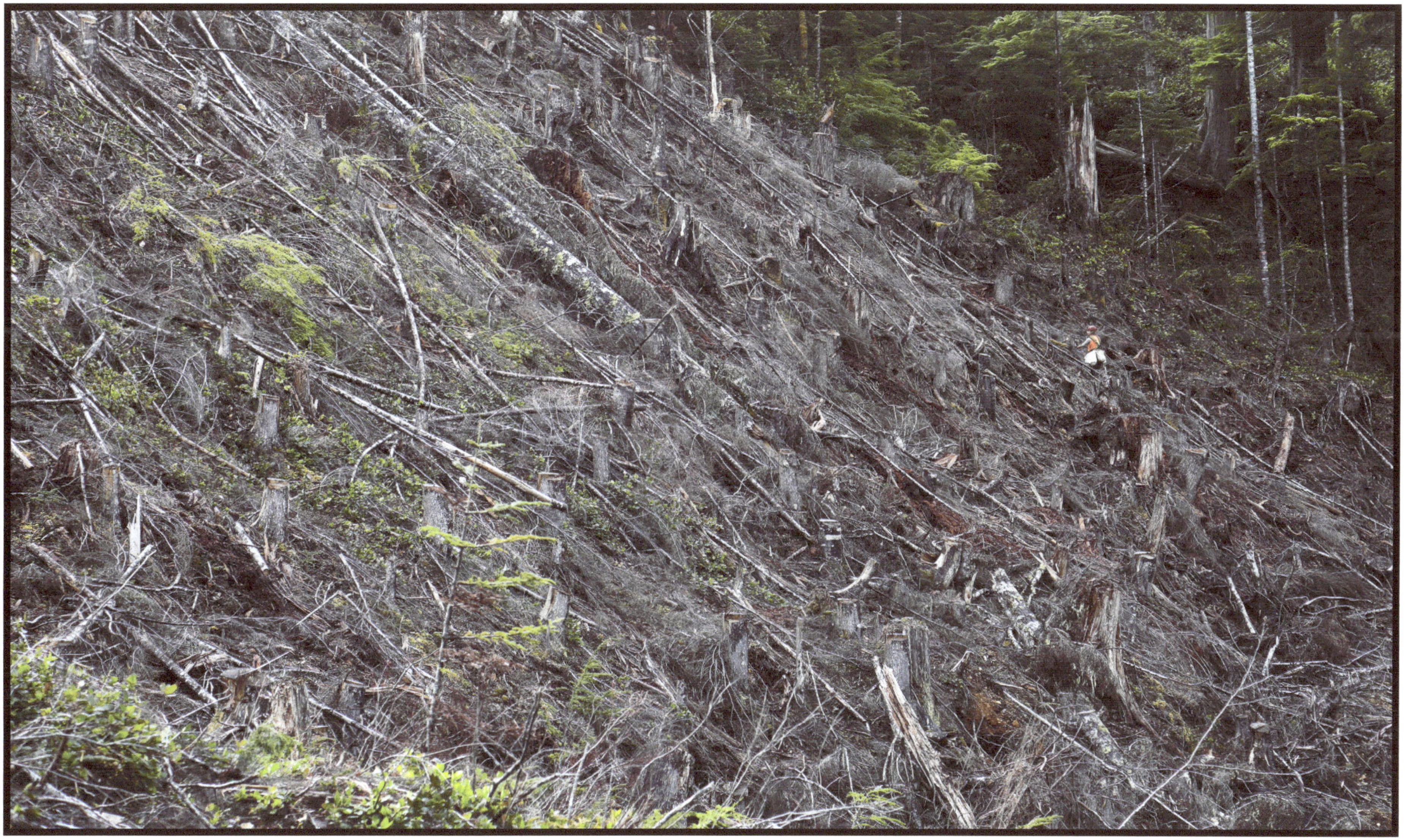

This block looks fairly clean.

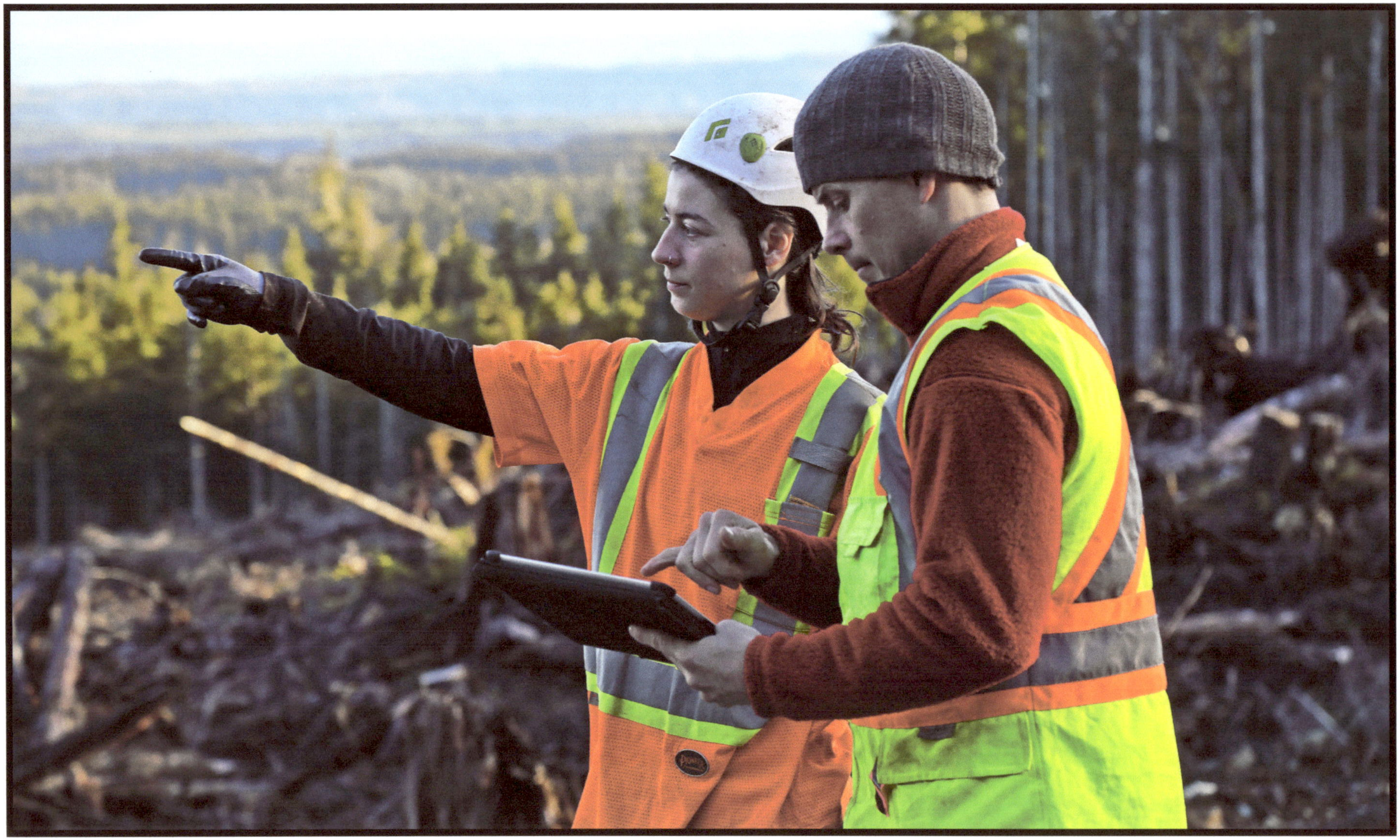

Liz, clarifying the boundary of her piece with Rene.

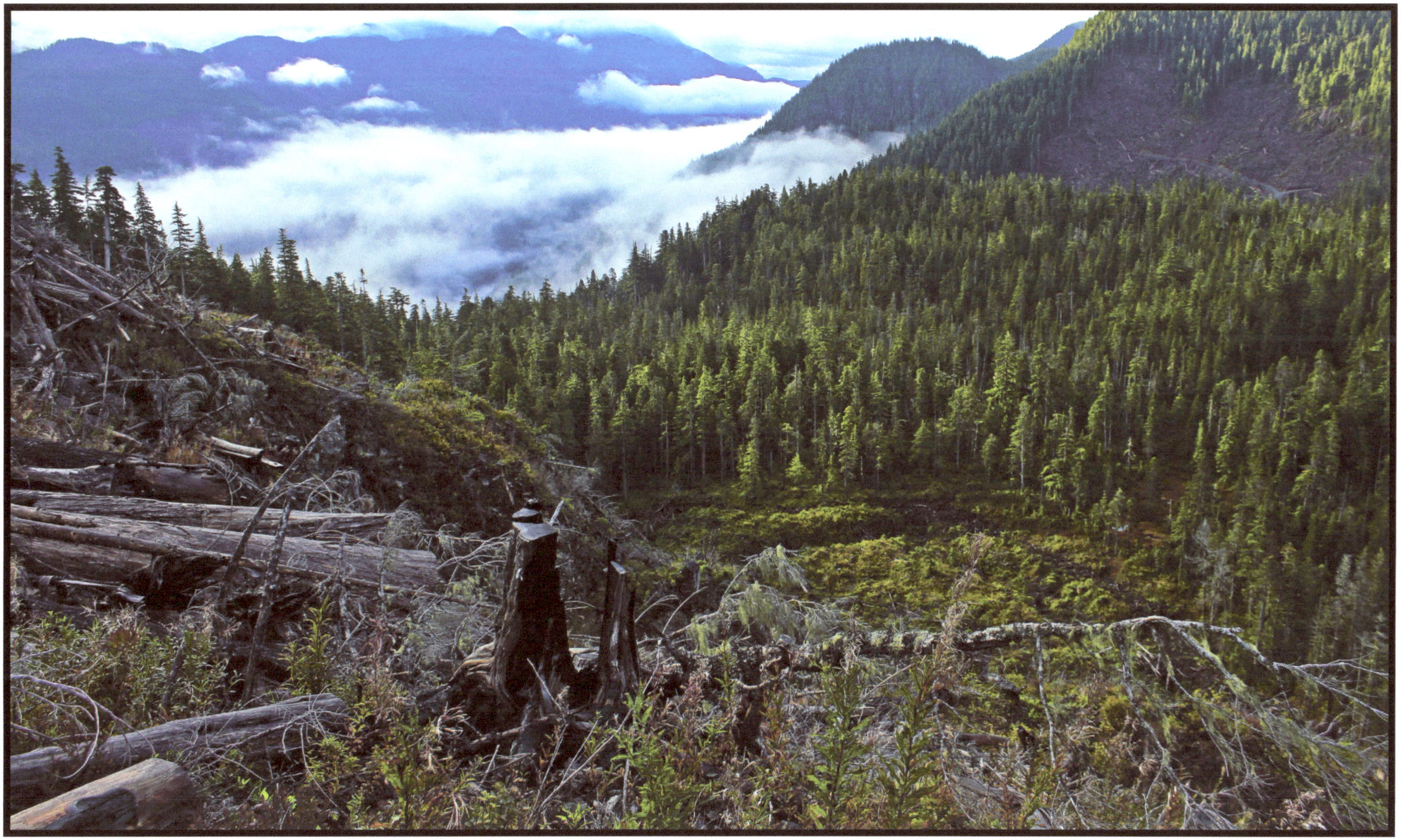

Here's another block that doesn't look too bad, until you realize that some of those mature trees at the bottom are over a hundred feet tall.

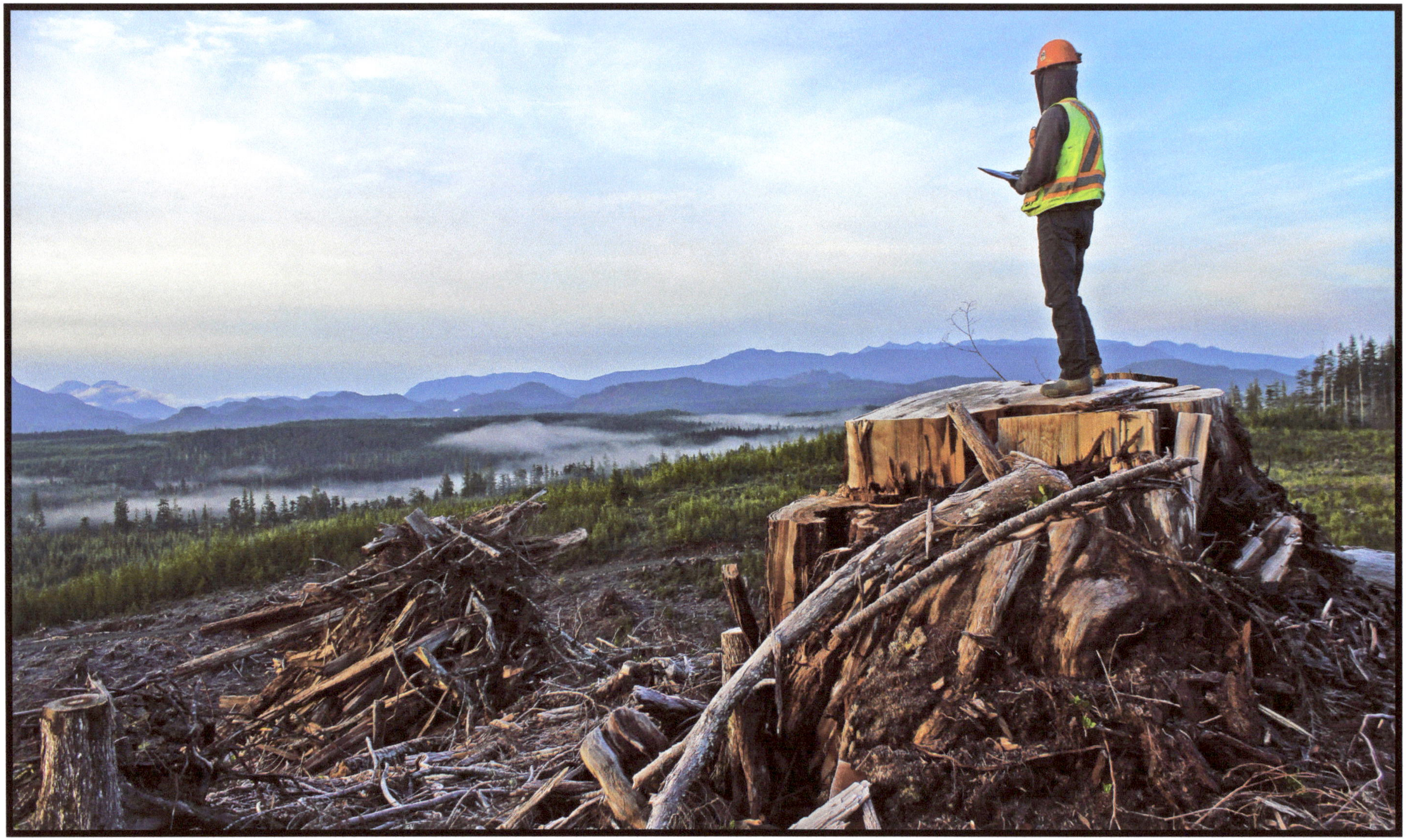

Getting oriented with the layout of the block.

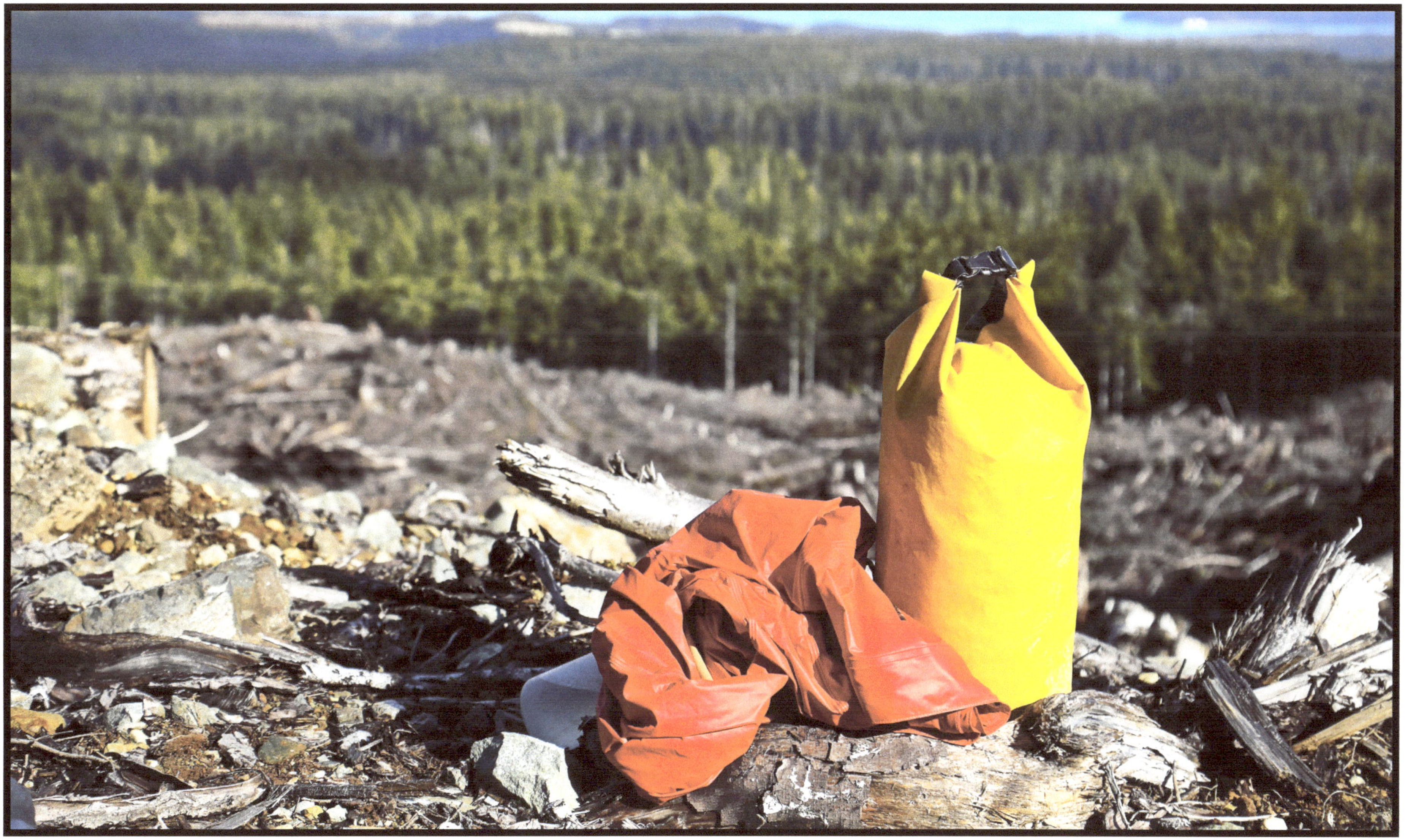

A "dry bag" is an essential piece of gear for a coastal planter. This waterproof sack will keep your stuff relatively dry. Planters use them to carry things like their lunch, tally book, antihistamines, spare gloves, and extra dry clothing.

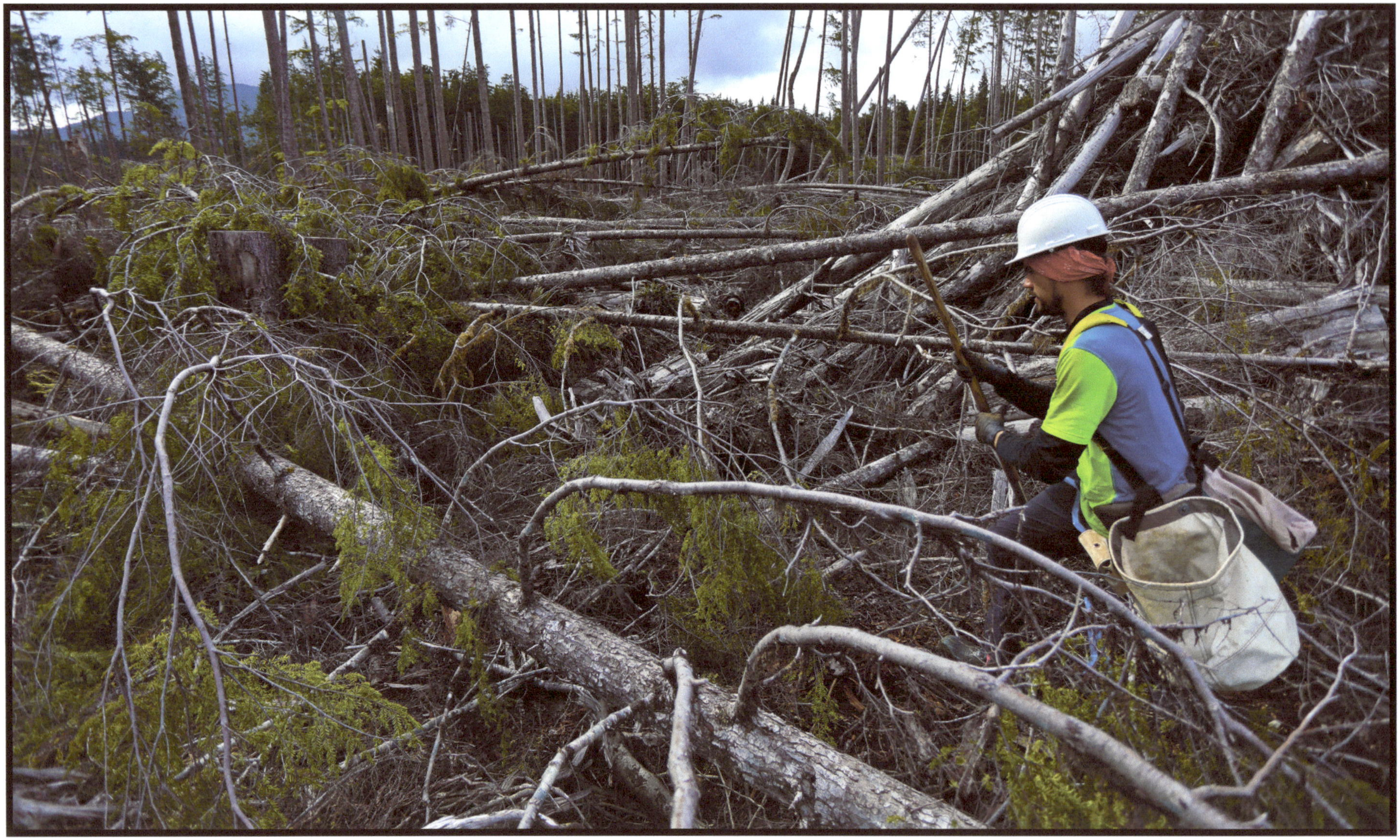

Don't let the slash get you down.

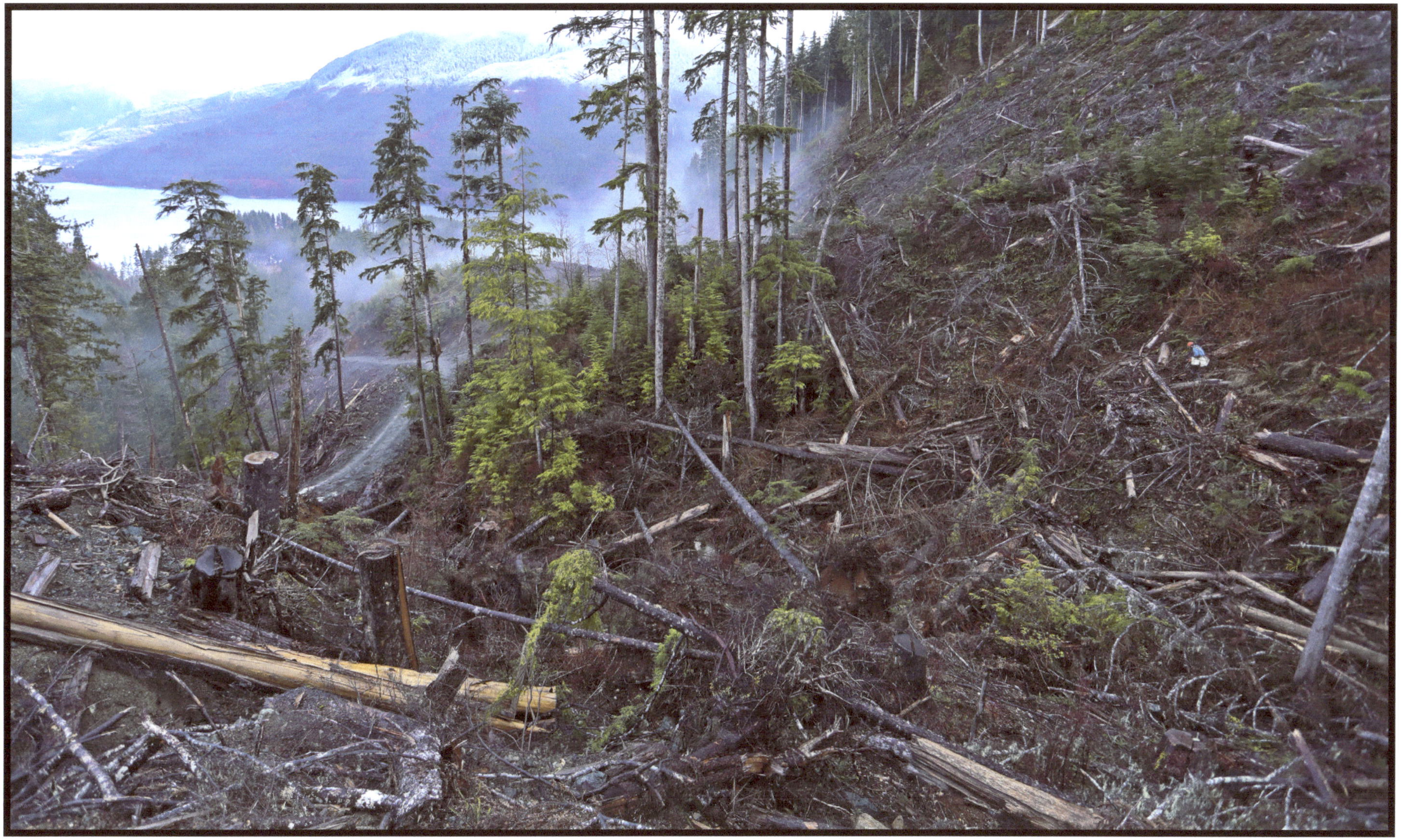

Find the planter. The high-vis helps.

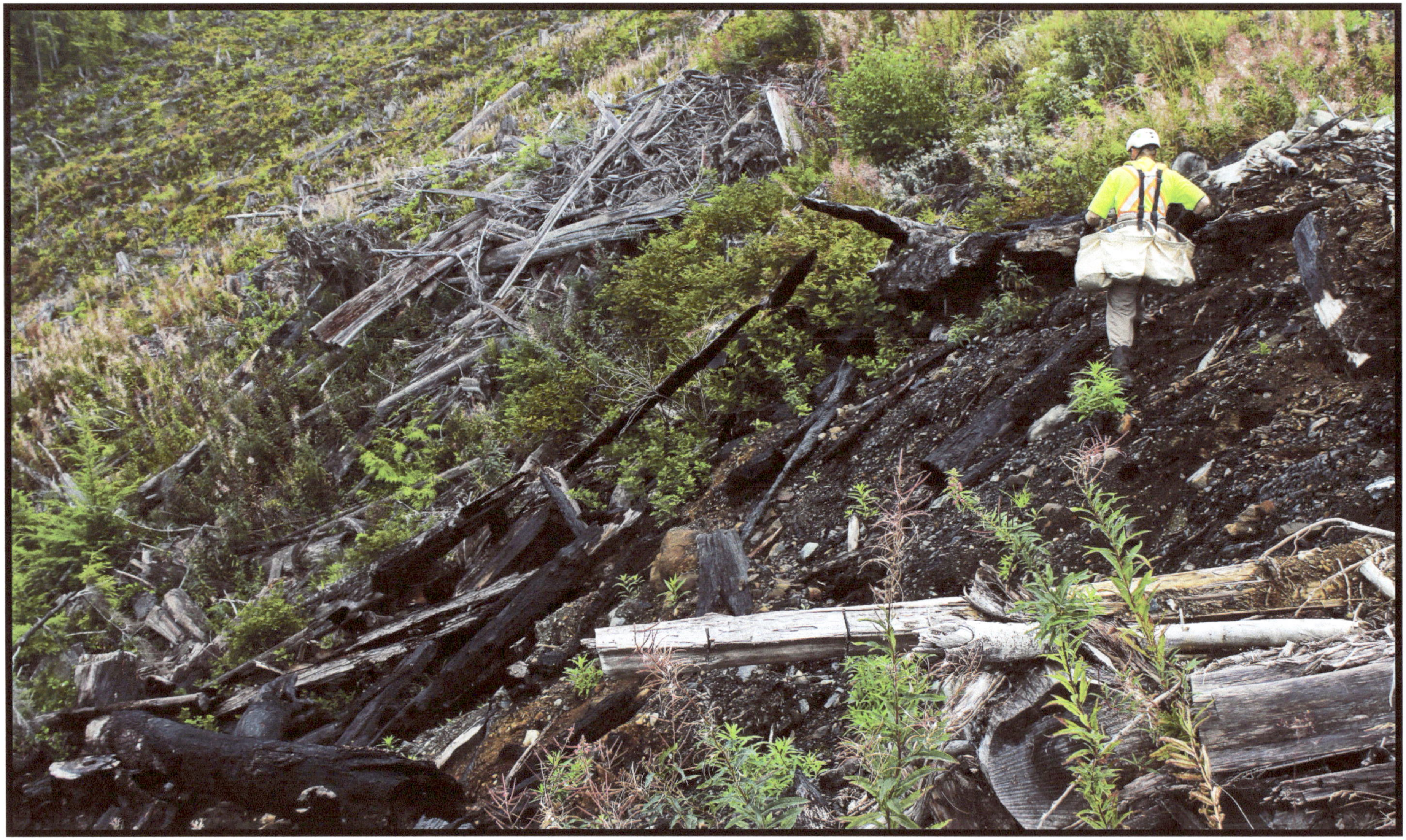

Sometimes, some of the slash on the blocks is piled then burned (assuming that the wood dries out enough to burn). Burning the piles allows us to plant a few extra trees in the block, rather than having the piles take up otherwise productive ground.

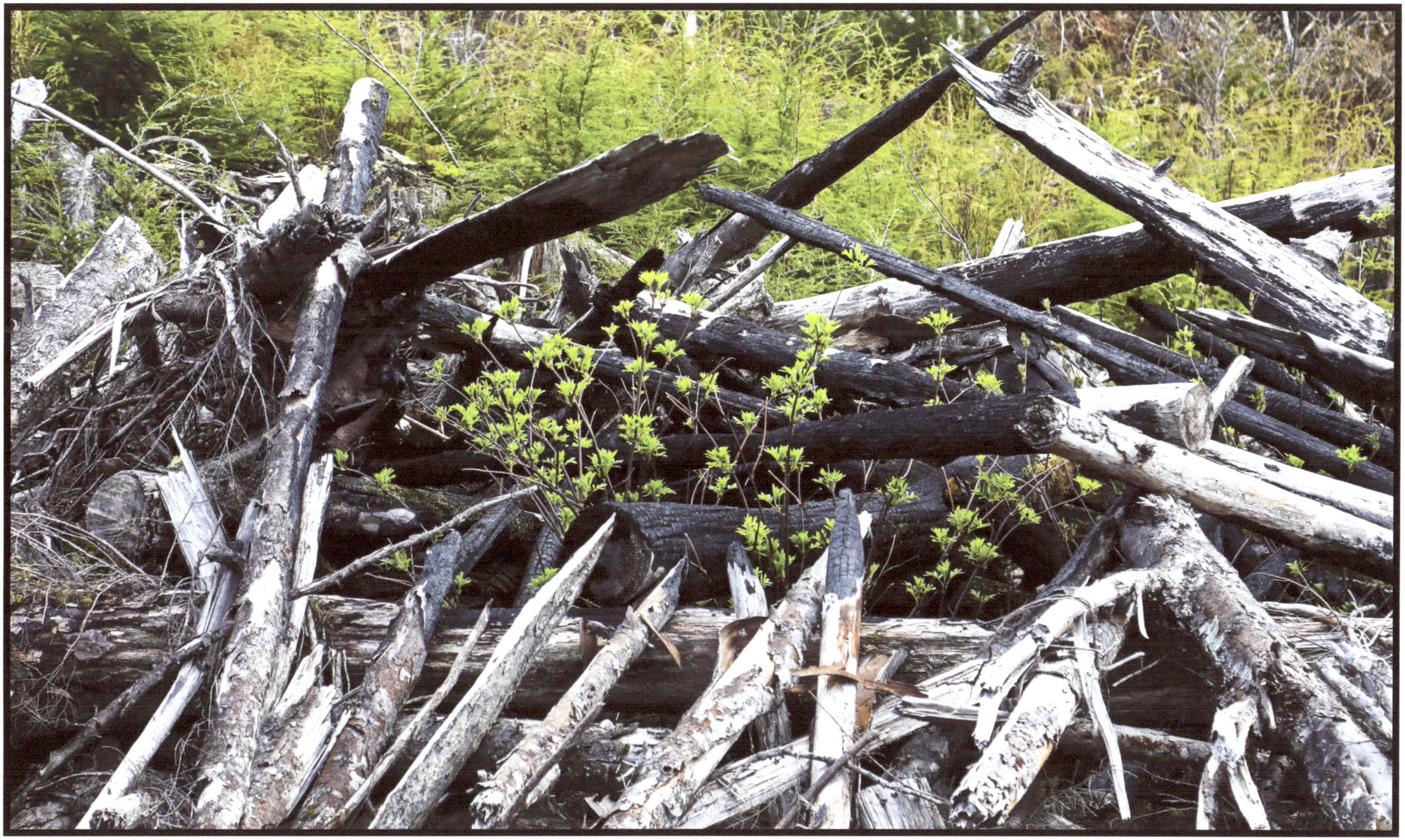

Of course, everything on the coast is harder. Sometimes, planting burns isn't all that it's cracked up to be.

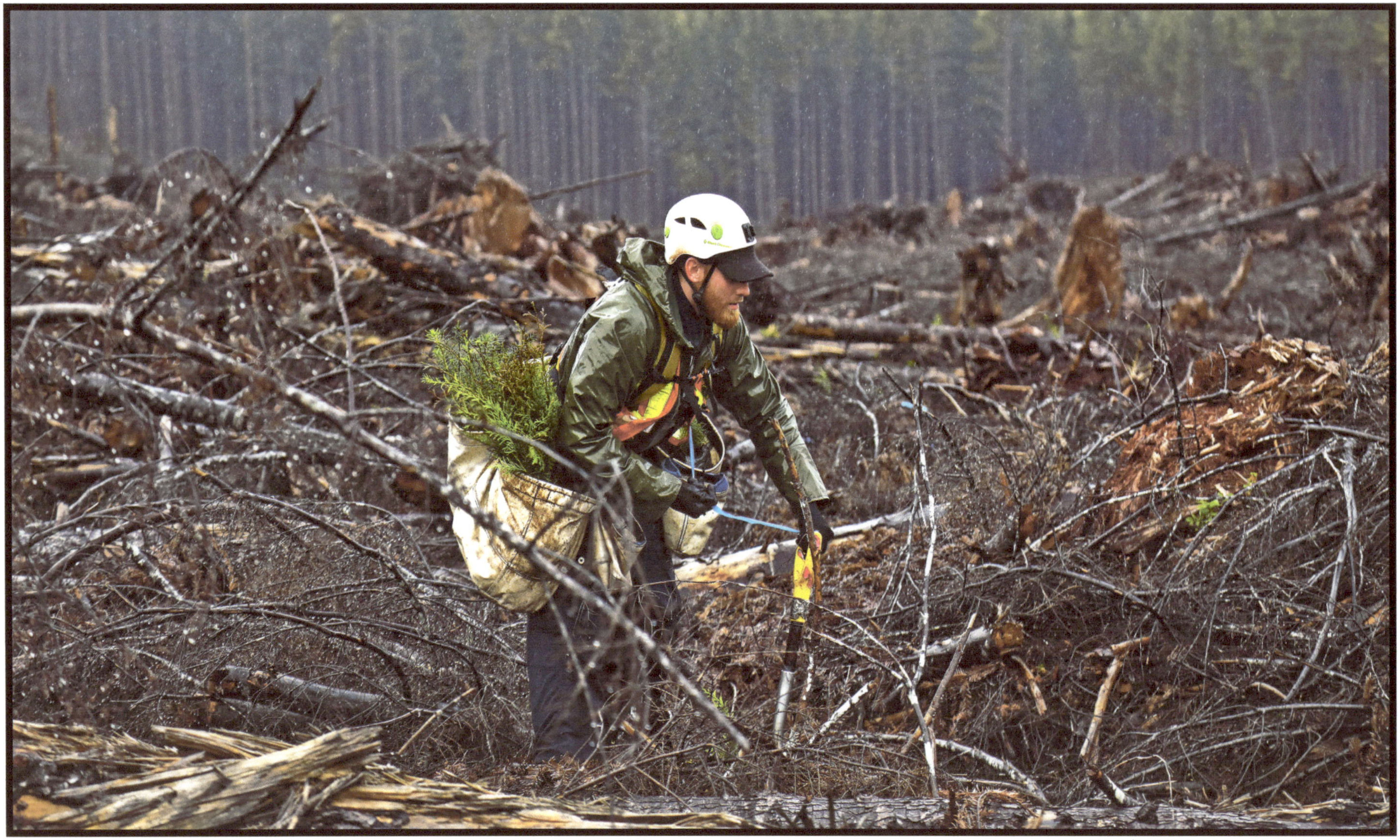

Andrew, on a surprisingly flat block. Andrew doesn't like flat blocks, because he finds them to be too boring.

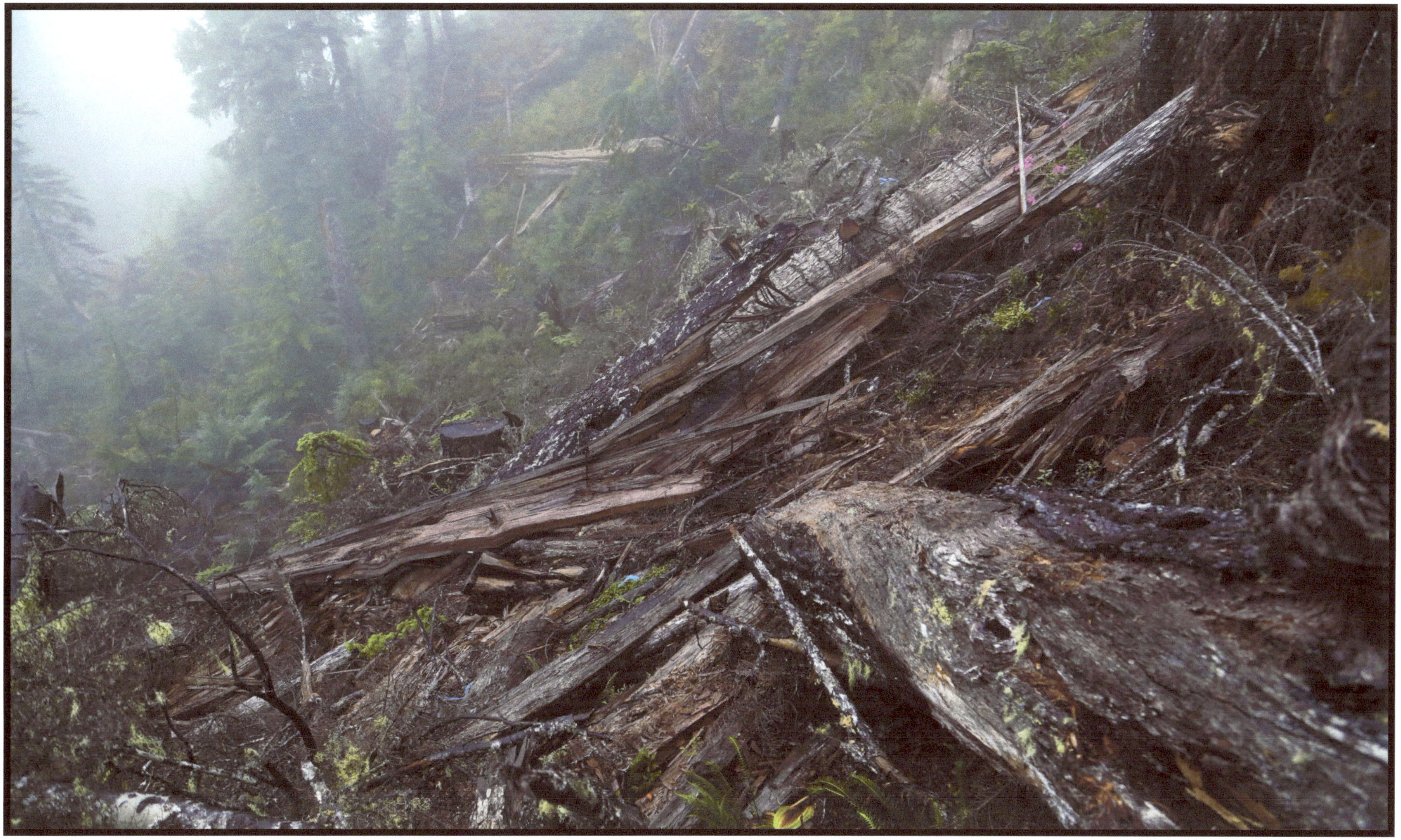

This misty block would be the type of ground that Andrew likes. If you look closely, you can see some pieces of blue flagging tape, which shows that this piece has already been planted.

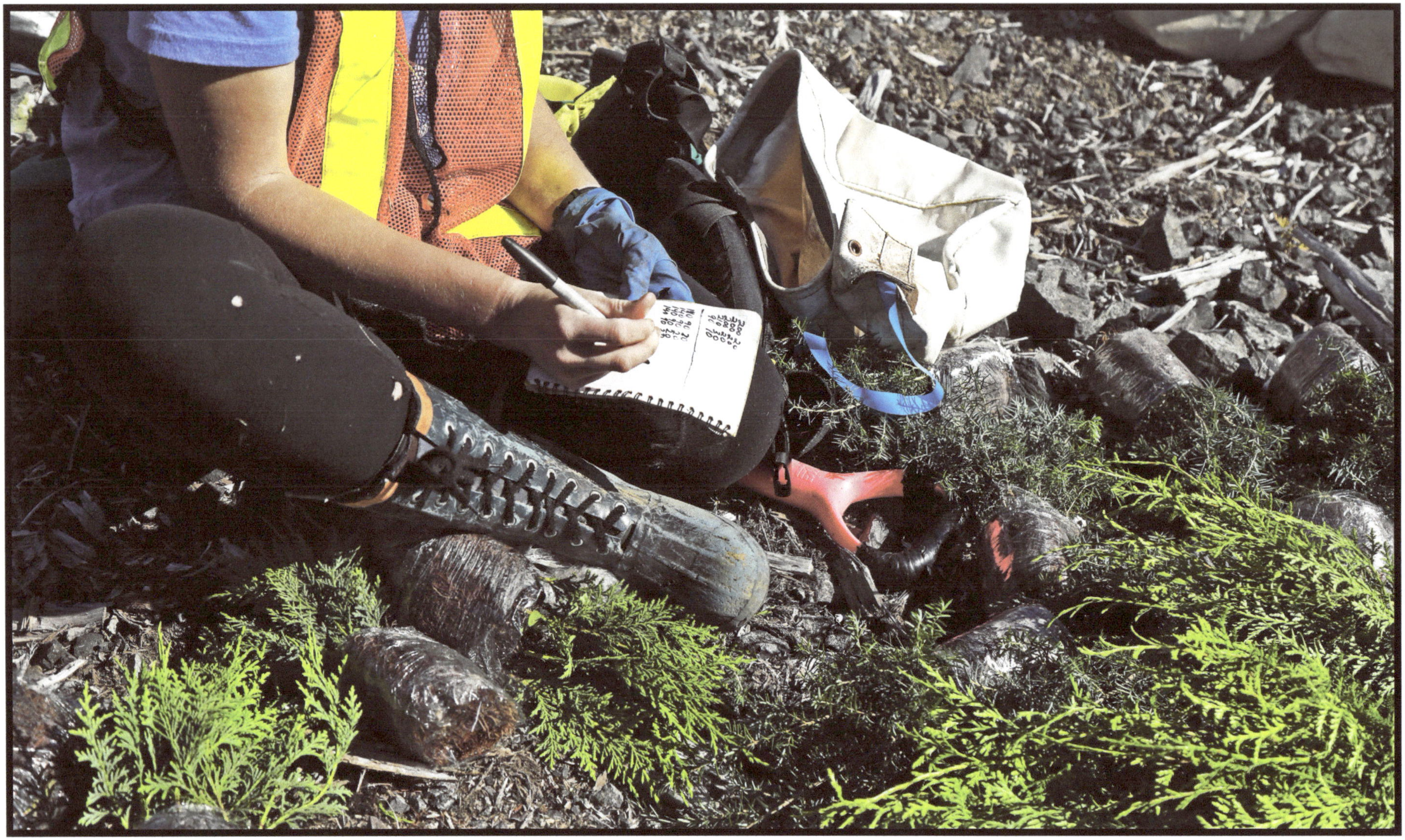

It's critical to have a waterproof tally book to keep track of the exact breakdown of what you've planted.

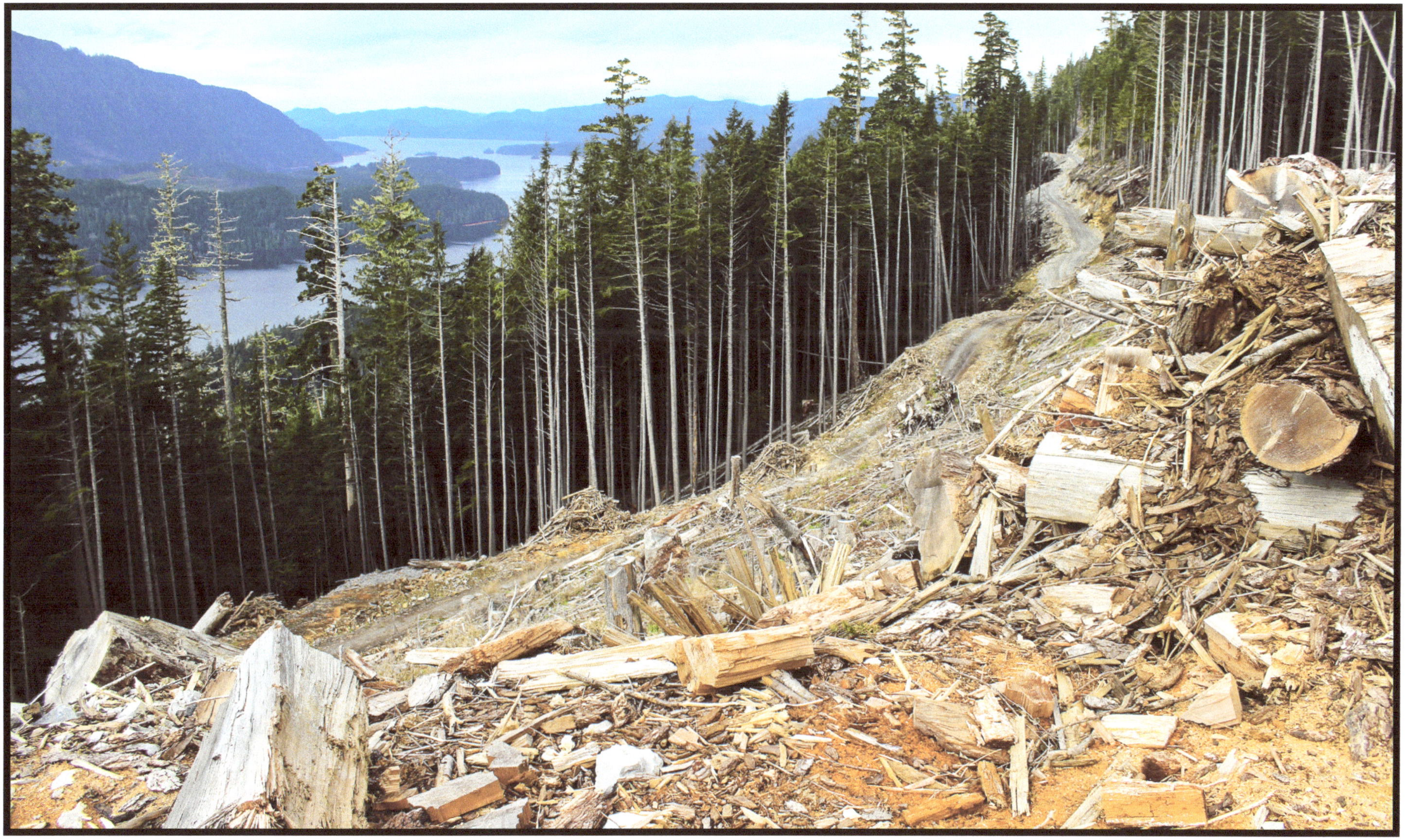

Another view from a block.

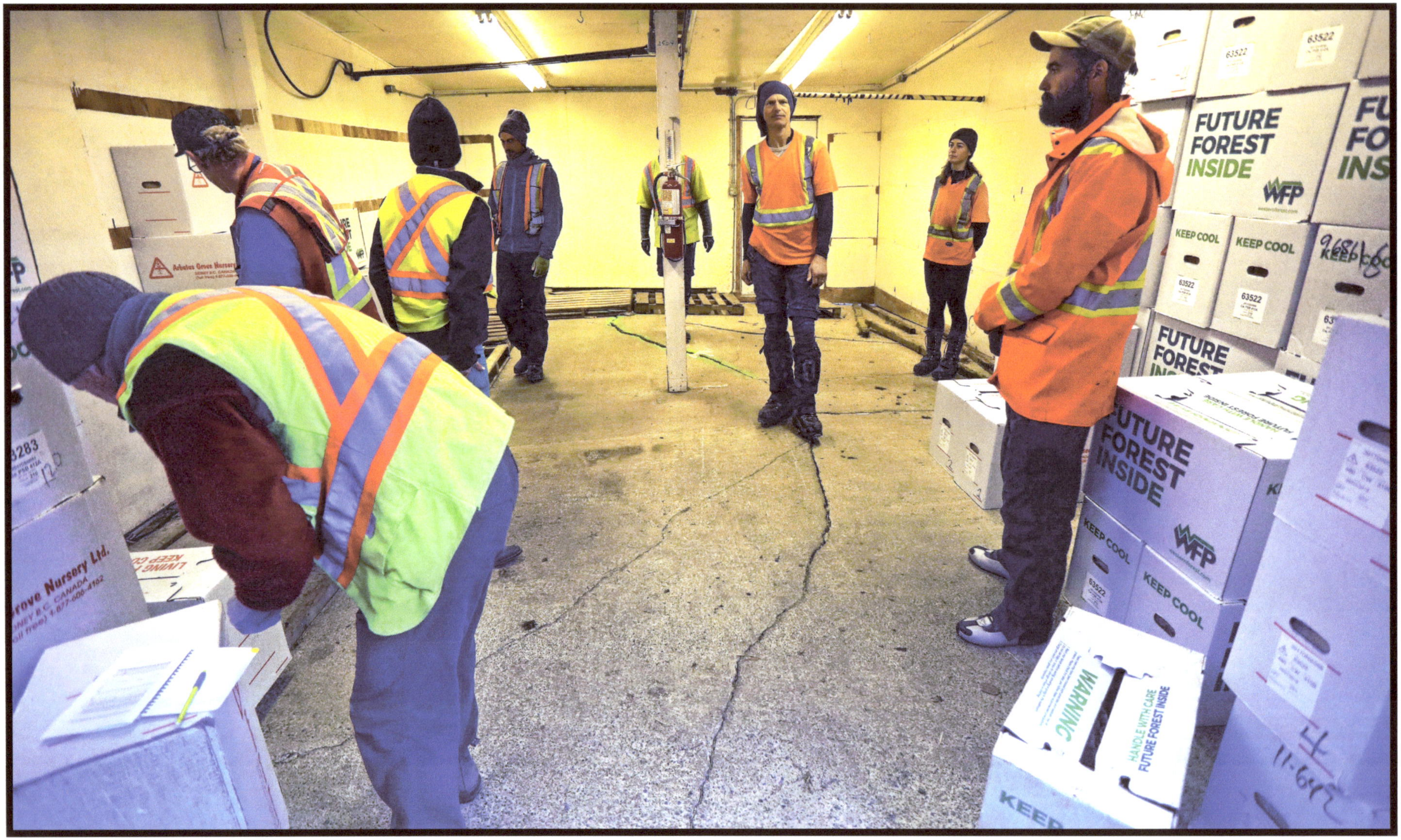

This cooler is almost empty. It's time to unload another shipment of trees.

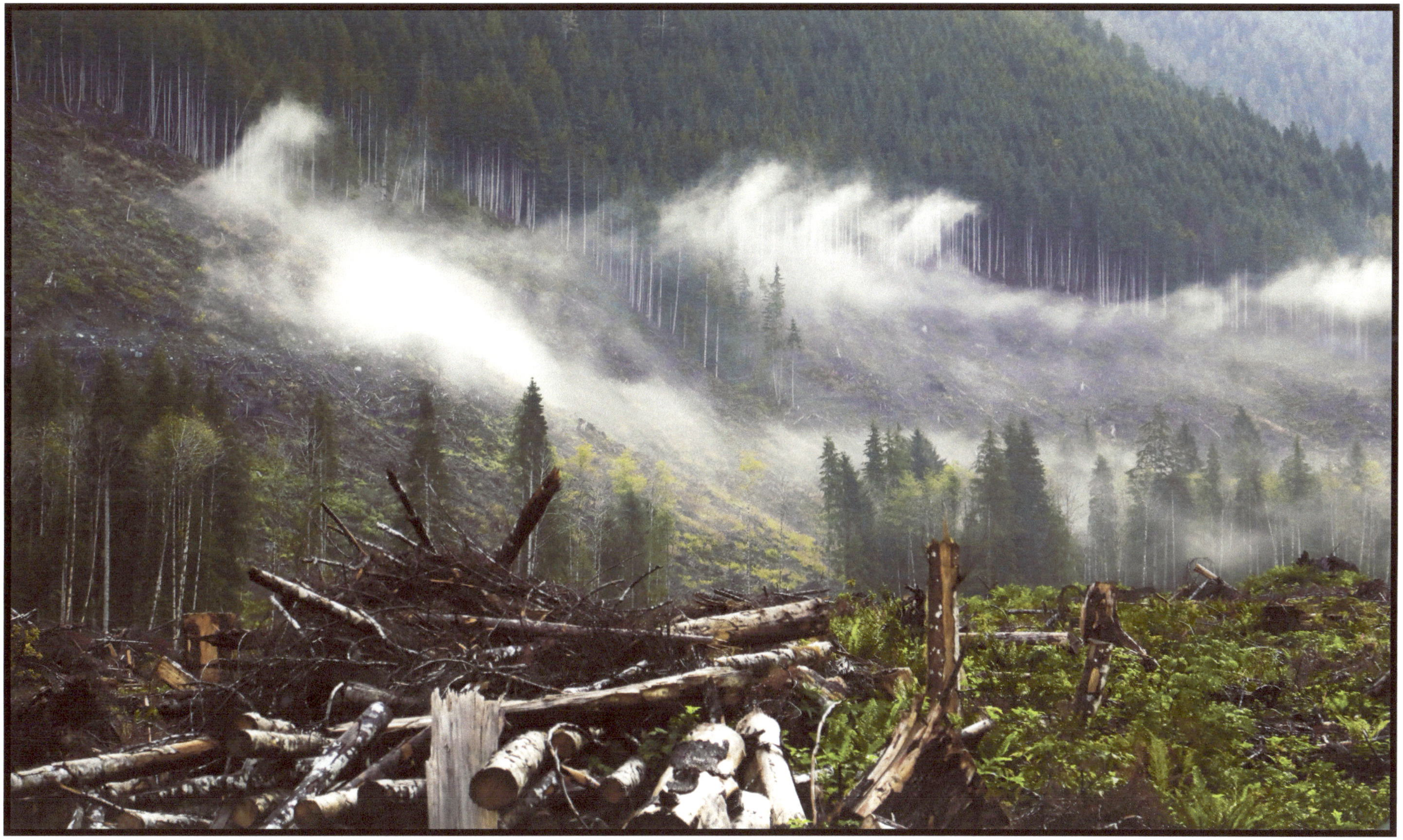

It's common to see low clouds moving through the valleys. Sometimes, when you're on a steep block, they'll blow past beneath you.

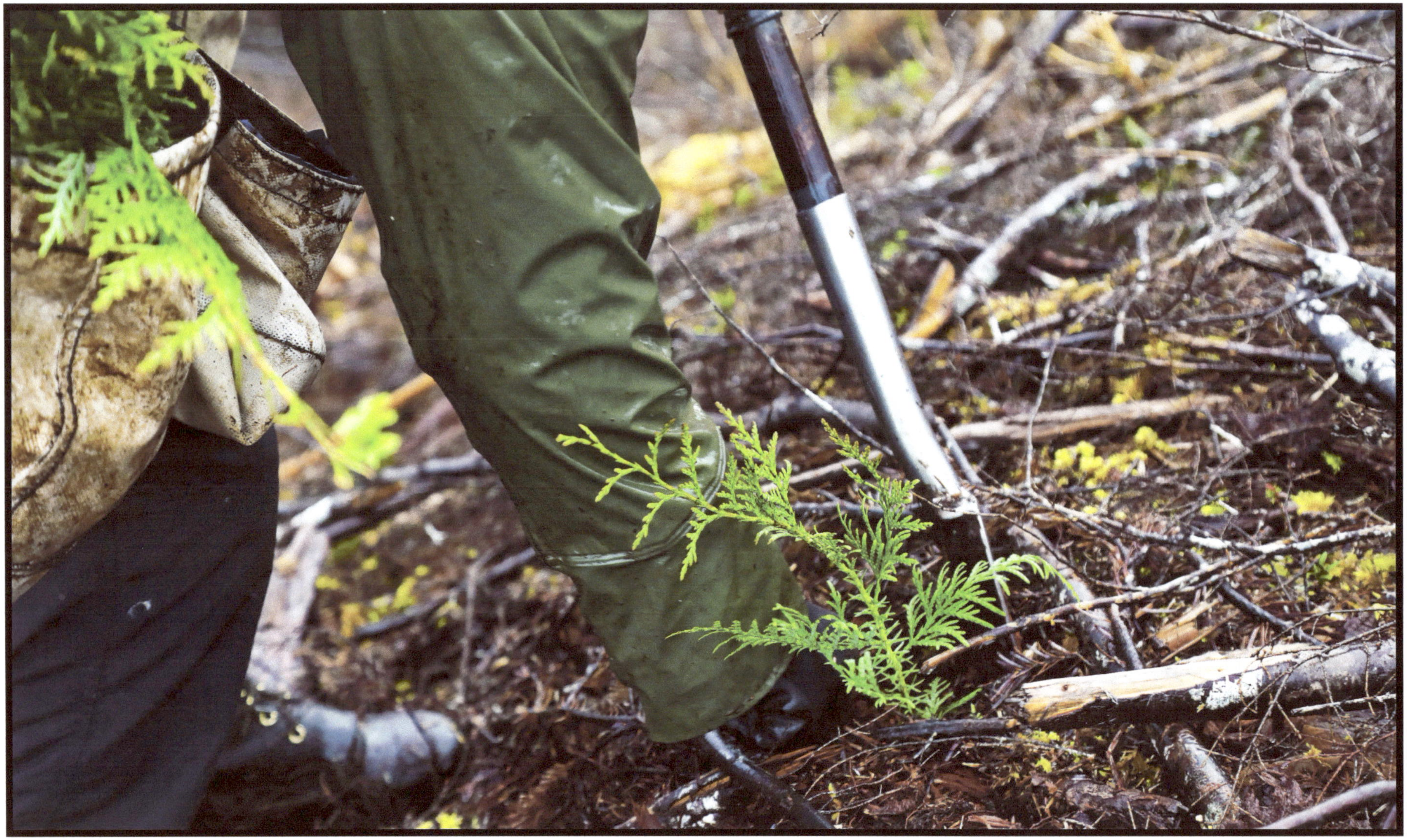

Closing the hole.

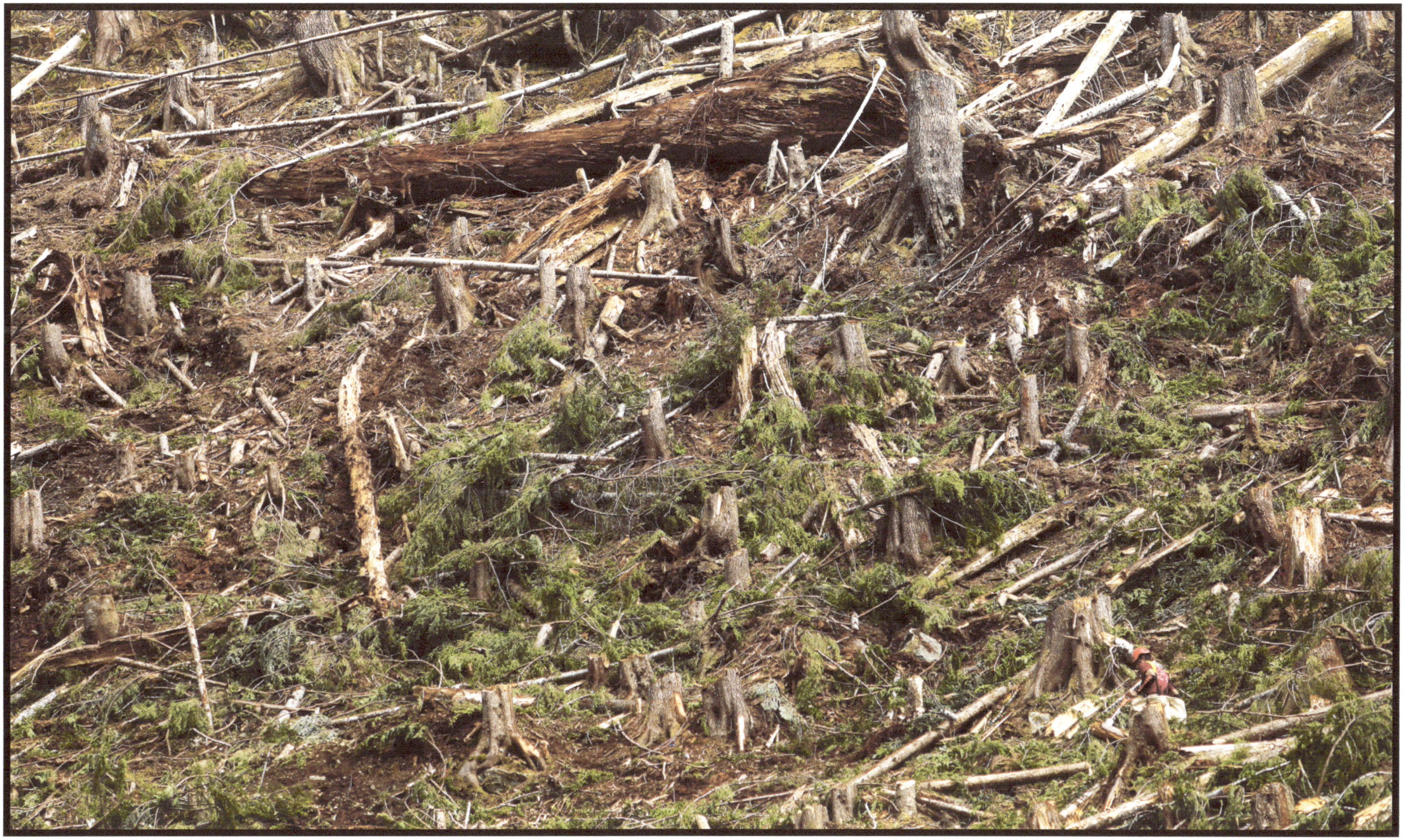

Find the planter.

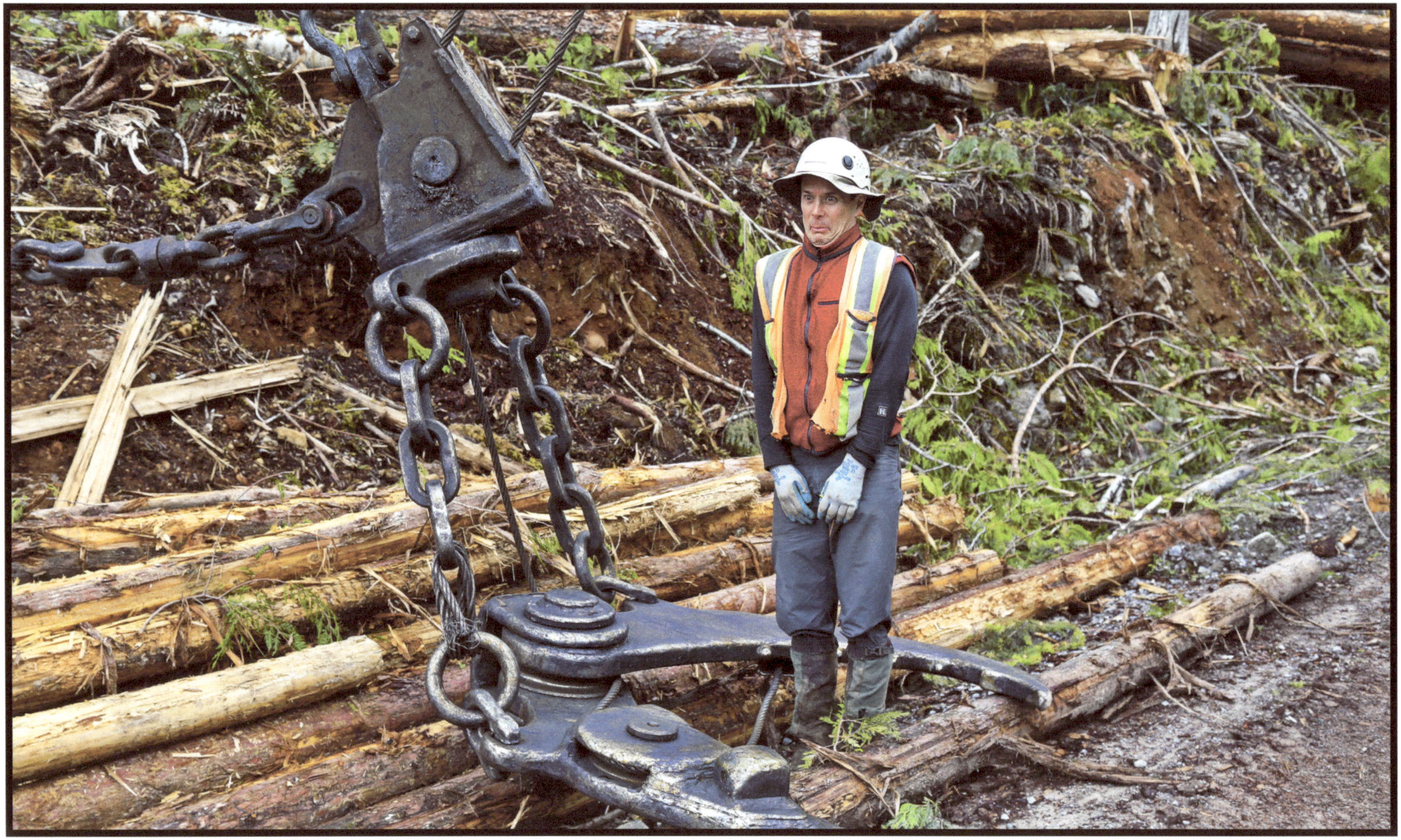

This piece of equipment is the grapple. It's used to pick fallen trees off a block and pull them out to roadside. Rene jumped into the photo to illustrate how large it is.

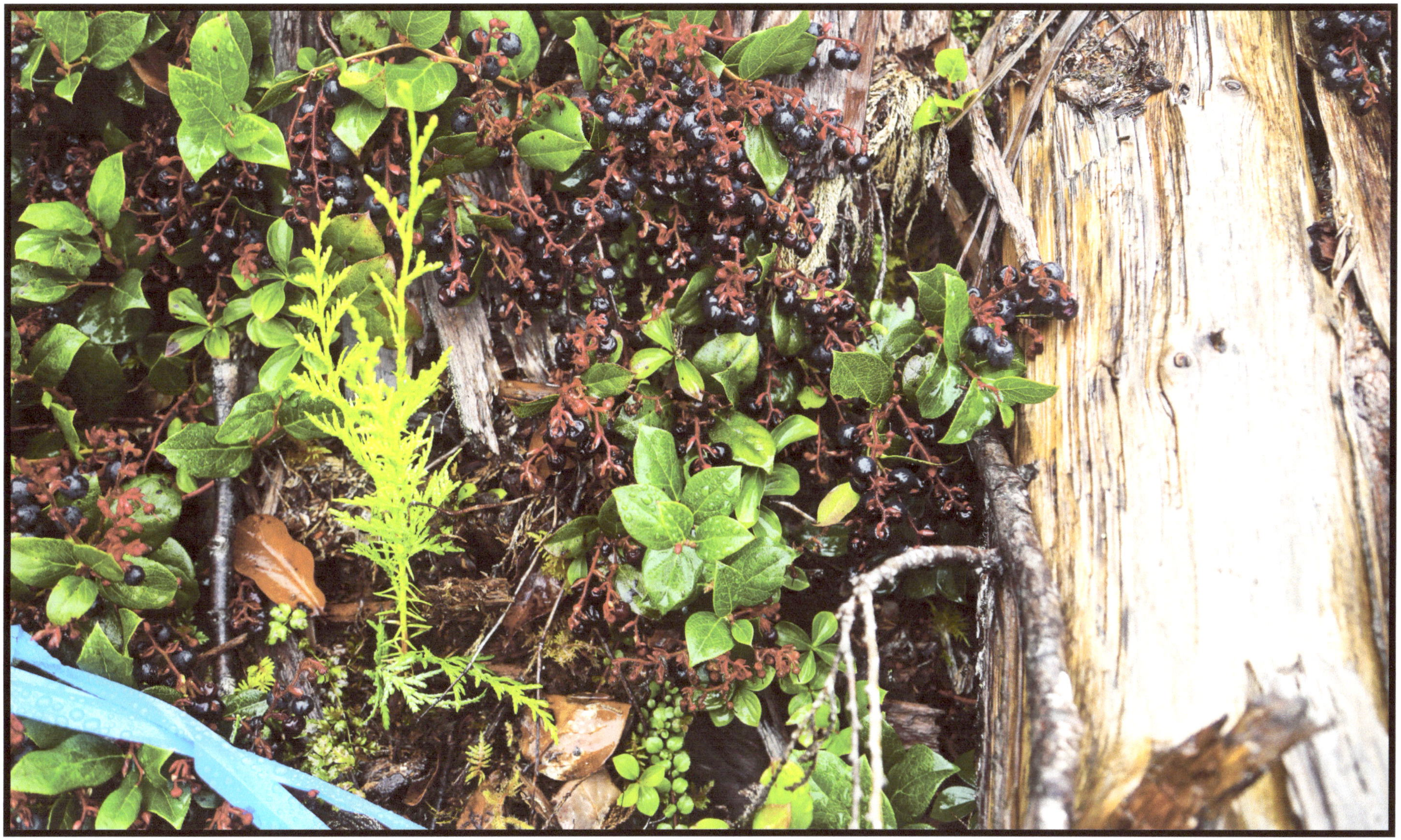

Salal is a very common plant on many coastal blocks. Salal holds its waxy shape for quite a while after being picked, so it is used commonly used in floral arrangements. There's a thriving industry of people who pick salal for florists, and salal from BC is shipped around the world. In the fall, salal plants are often heavily laden with ripe berries. These berries are edible.

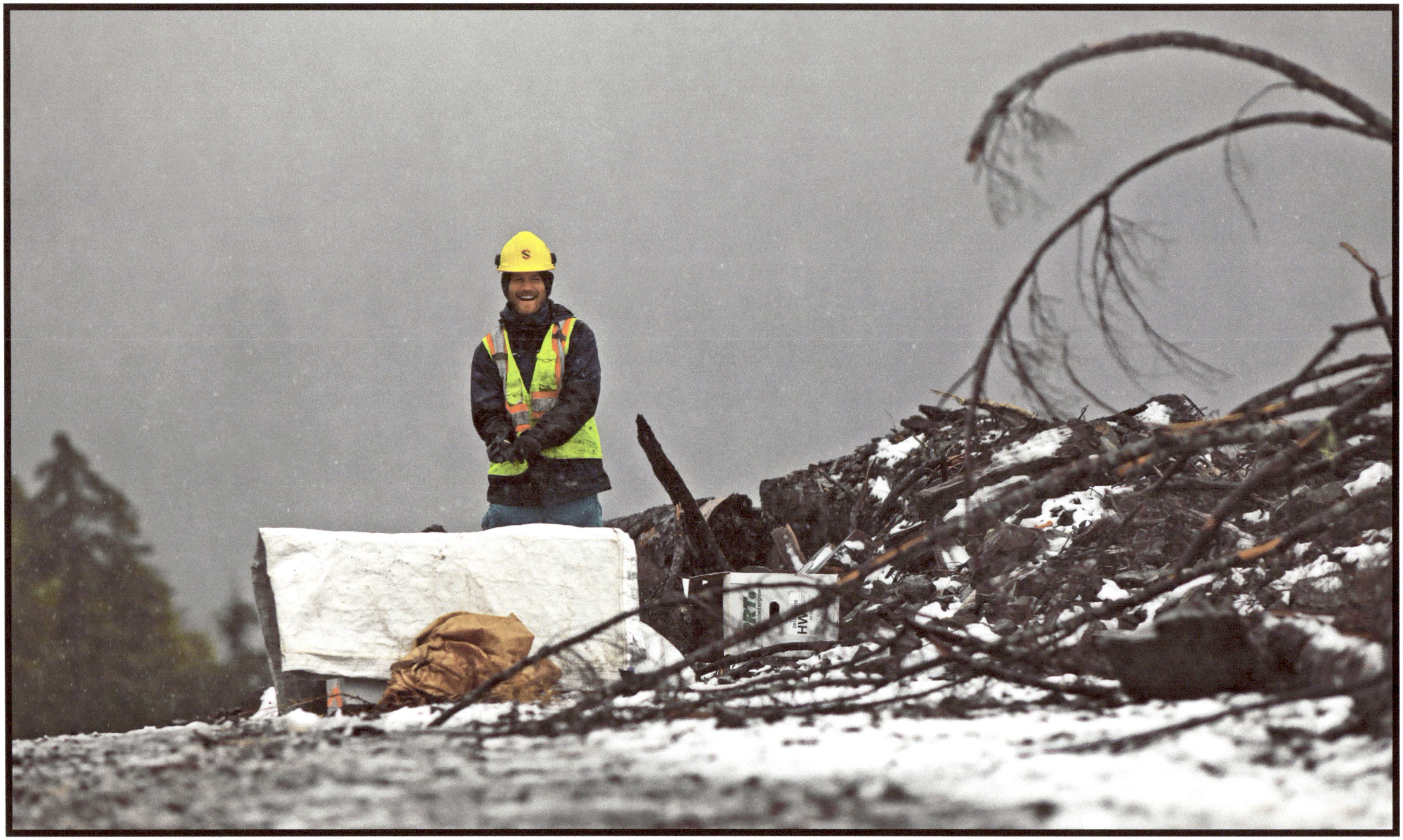

A chilly spring morning.

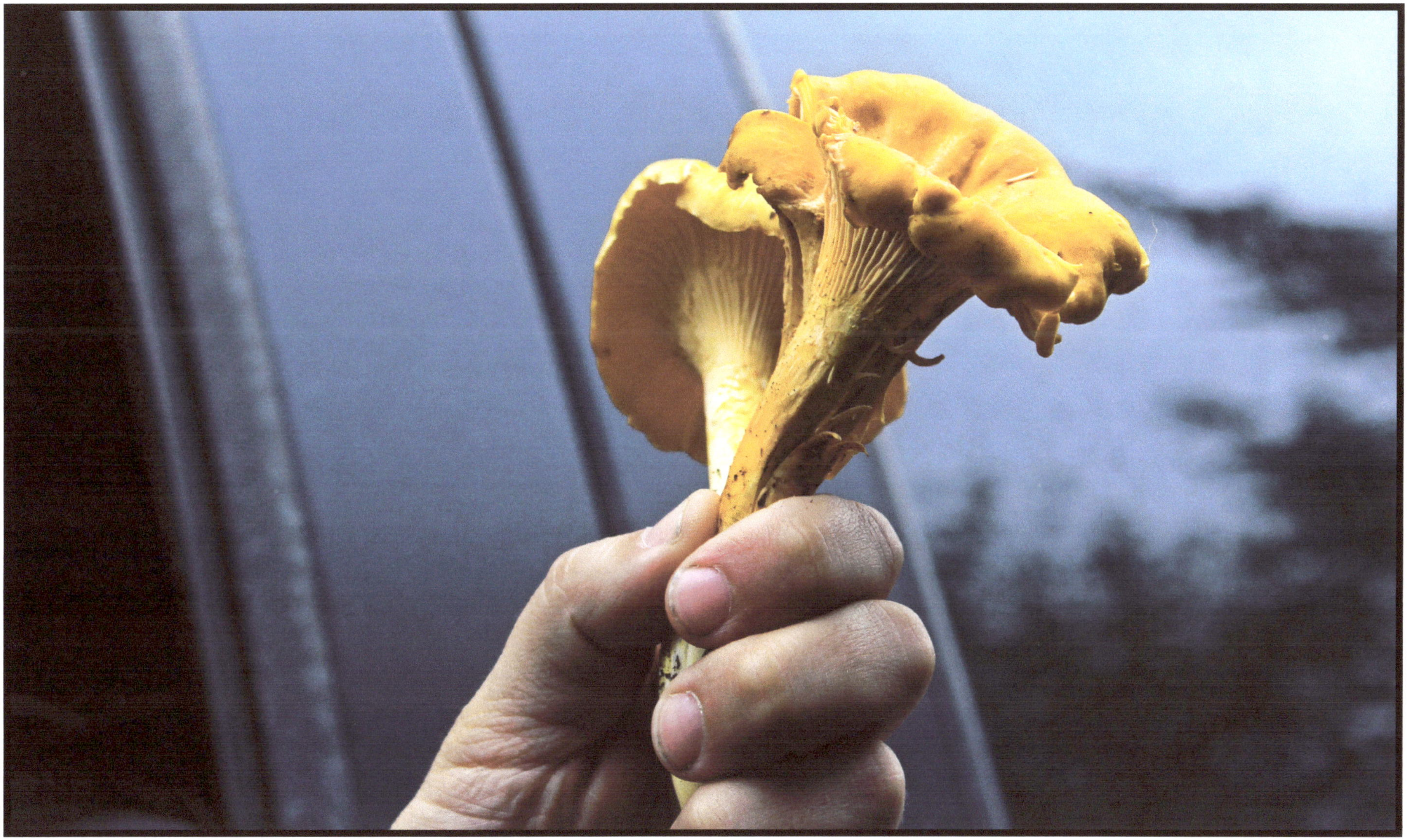

It's often possible to find a lot of different mushrooms and other types of fungi on the blocks. Although many are toxic, a few are edible. The mushrooms shown here are chanterelles, which are very tasty. Morels are also common in some areas. Before trying to eat any mushrooms, it's important to get a proper field guide that can help you identify which ones are toxic. Inexperienced foragers can be tricked by false morels or false chanterelles.

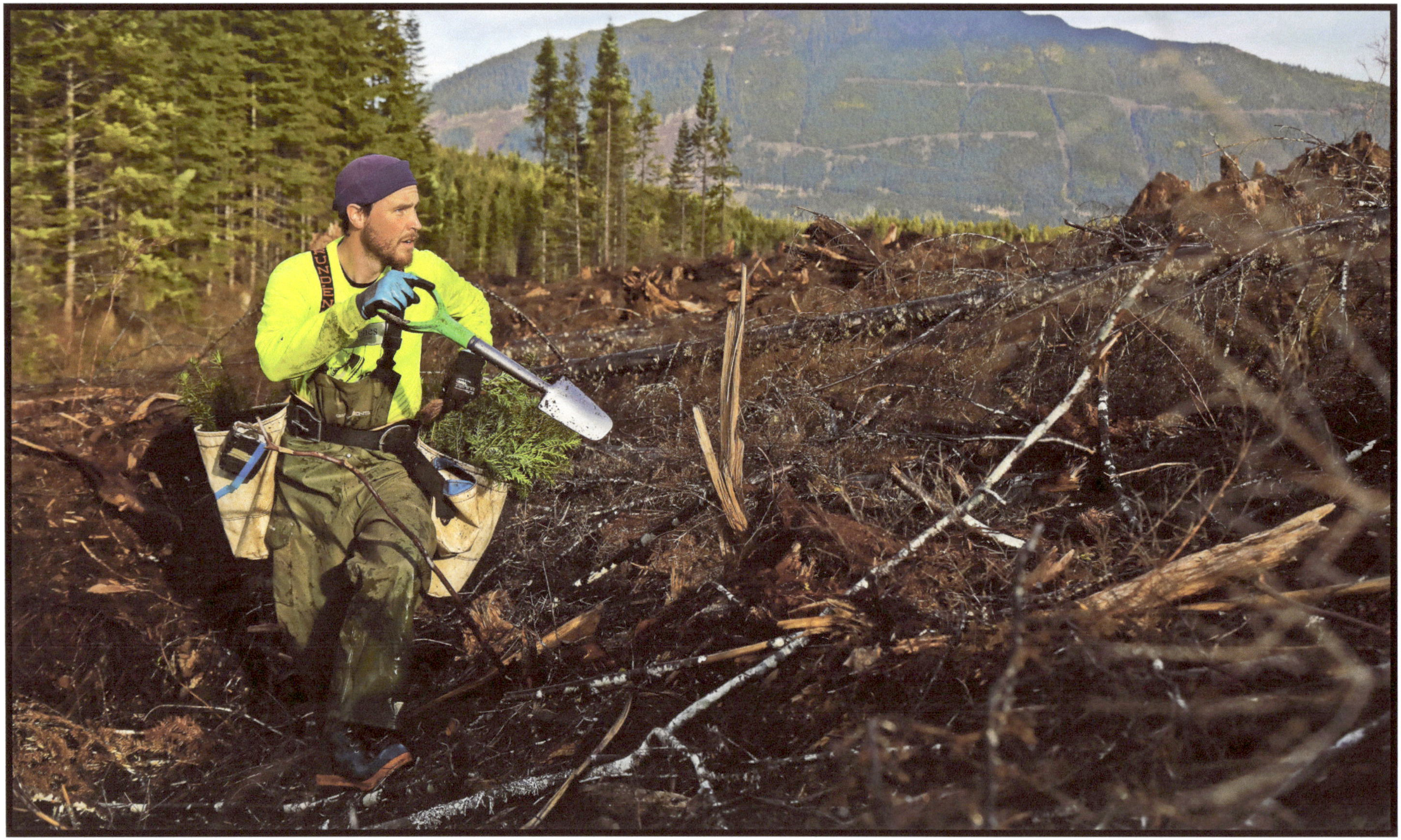

The Lars Zergun.

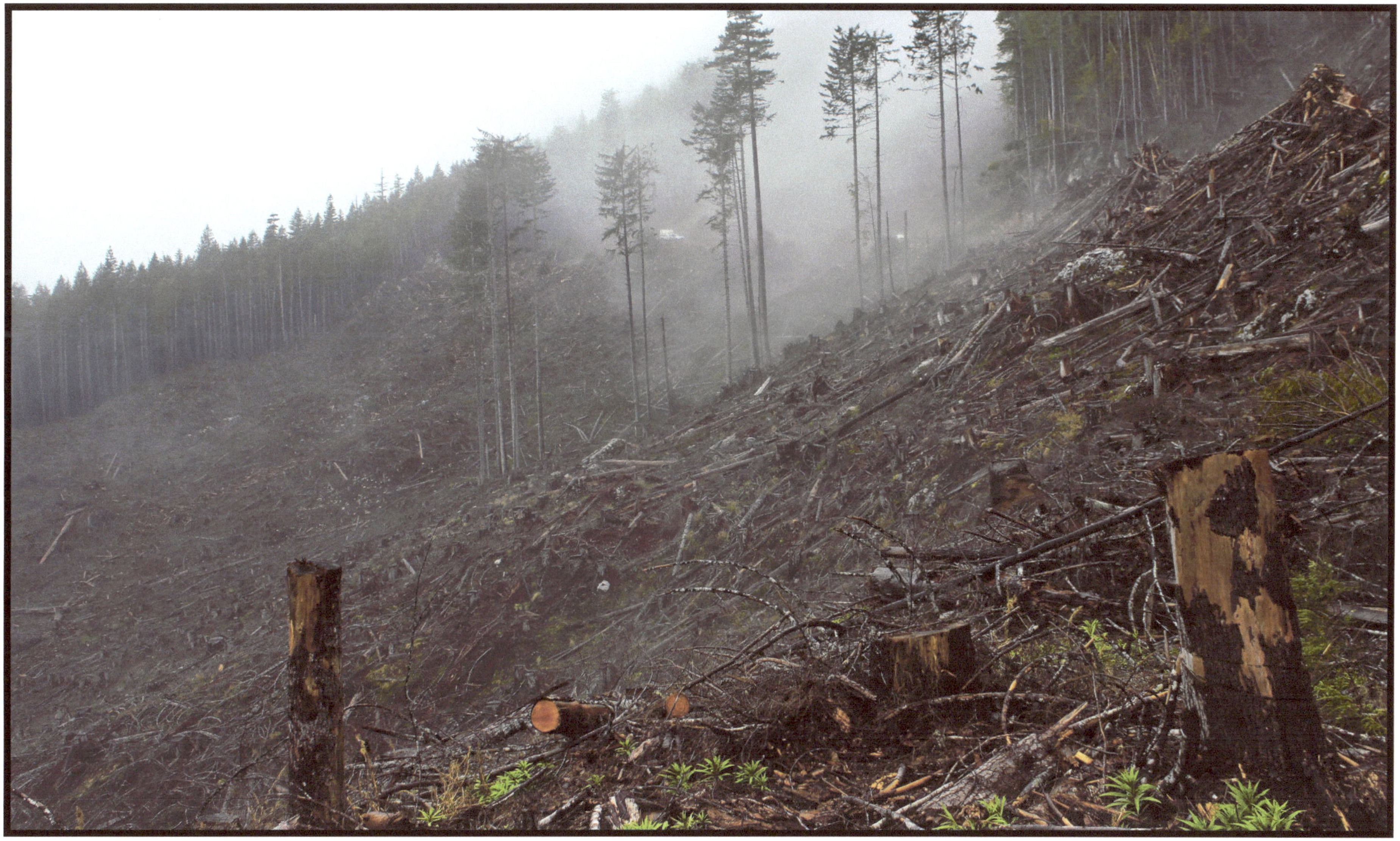

Another spring block.

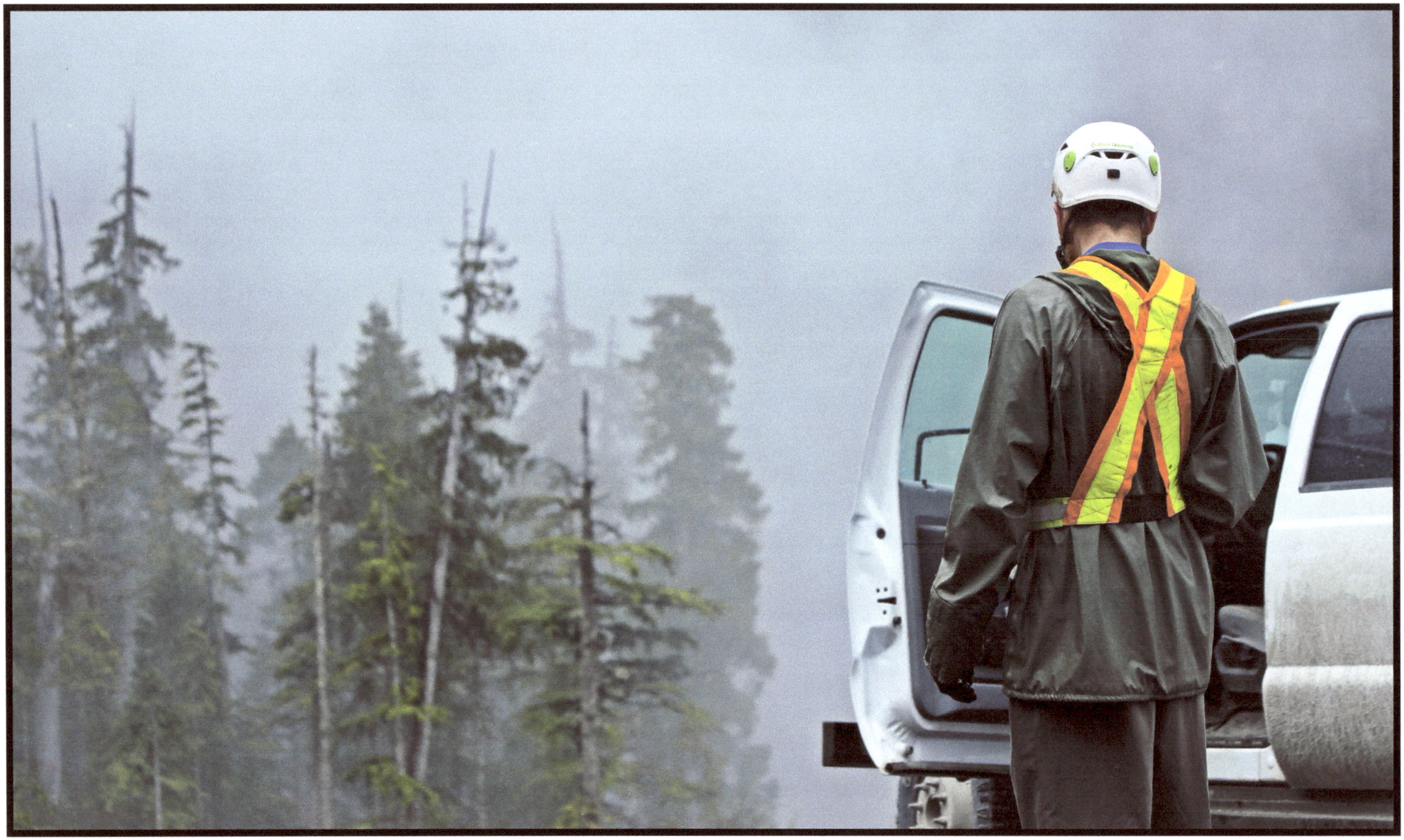

Mist is a constant presence throughout the spring season.

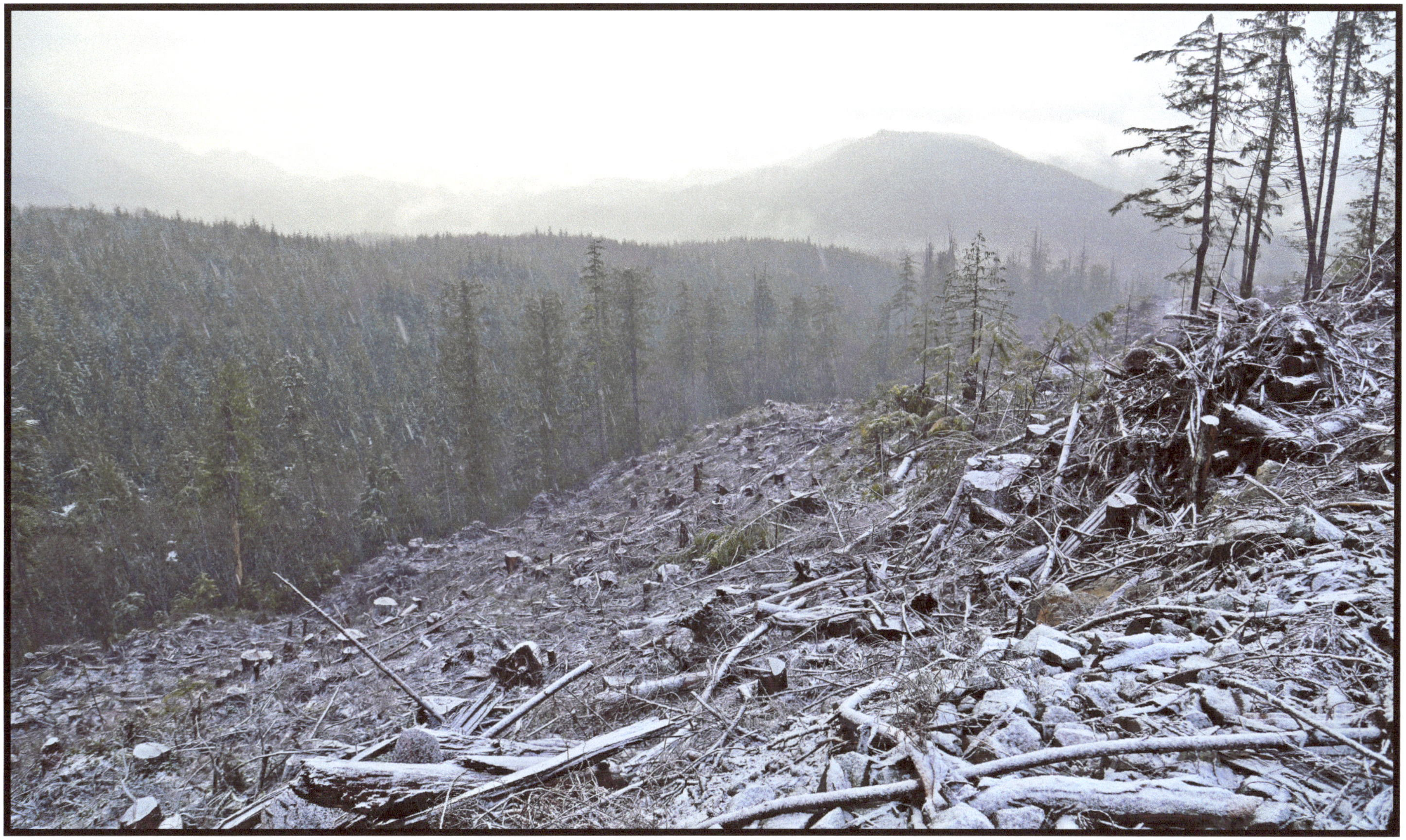

Ah, a change in the weather. Some freshly falling snow. Luckily, it's not that deep yet, so we should still be able to plant for a while.

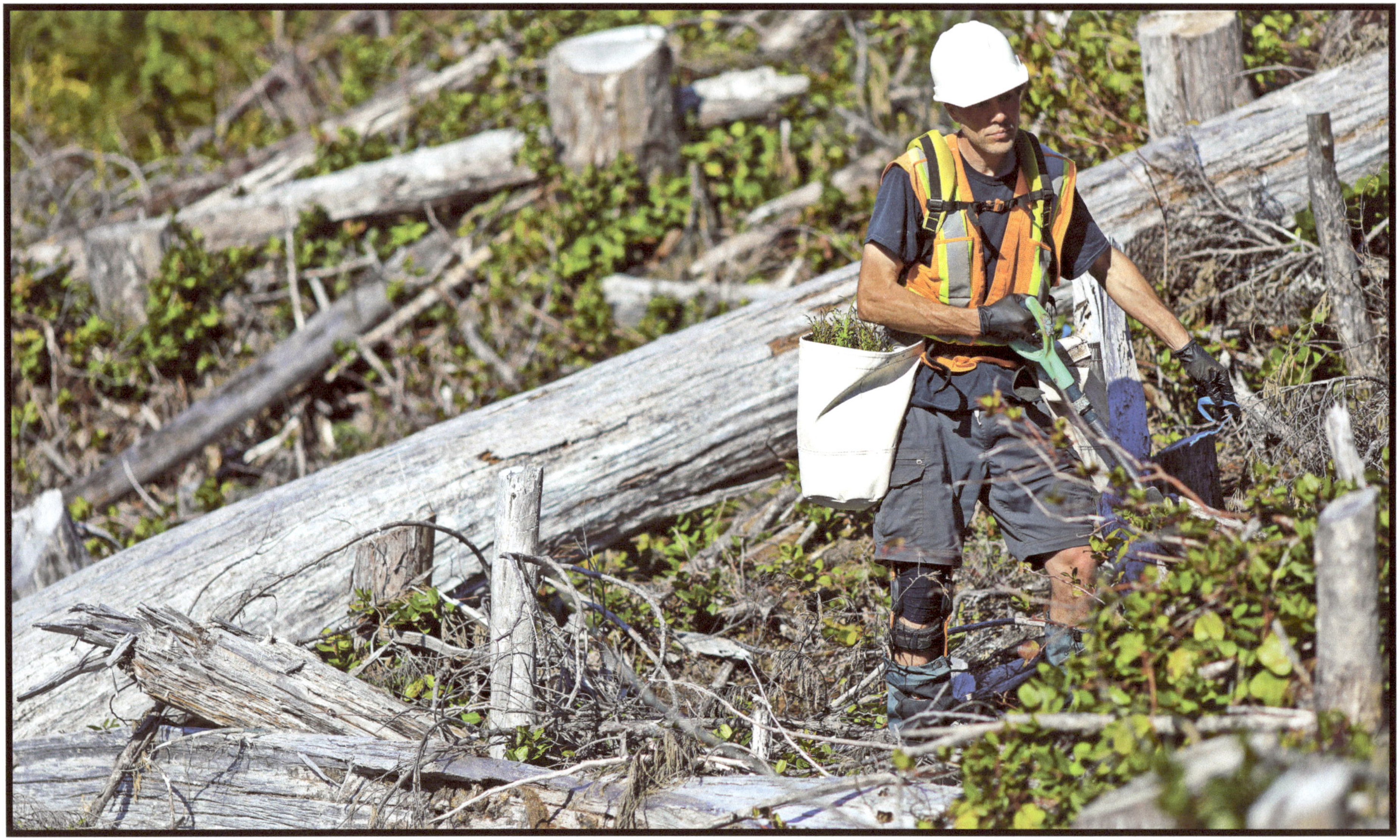

Rainier lives in BC, and you probably won't find too many people that enjoy all of the crazy aspects of planting more than he does.

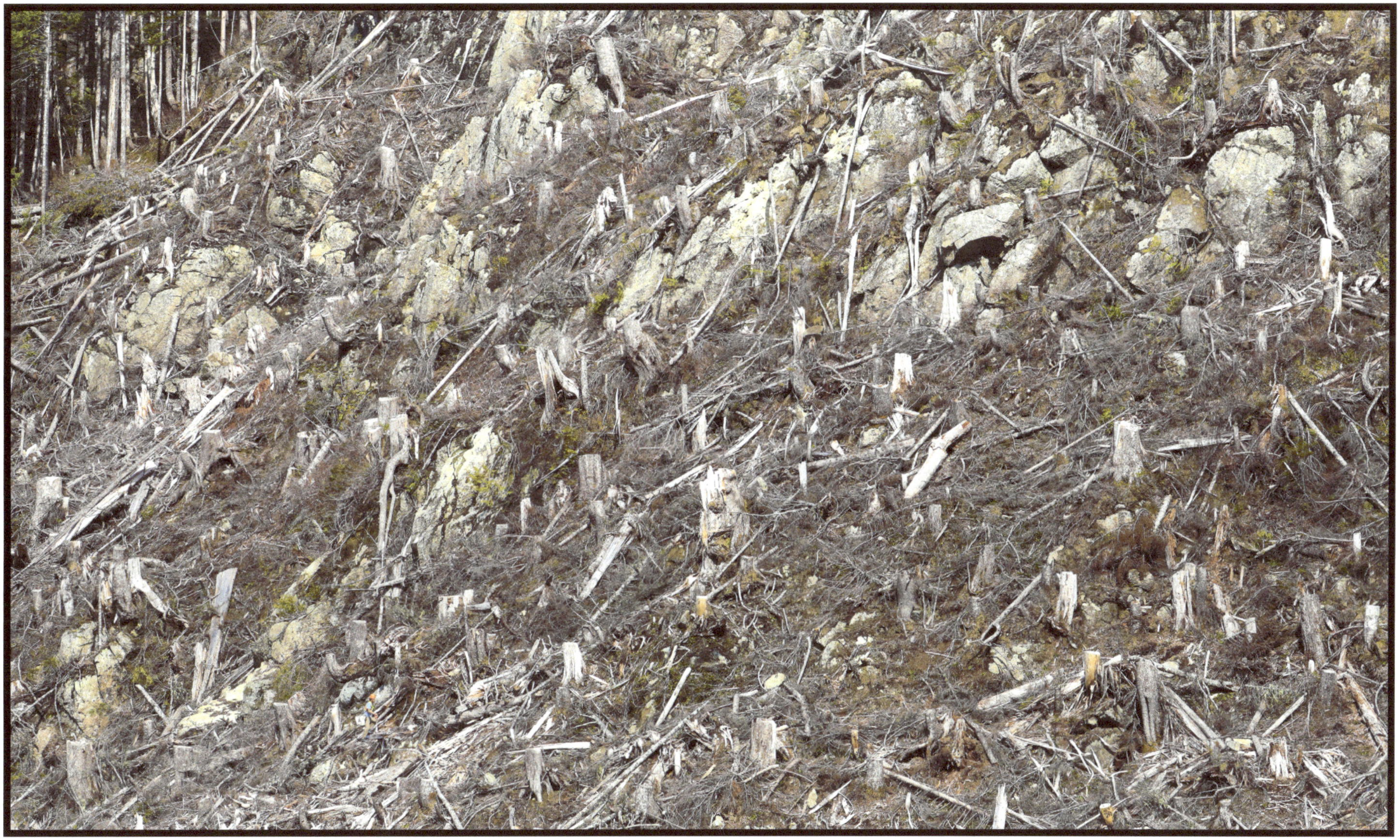

Find the planter.

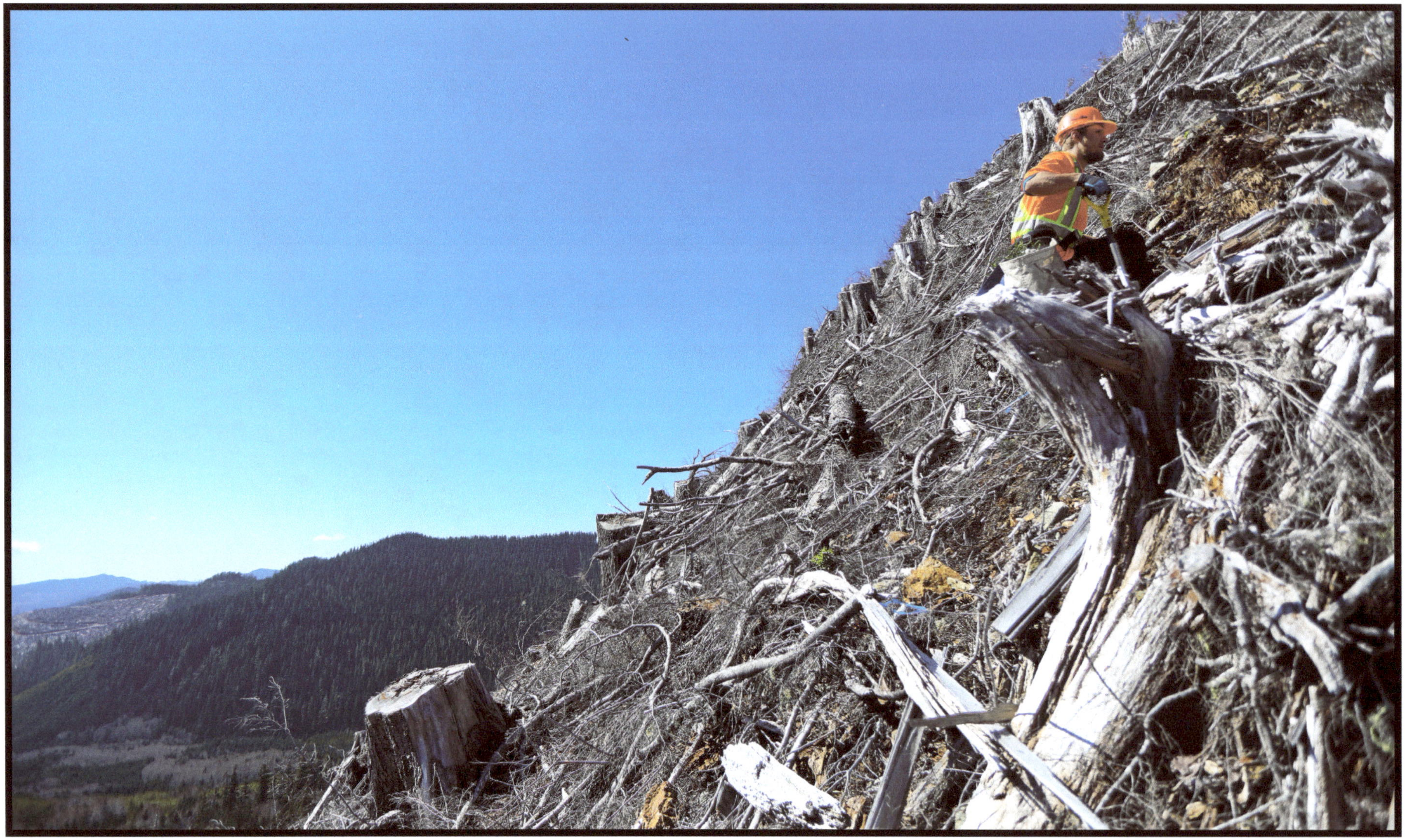

This is one of the few photos in this book that I feel truly illustrates how steep some of our blocks are. Evan says that he has always been drawn to the mountains and to being outdoors, and that's one of the reasons why he loves the challenge of steep coastal blocks. When he isn't planting, Evan spends a lot of time mountain biking, snowboarding, and playing hockey.

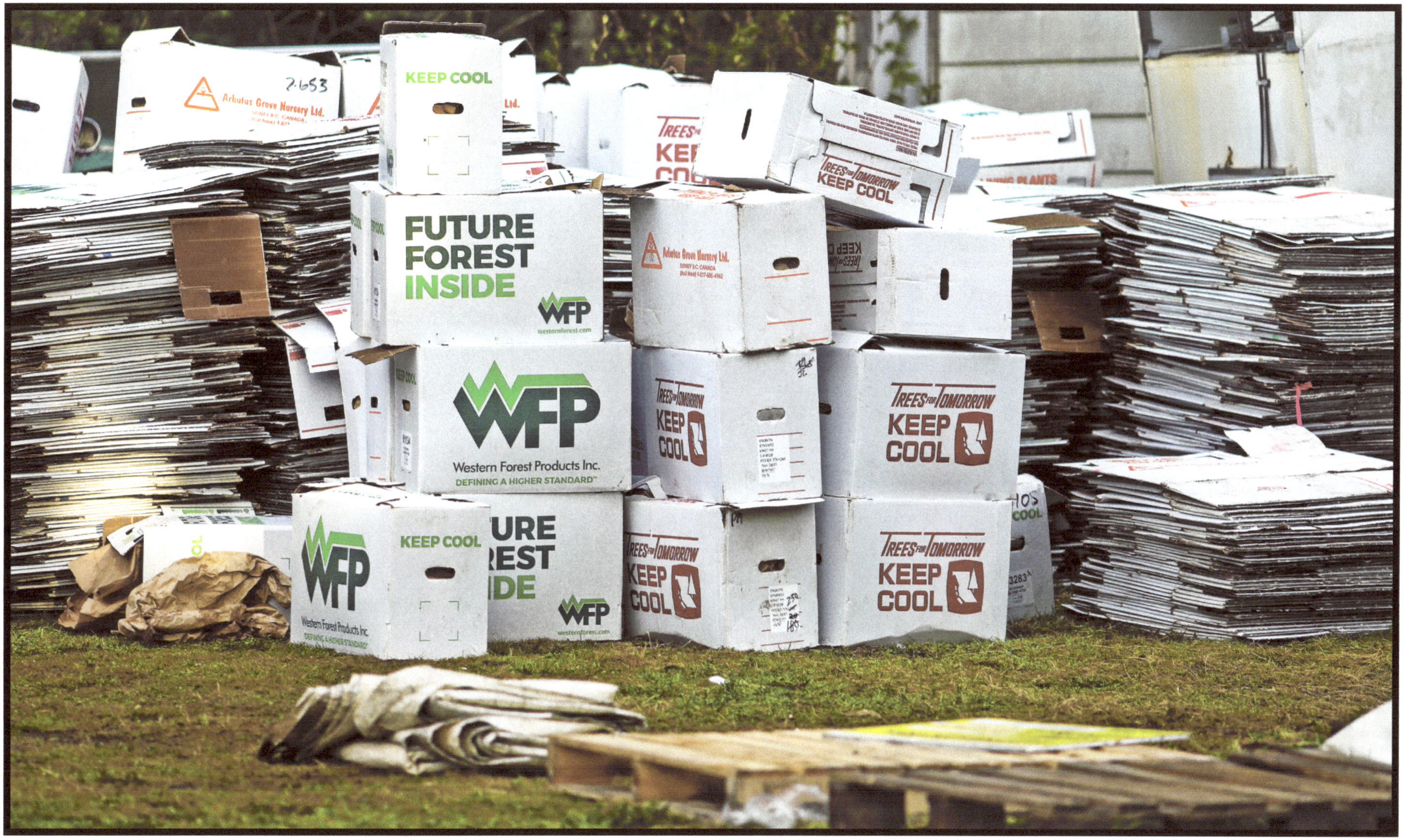

Piles of empty tree boxes. It looks like we've planted a lot of trees so far.

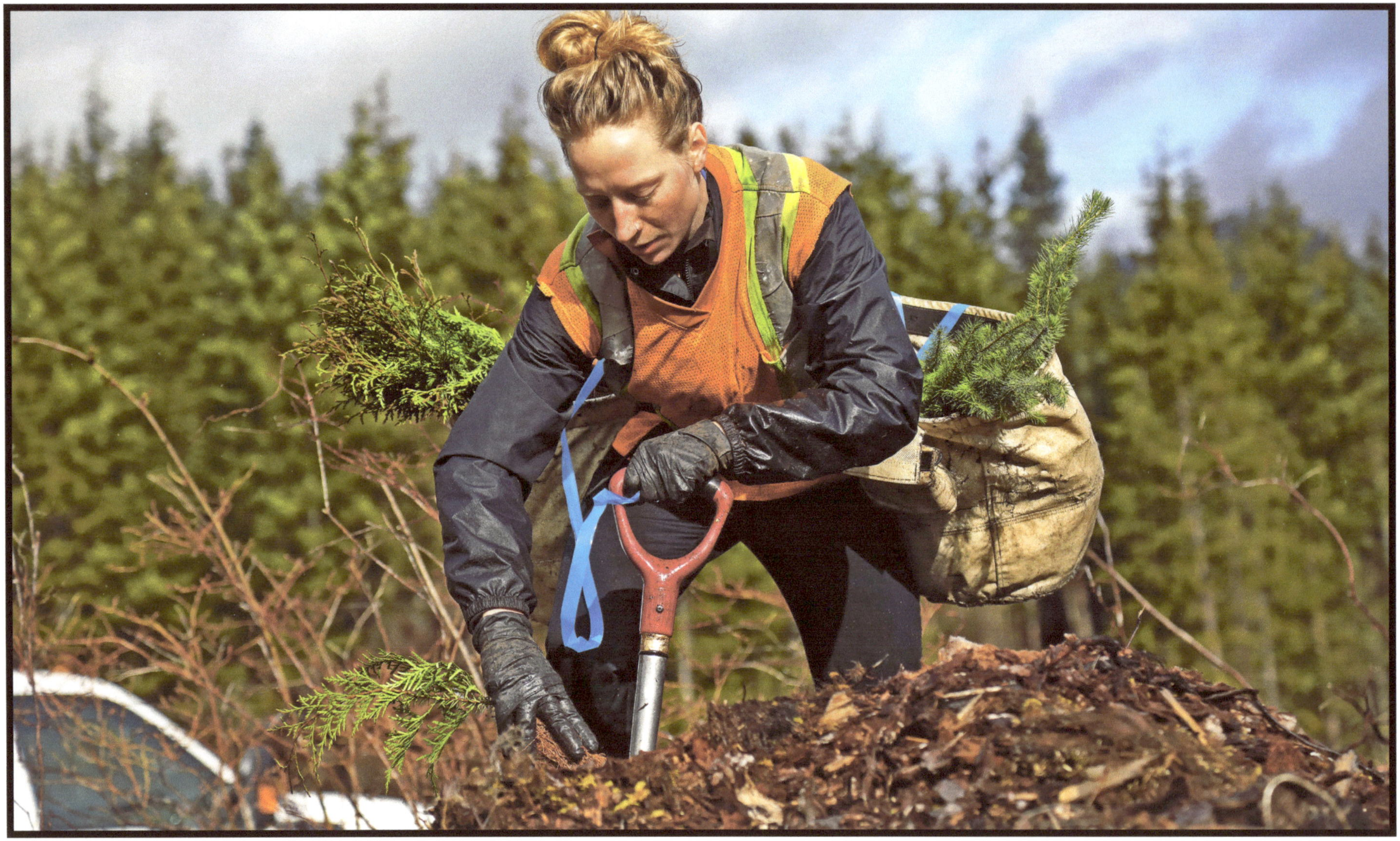

These are some fairly large seedlings that we're planting.

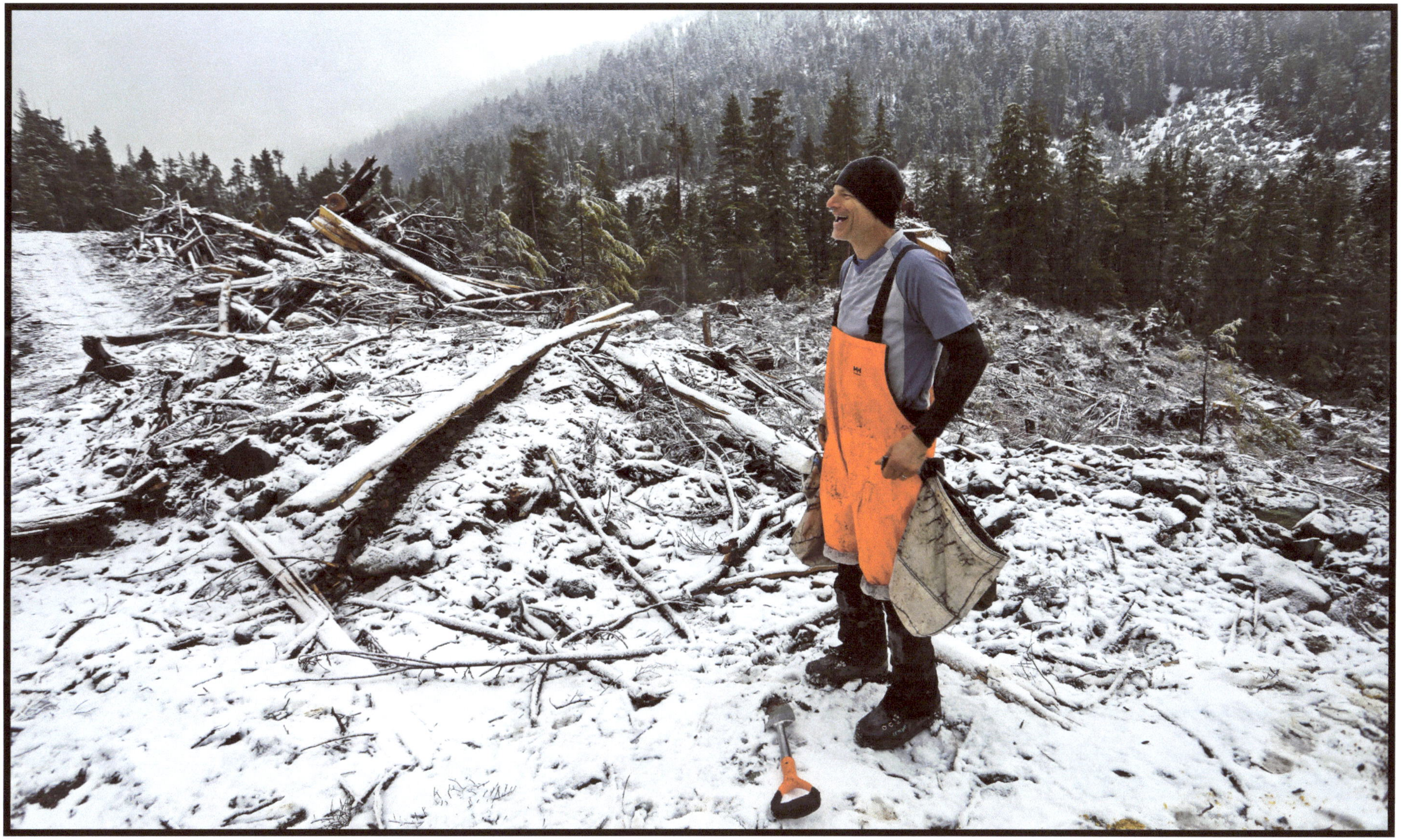

Wayne is laughing at the snow. Wayne is always laughing or smiling.

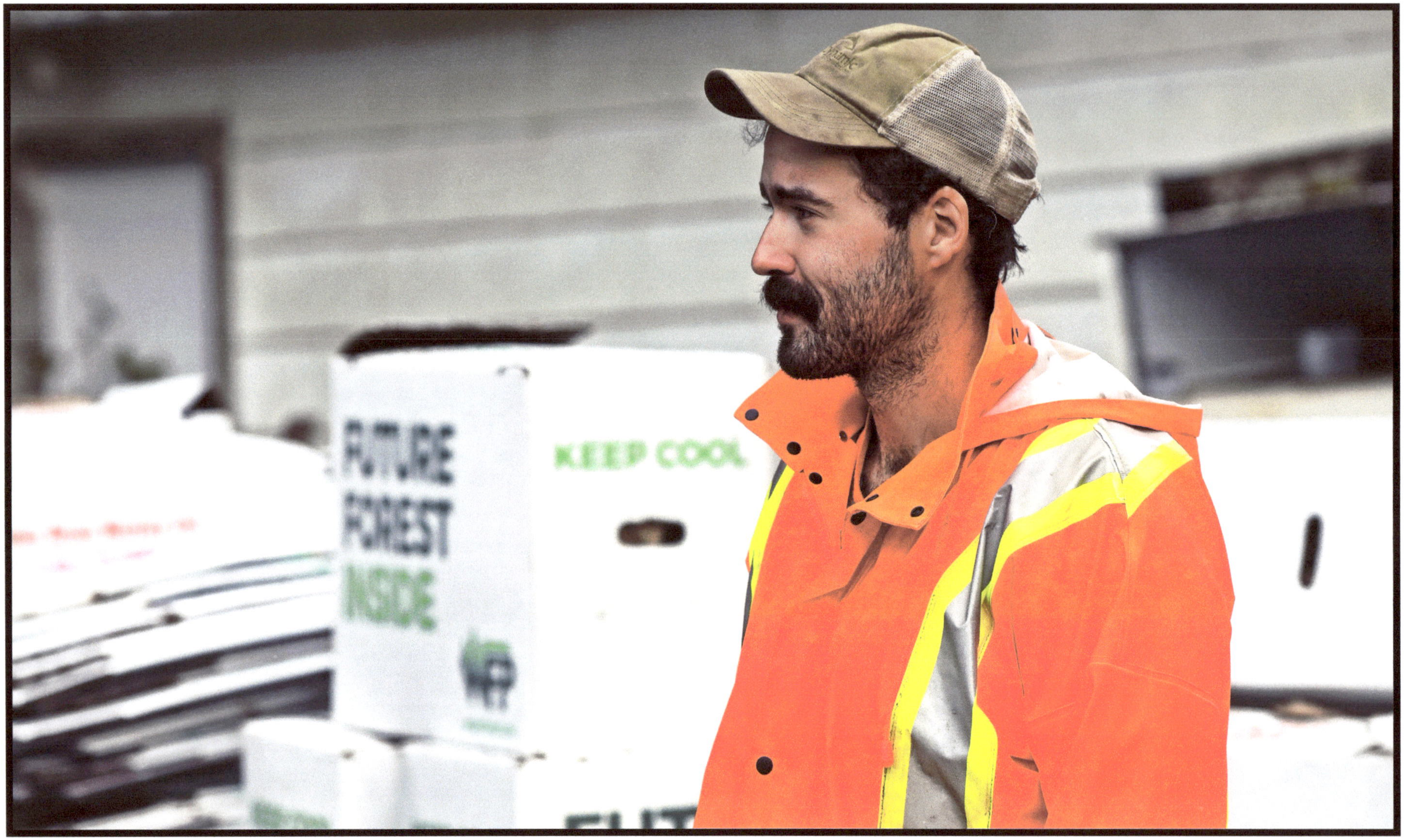

Kiernan is from Saskatchewan. The coastal slopes are very much unlike the ground he's used to back home, but he still loves his work. Kiernan enjoys working on projects involving wood-working and machining.

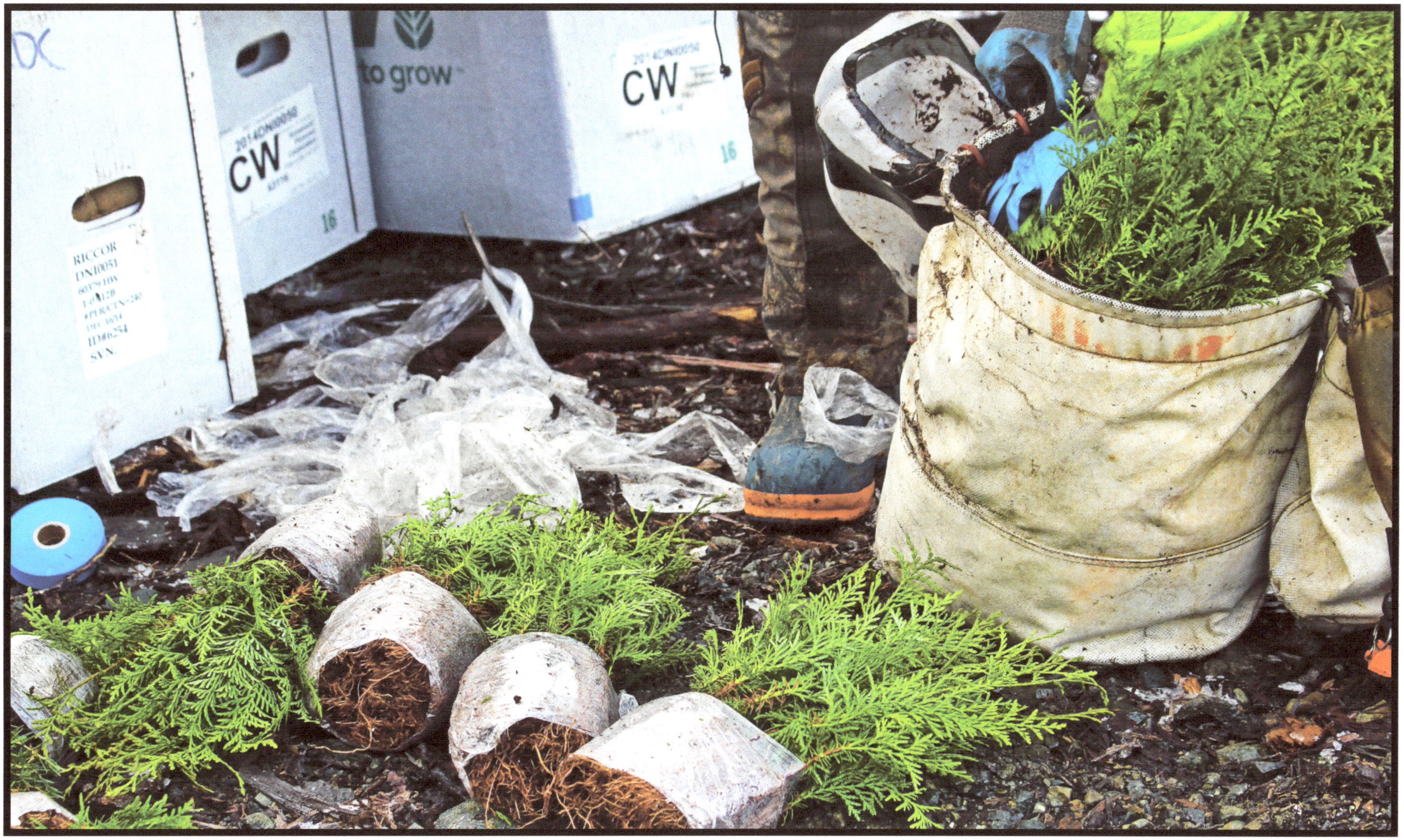

Bagging up with a load of red cedar.

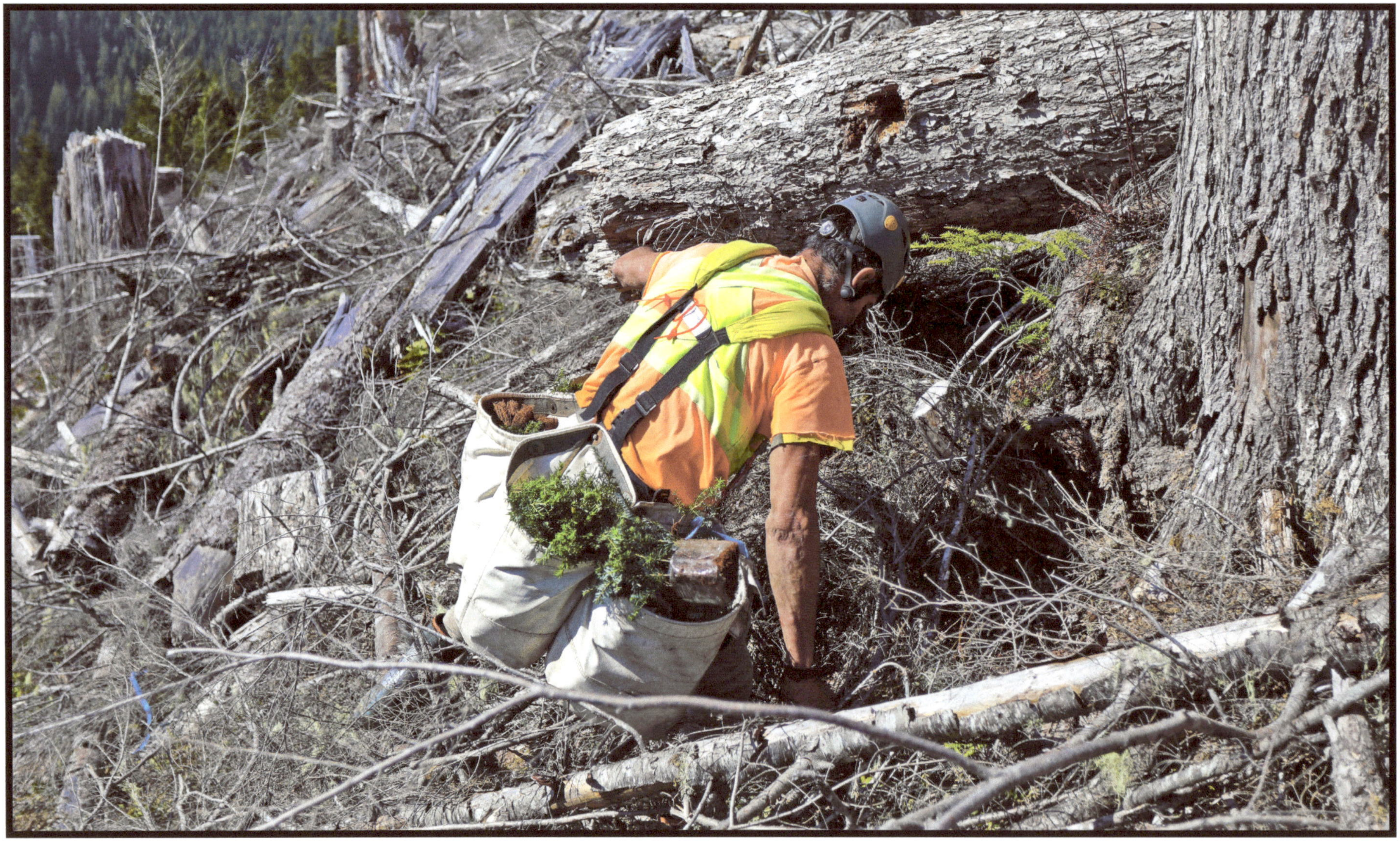

Simon, easily navigating his way through a slash-filled block.

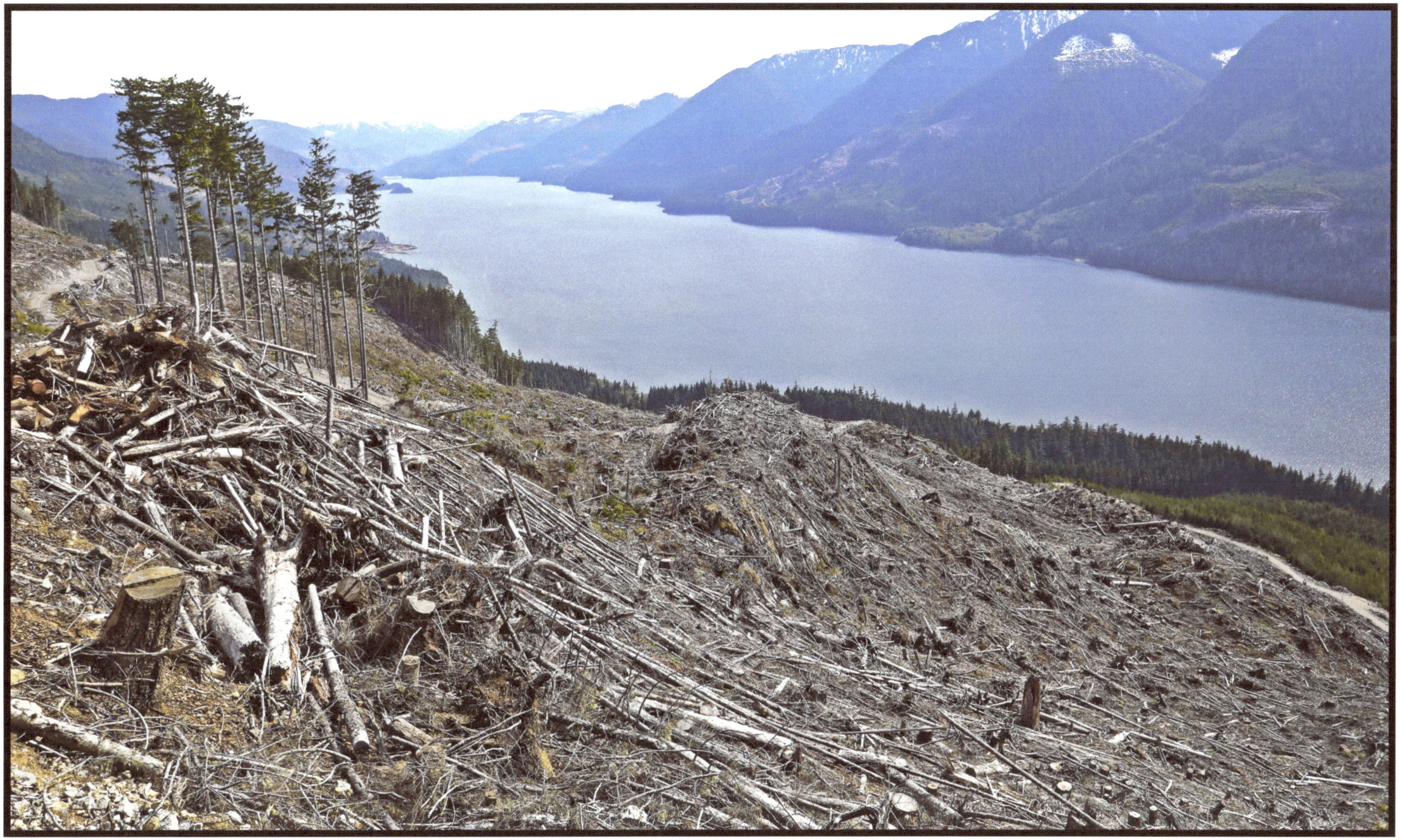

Looking back toward Port Alice.

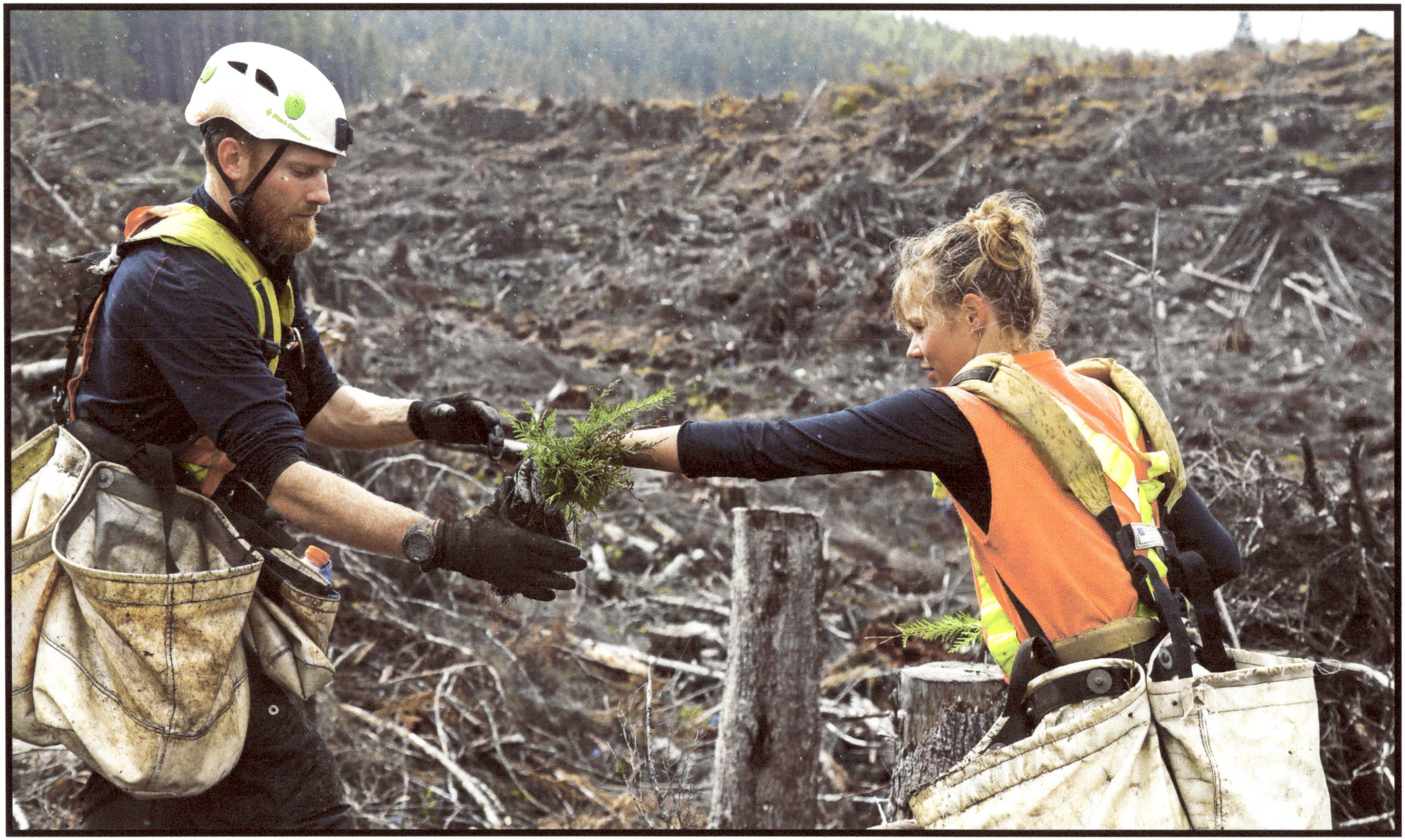

Although it's hard to see without being able to zoom in, this photo does a great job of capturing the falling raindrops in the air.

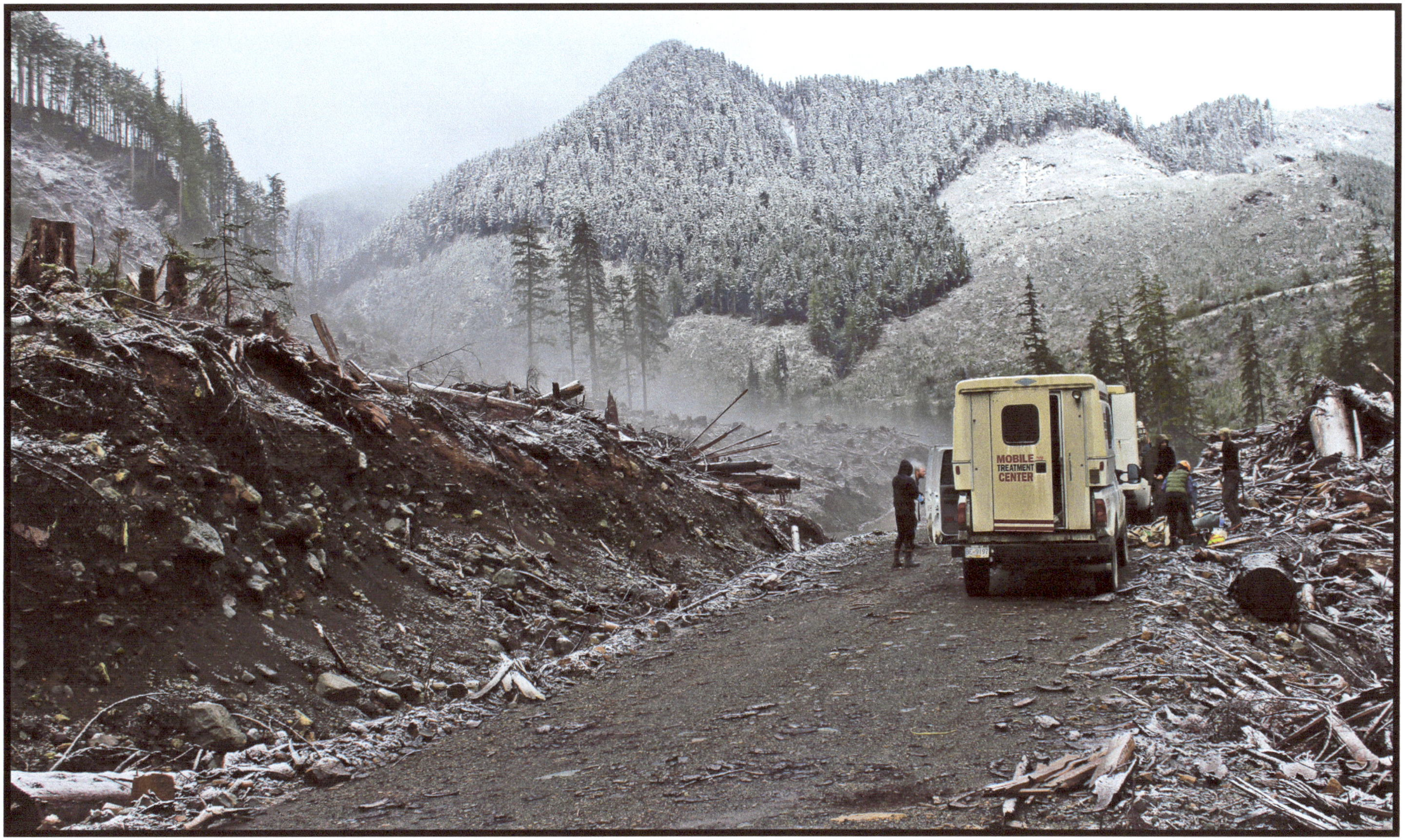

Pushing the limits at the base of the snow line.

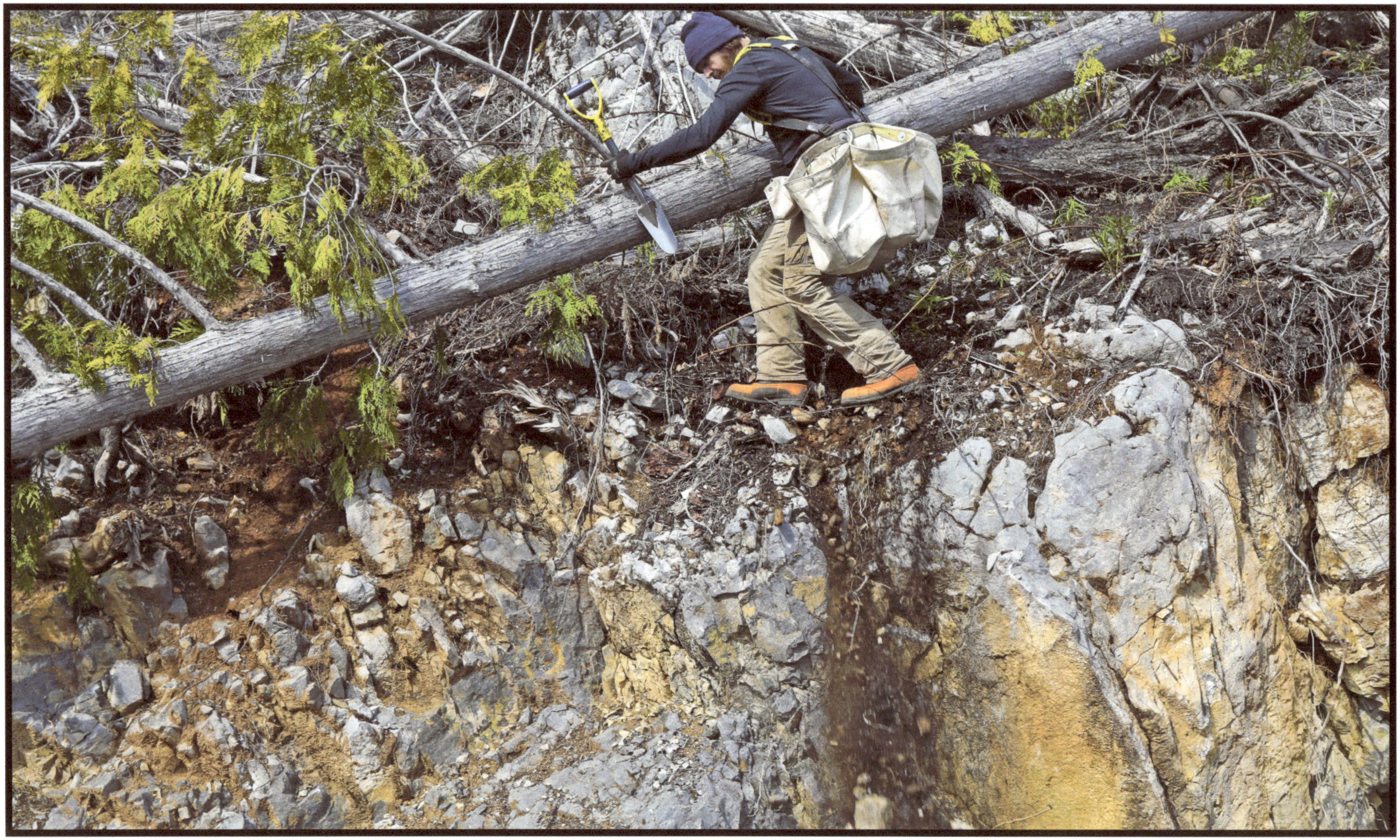

It's important to watch your footing.

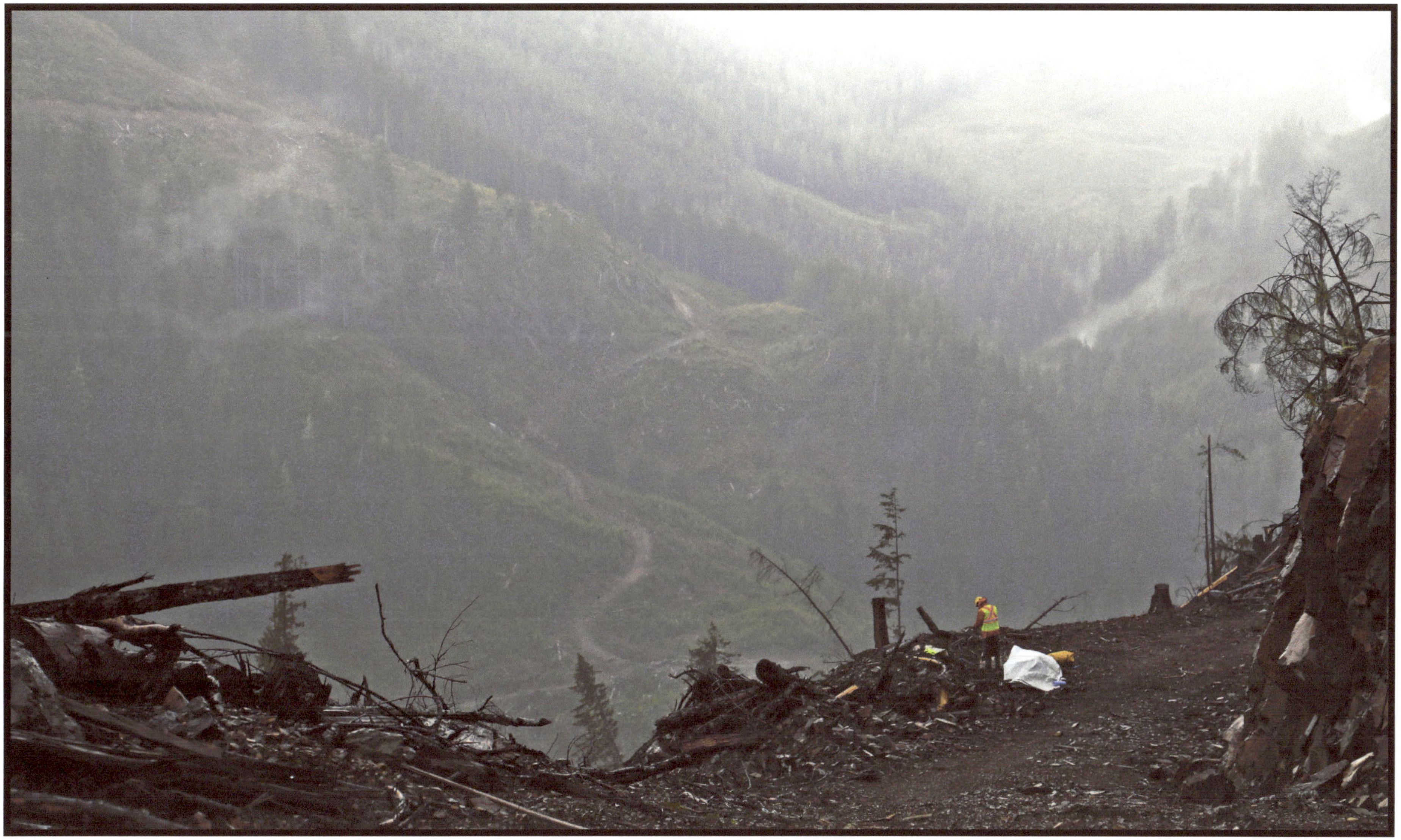

The roads in the block on the other side of the valley help to show the size of this area.

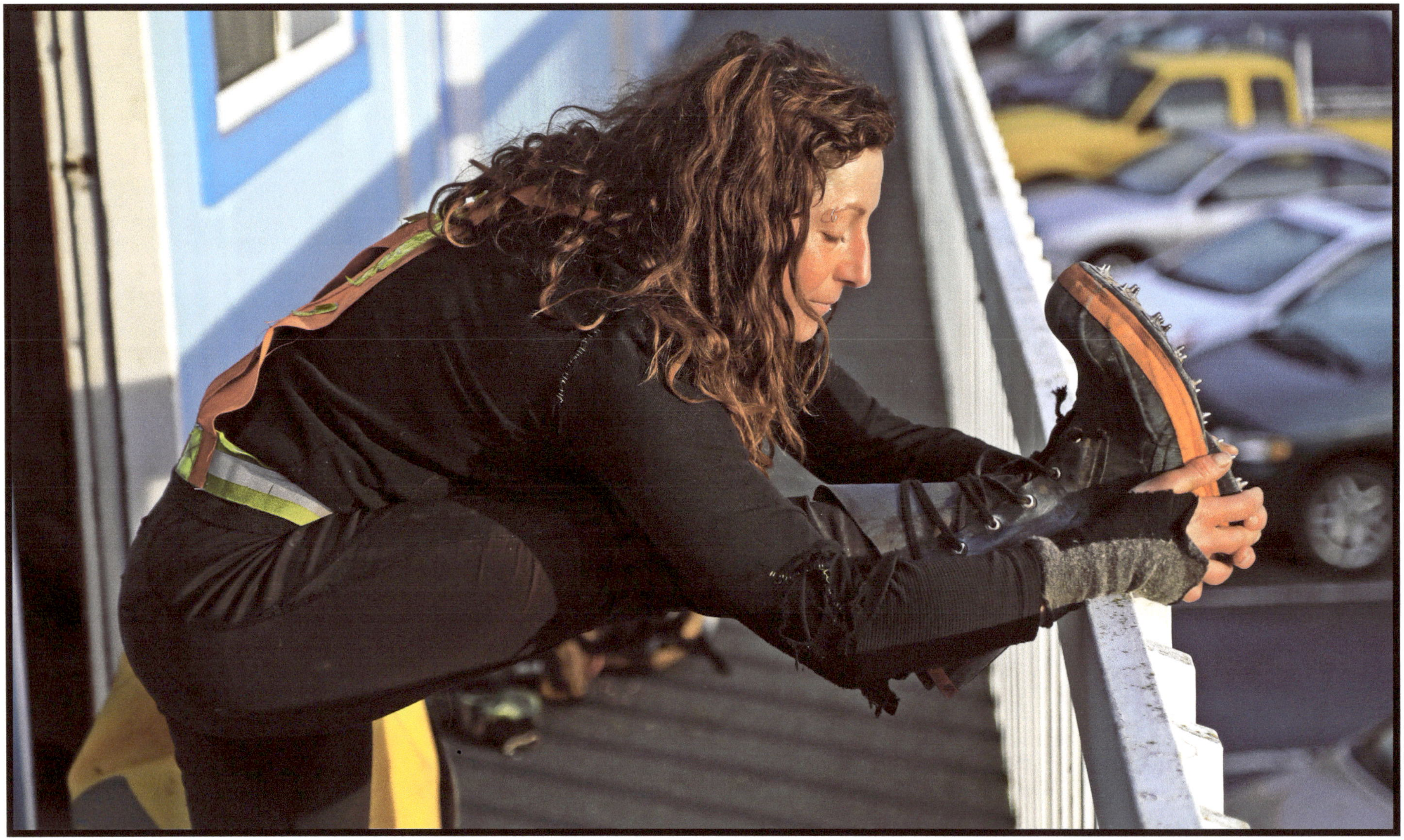

It's important to stretch your muscles in the morning, before starting to work.

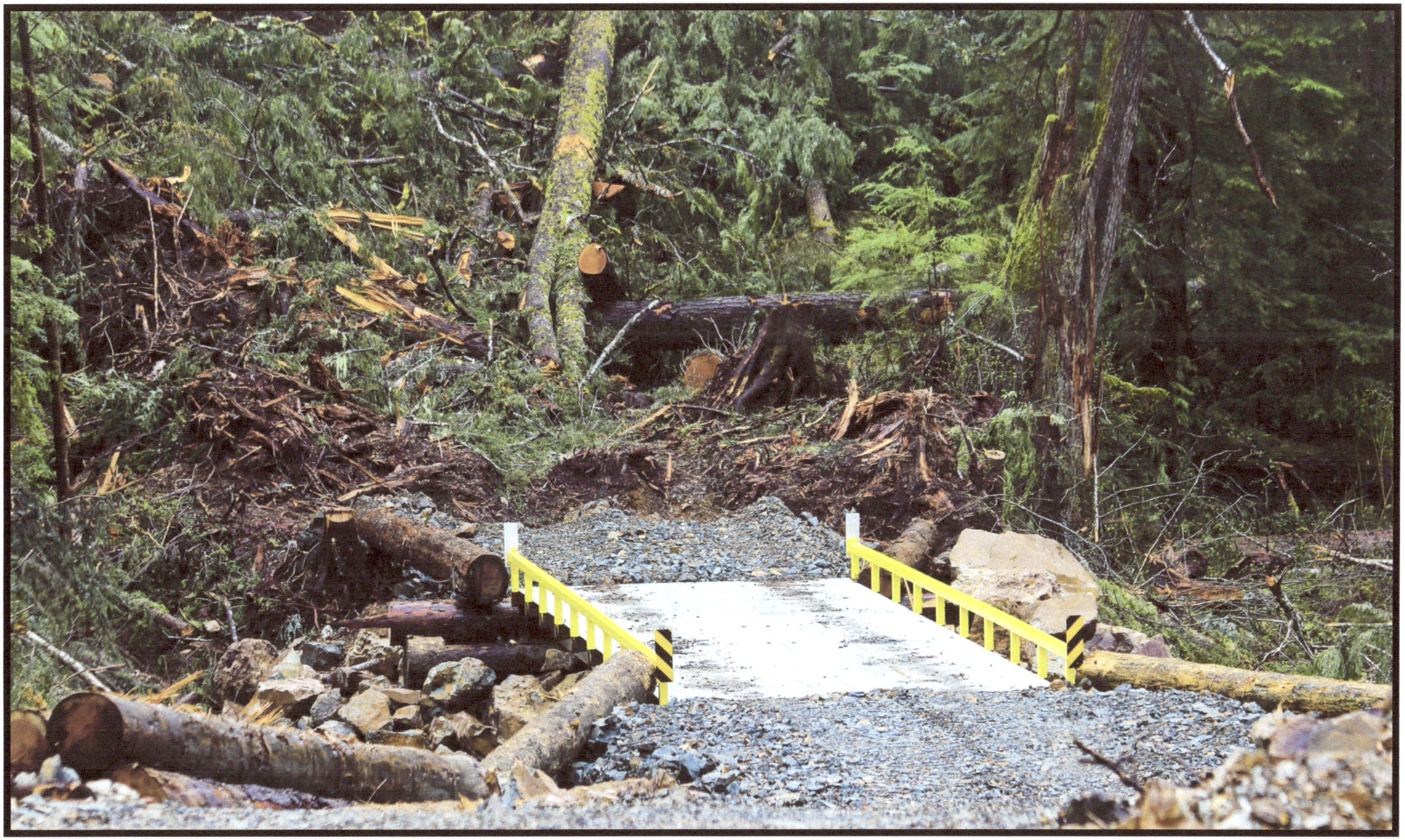

A bridge to nowhere.

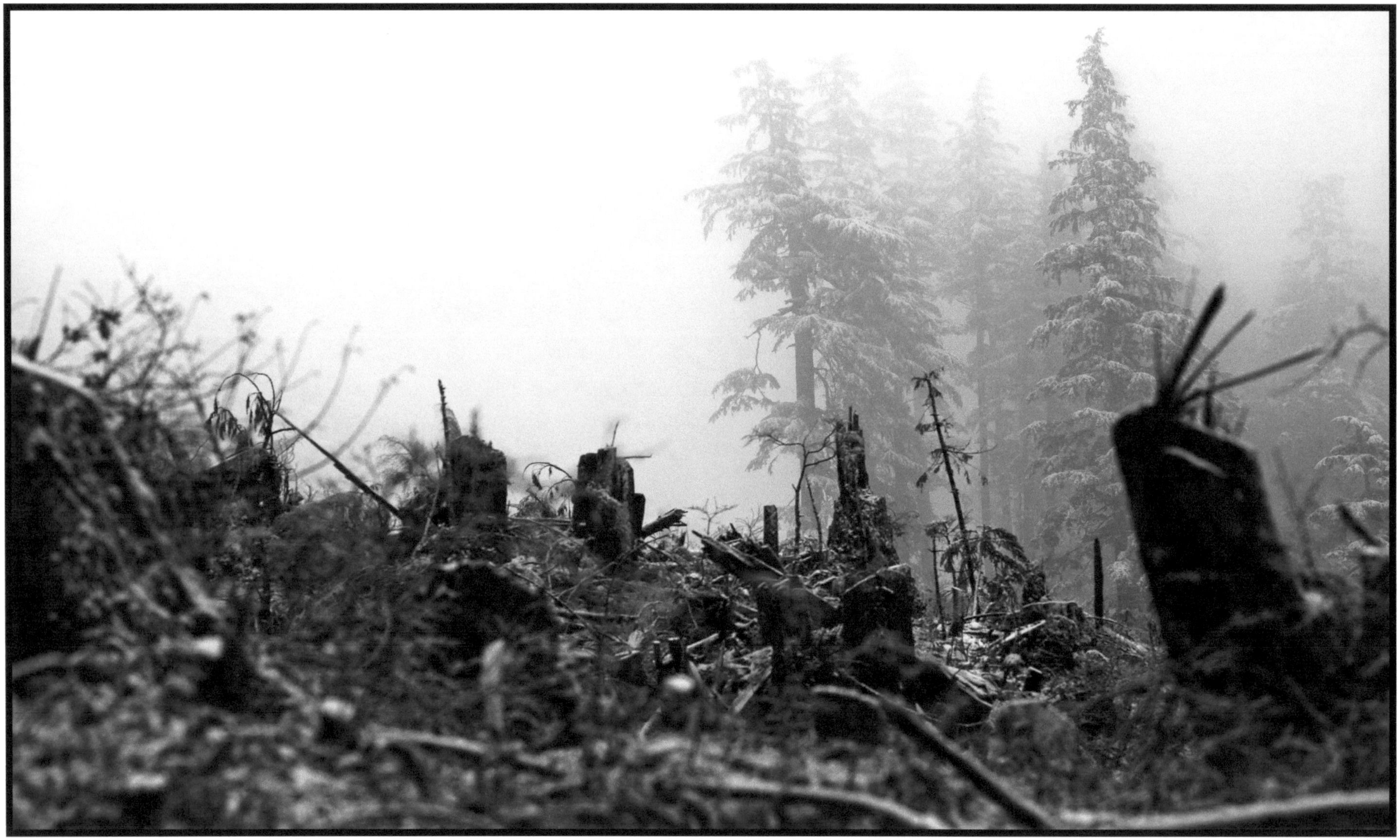

A typical grey day in late February.

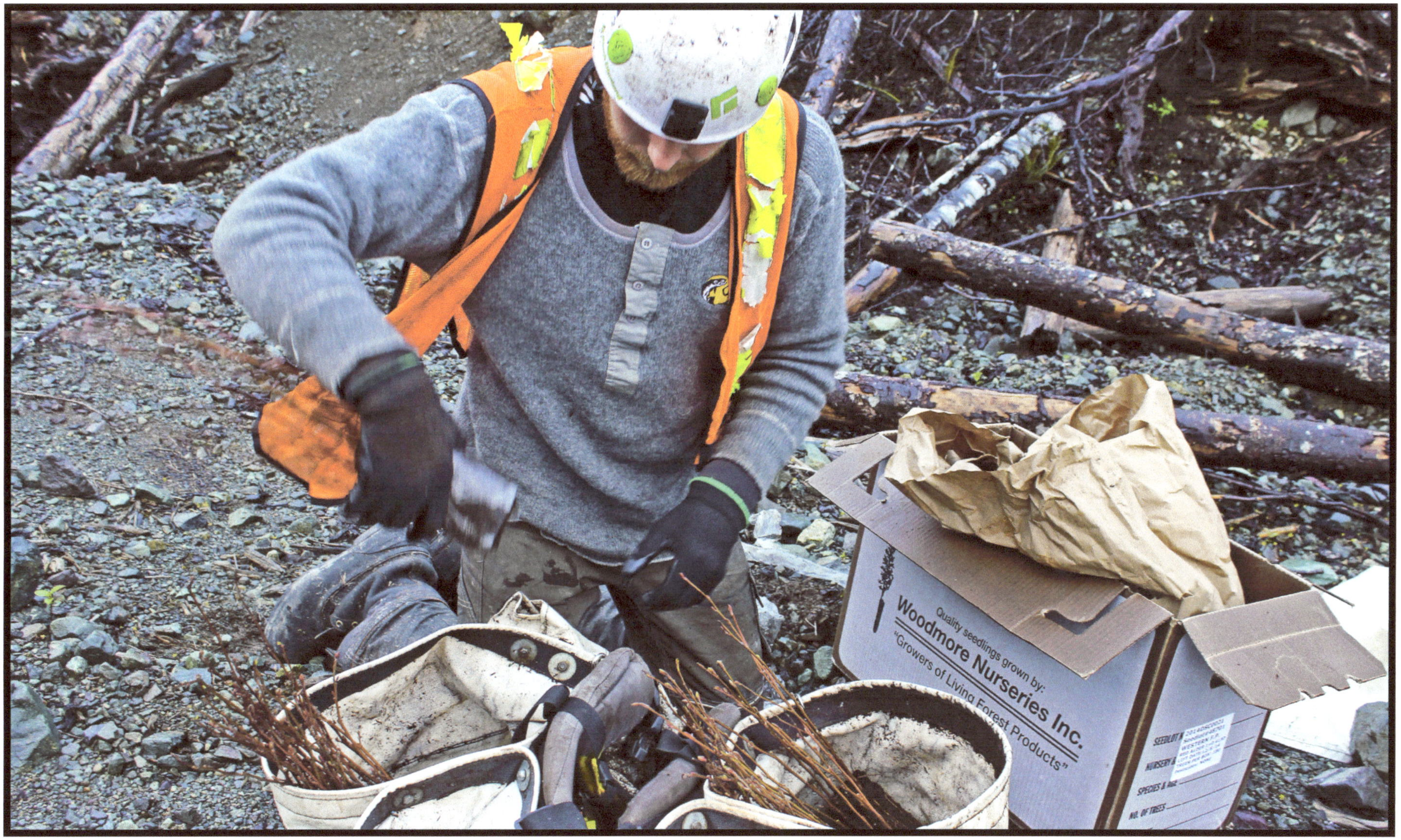

Bagging up with a load of red alder, a deciduous species that is sometimes planted at lower elevations.

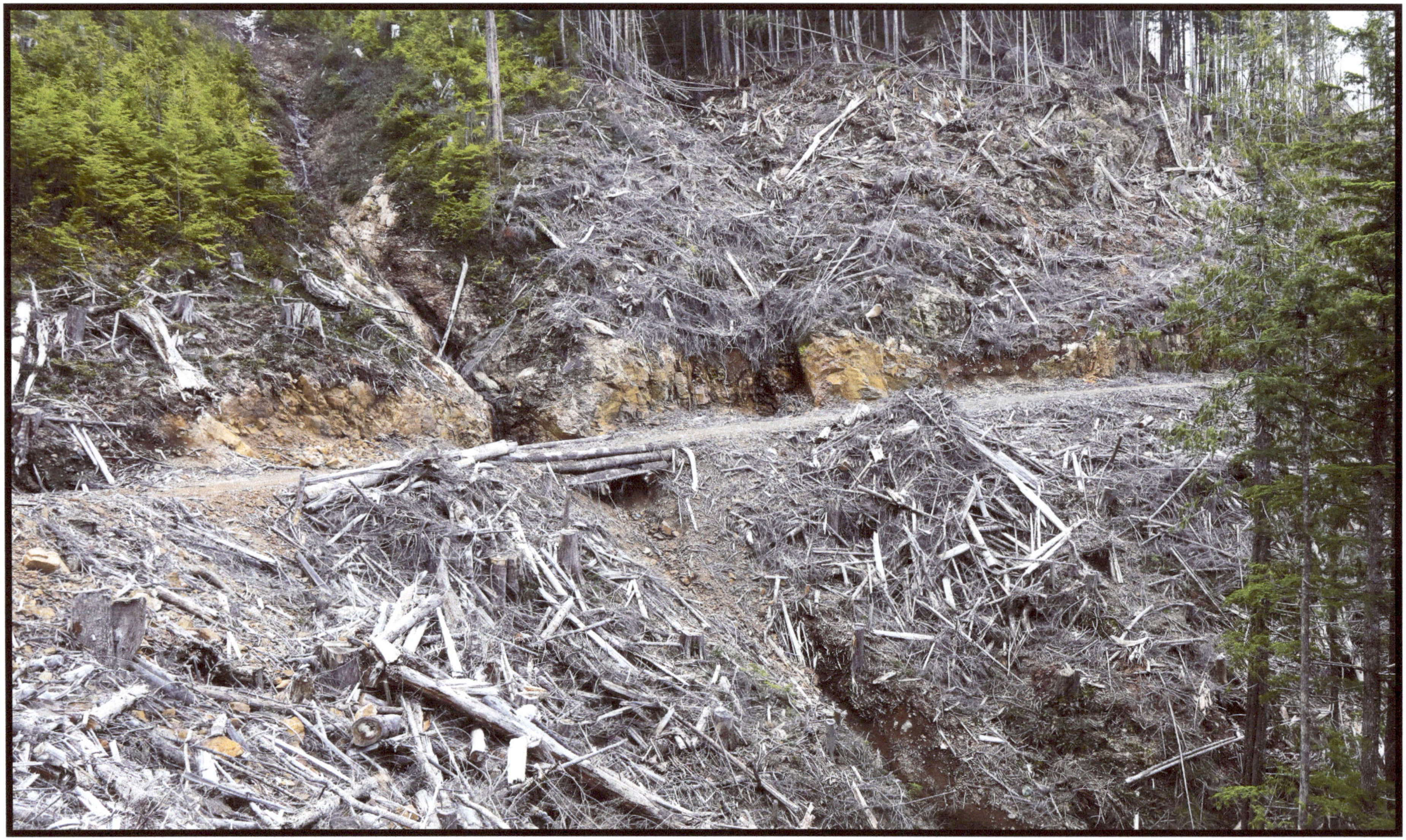

The piece above the road took me an entire day to plant. I wish there was a truck in this photo to show the scale. The main face above the road in the center of this photo was about twenty feet high. It was almost impossible for me to climb up into this piece in the morning, even in the area where it appears to dip close to the road. The foreman helped by tossing trees up to me all day, while I stayed up in the piece. I often think that it would be helpful if each of our trucks carried an extension ladder, to help us get up into steep pieces.

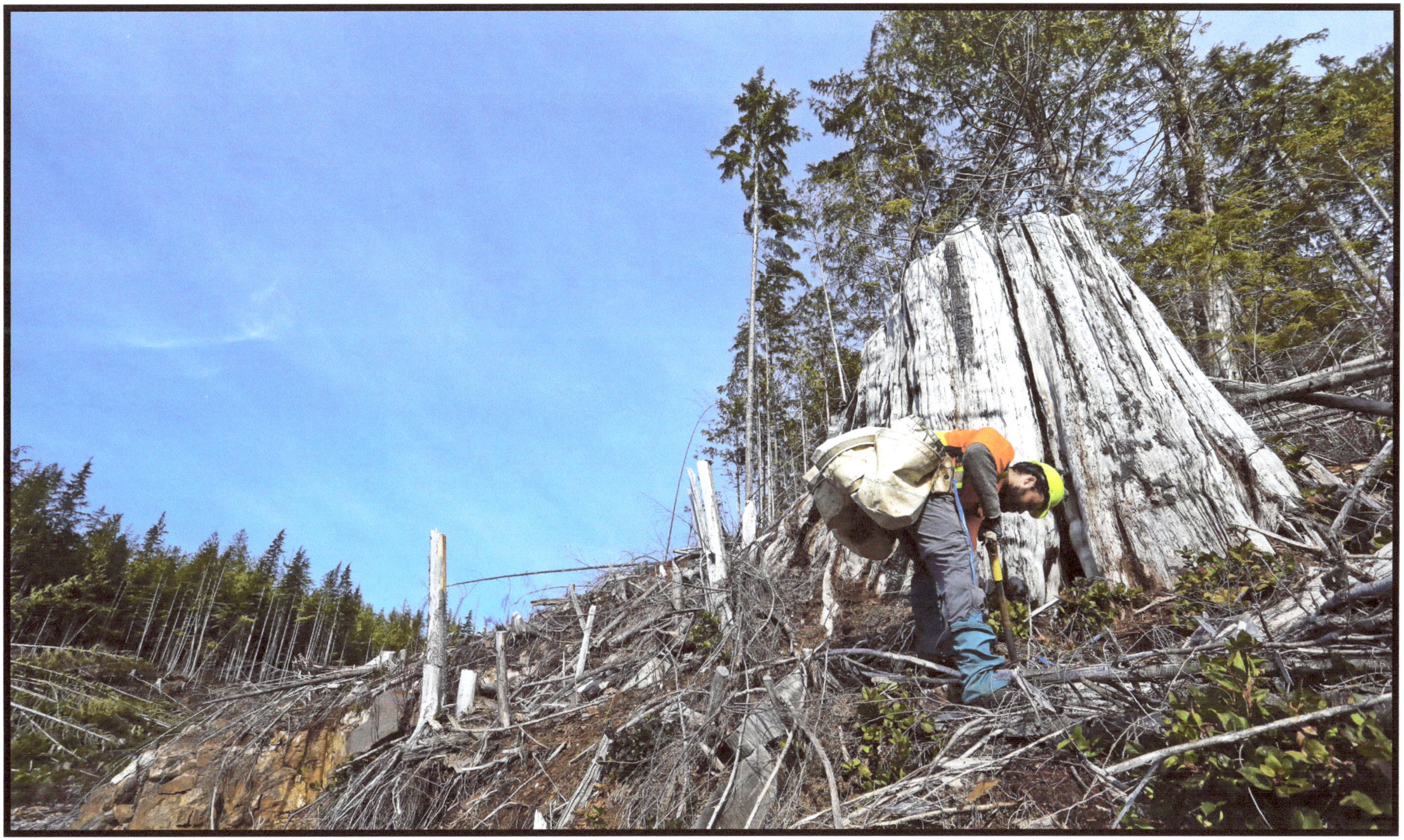

Kiernan, planting beside an old cedar stump.

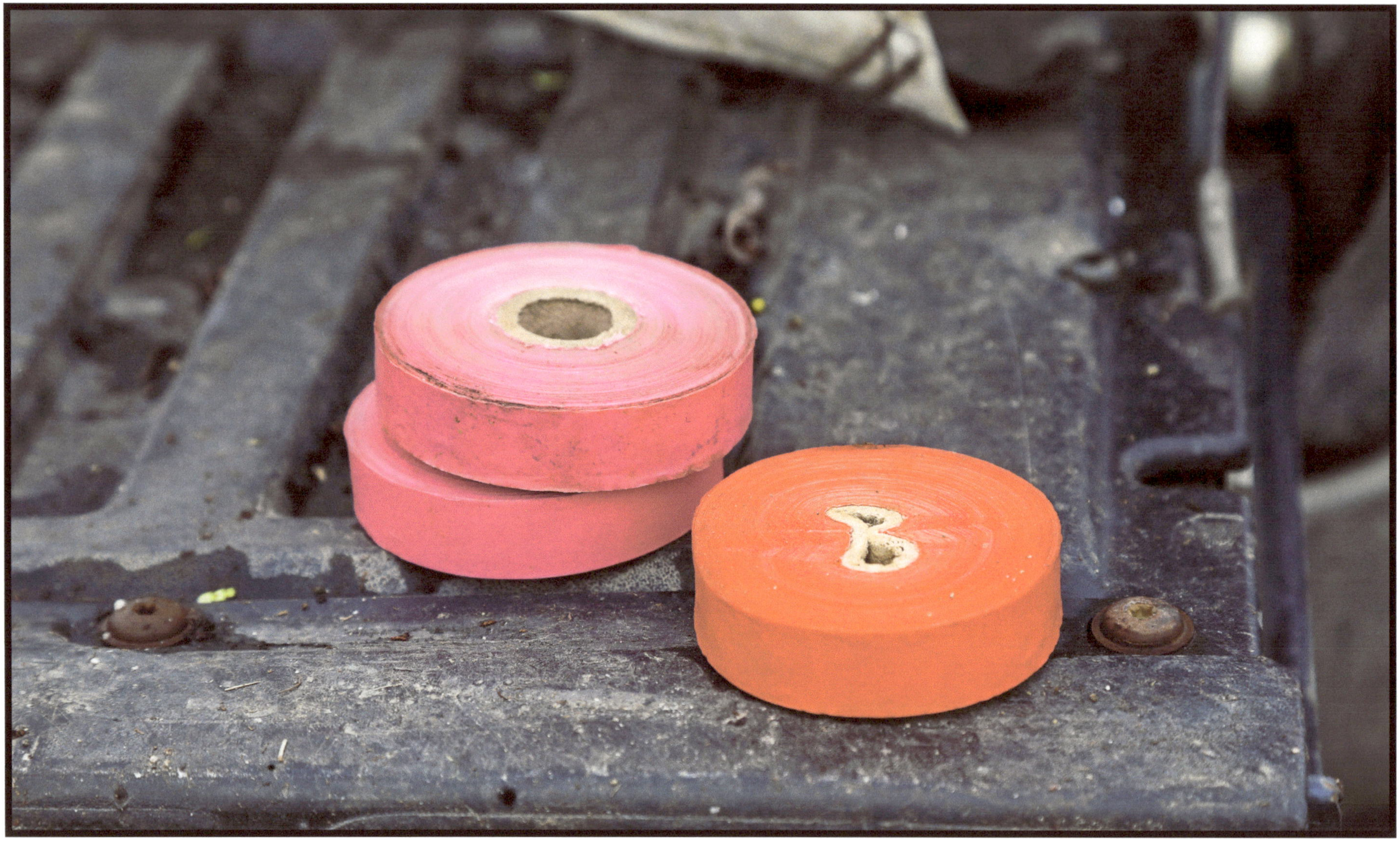

With the heavy slash and vegetation, it's difficult to see where new seedlings have been planted, which makes it hard for us to manage our spacing and planting densities. We typically throw small pieces of flagging tape on the ground beside the seedlings as we work our way through a block, to help identify which areas have already been covered. Planters have an almost universal preference for blue flagger, although some other colours are used occasionally. These orange and pink rolls will probably sit in the back of the truck until all of the blue ribbon is gone.

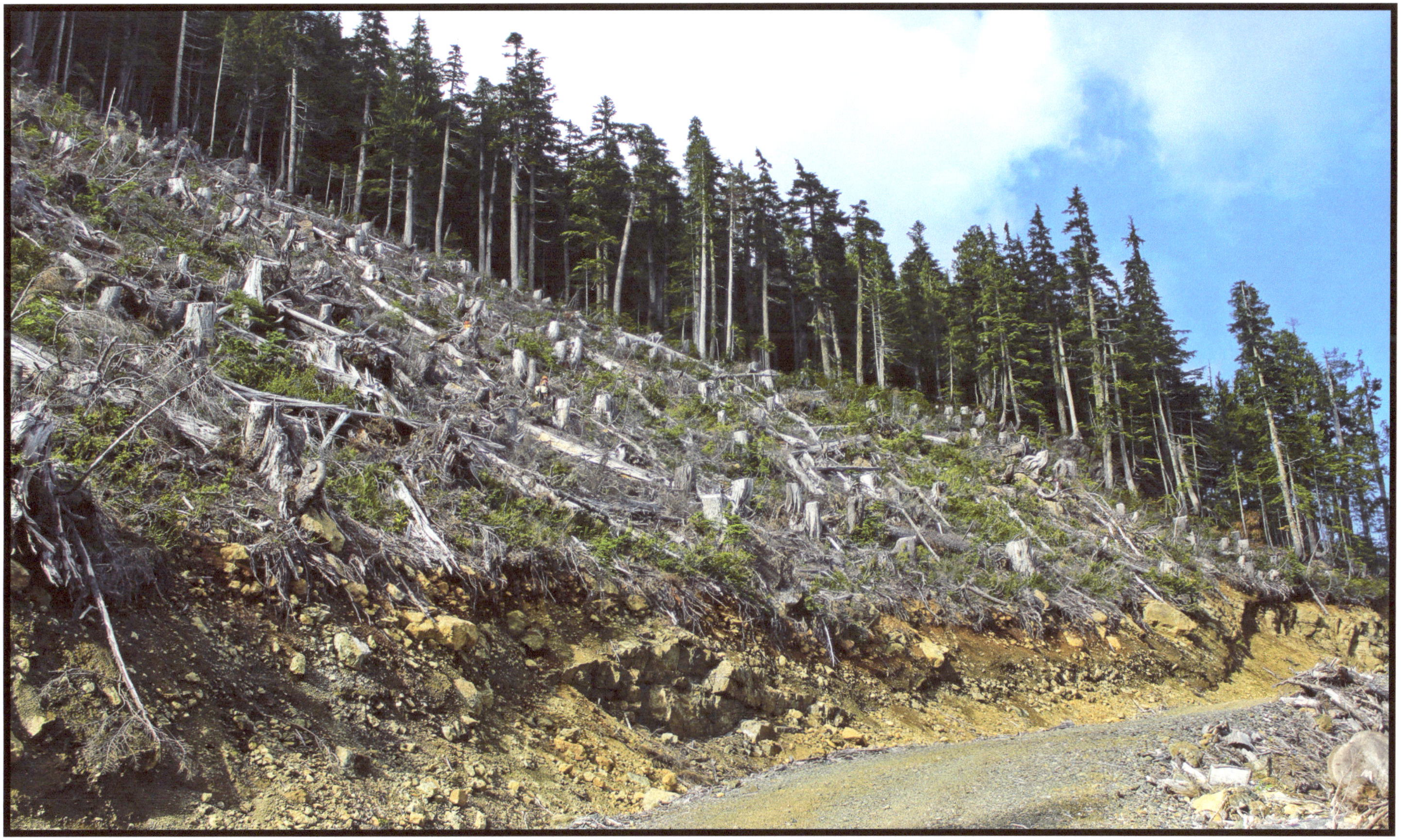

Find the planters.

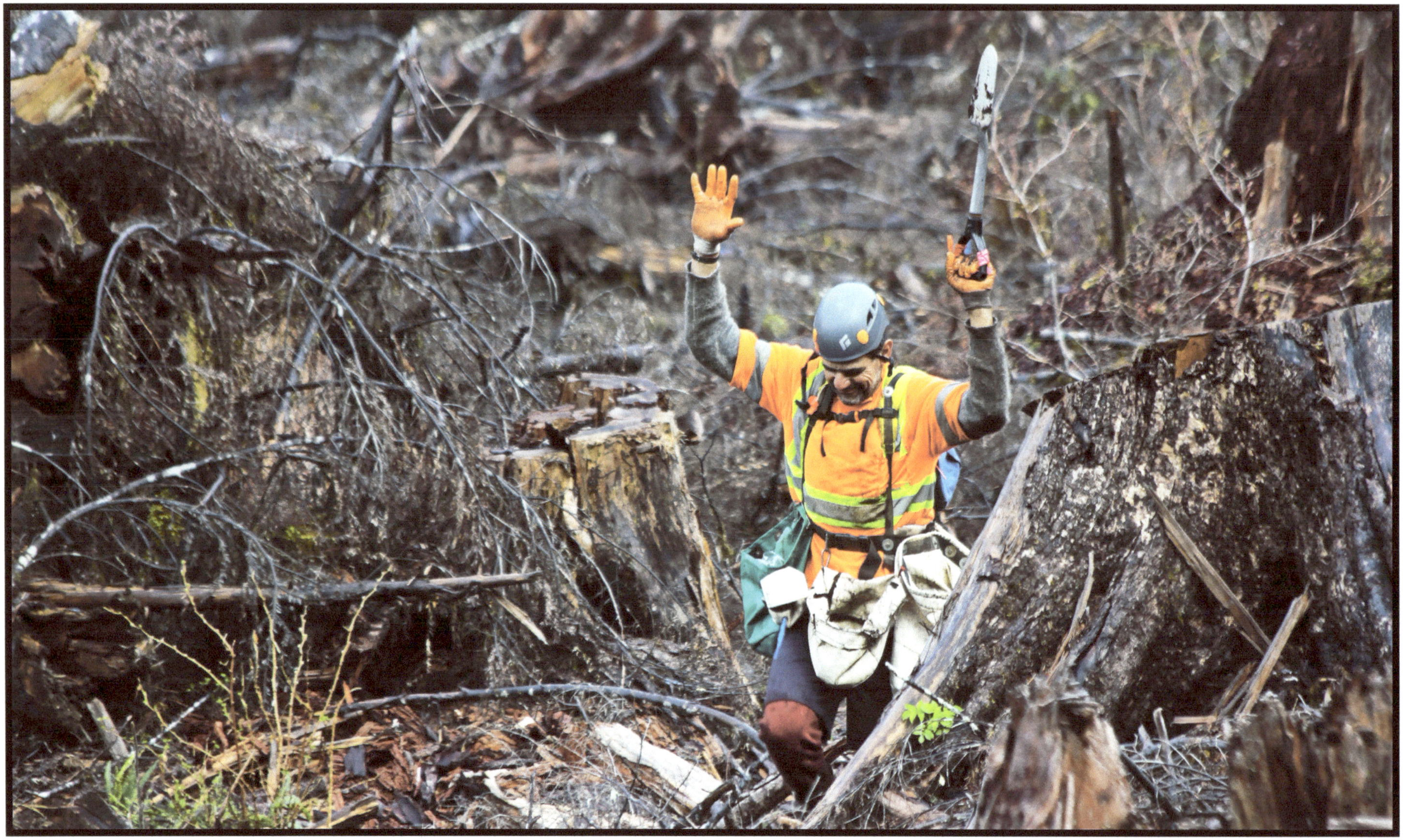

Simon.

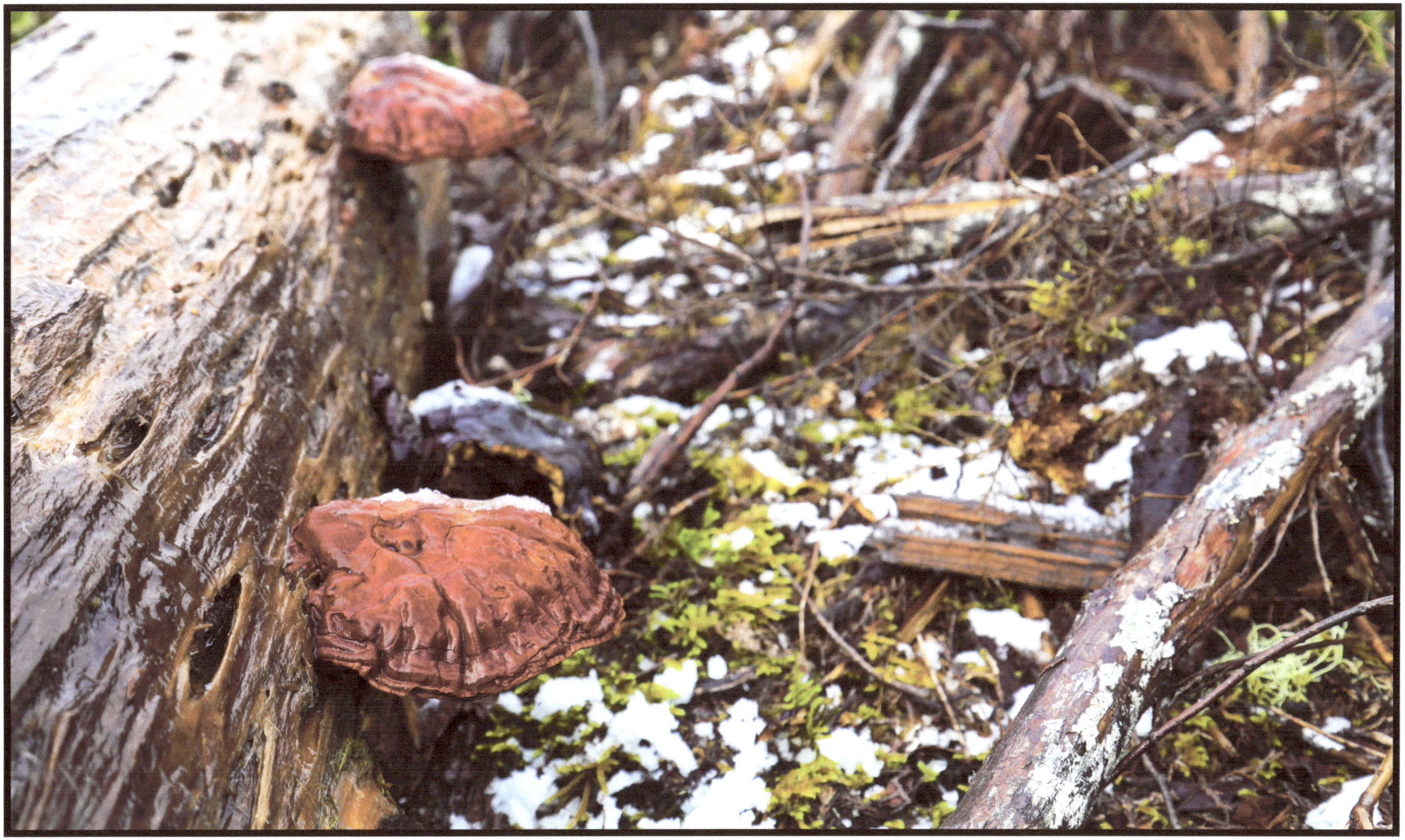

Some interesting fungal conks.

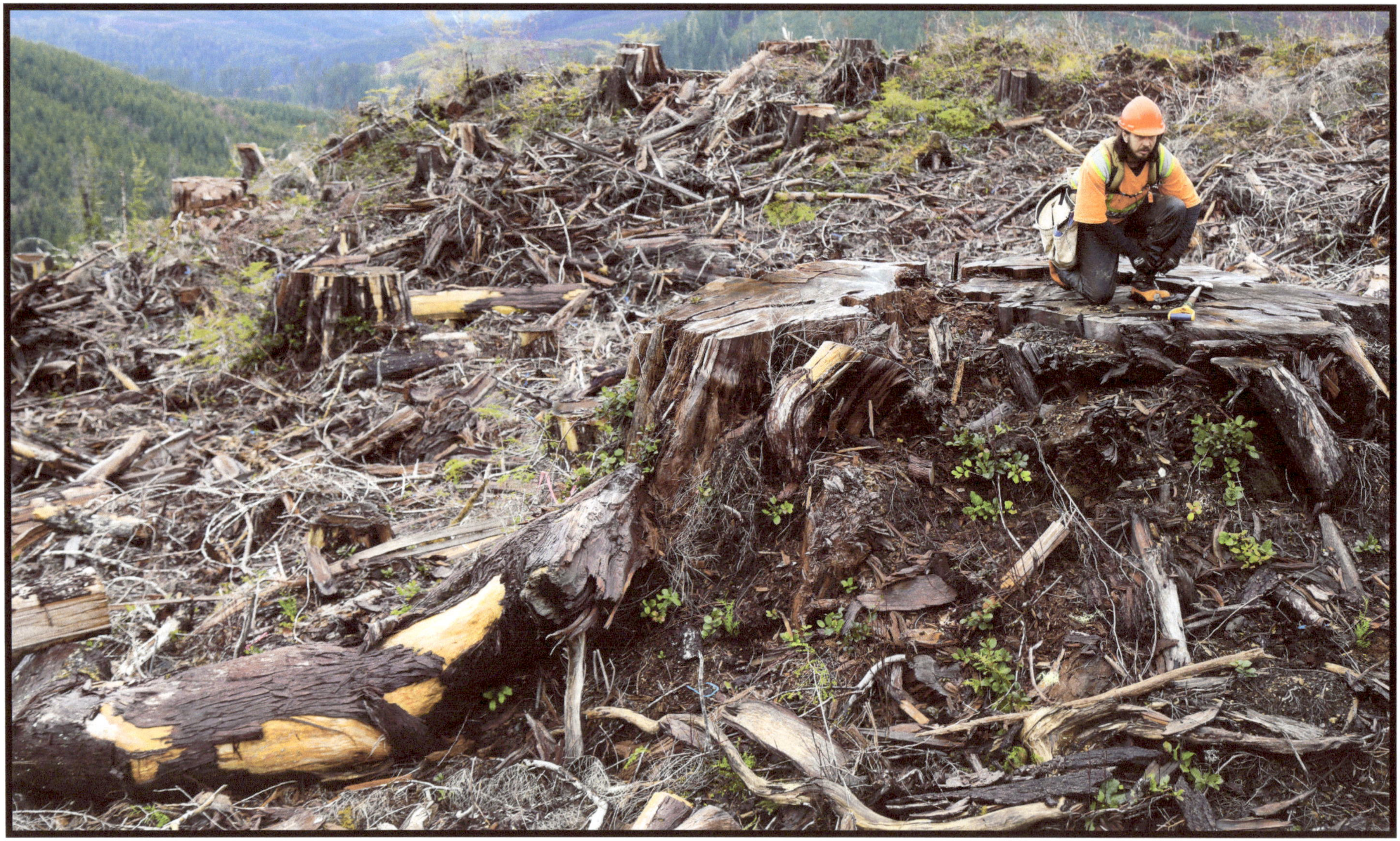

Stopping to tie up my boot lace on top of an old cedar stump. This must have been a very large tree.

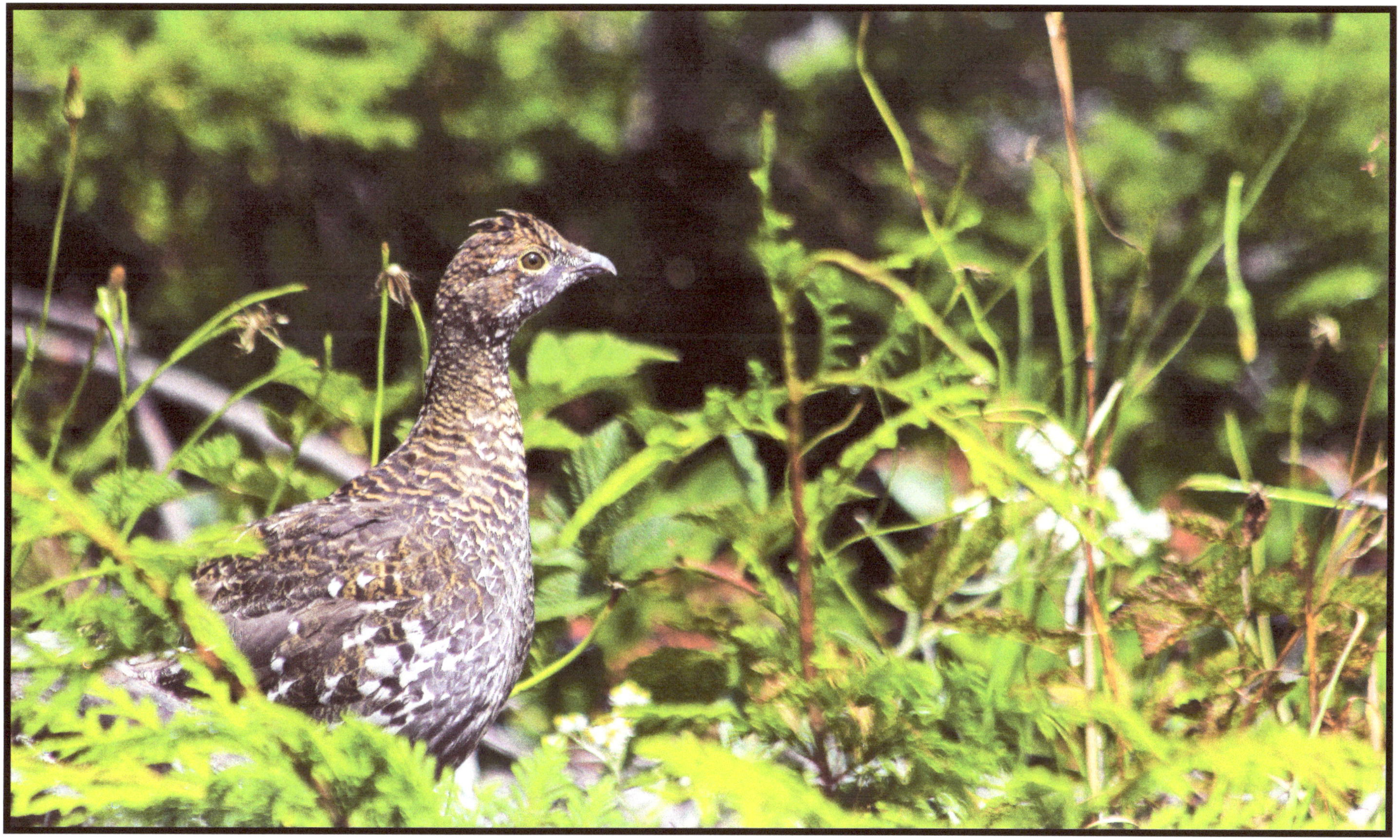

This bird probably hopes that I can't see it.

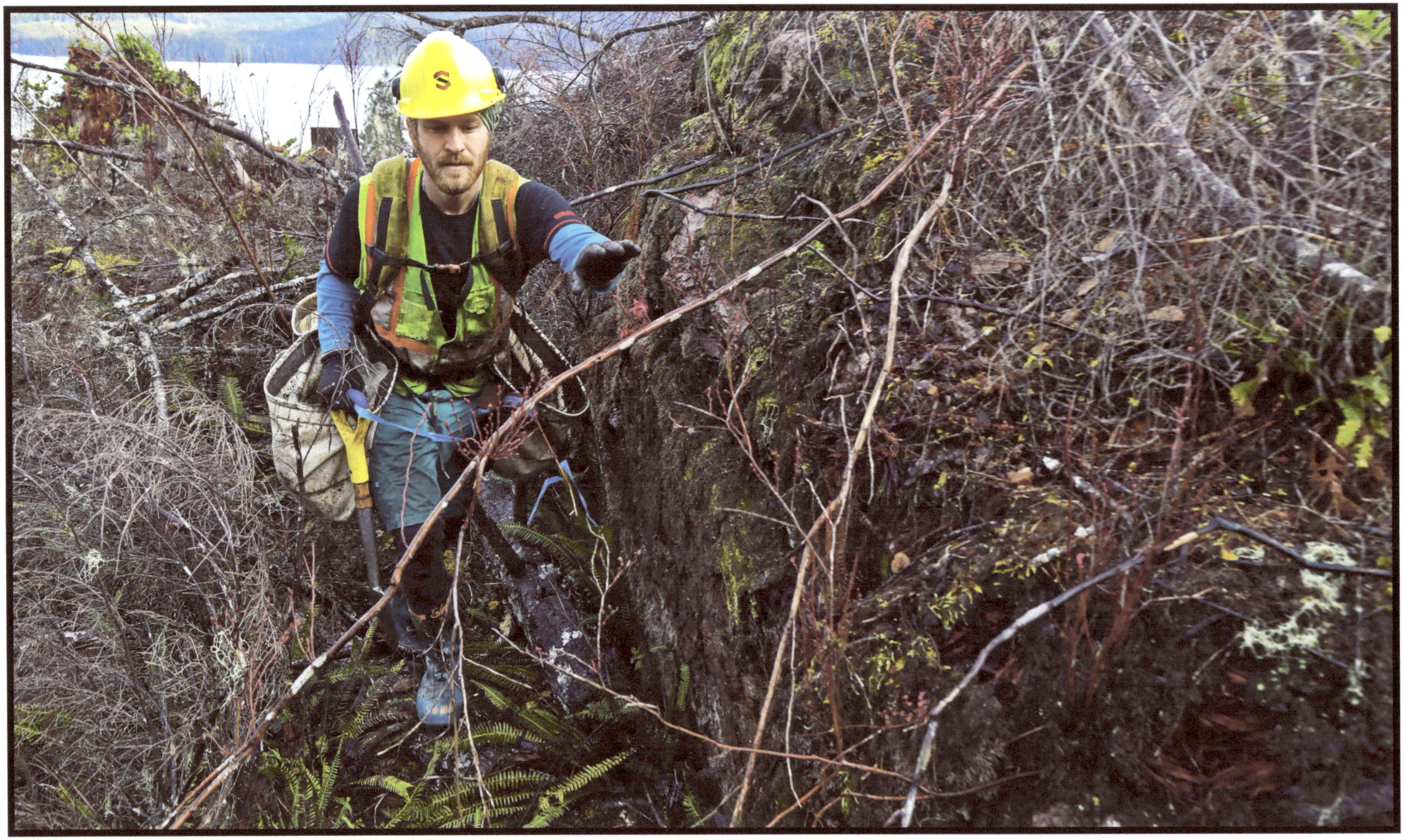

To someone who hasn't planted on the coast, this ground may look pretty heinous. However, this is pretty average.

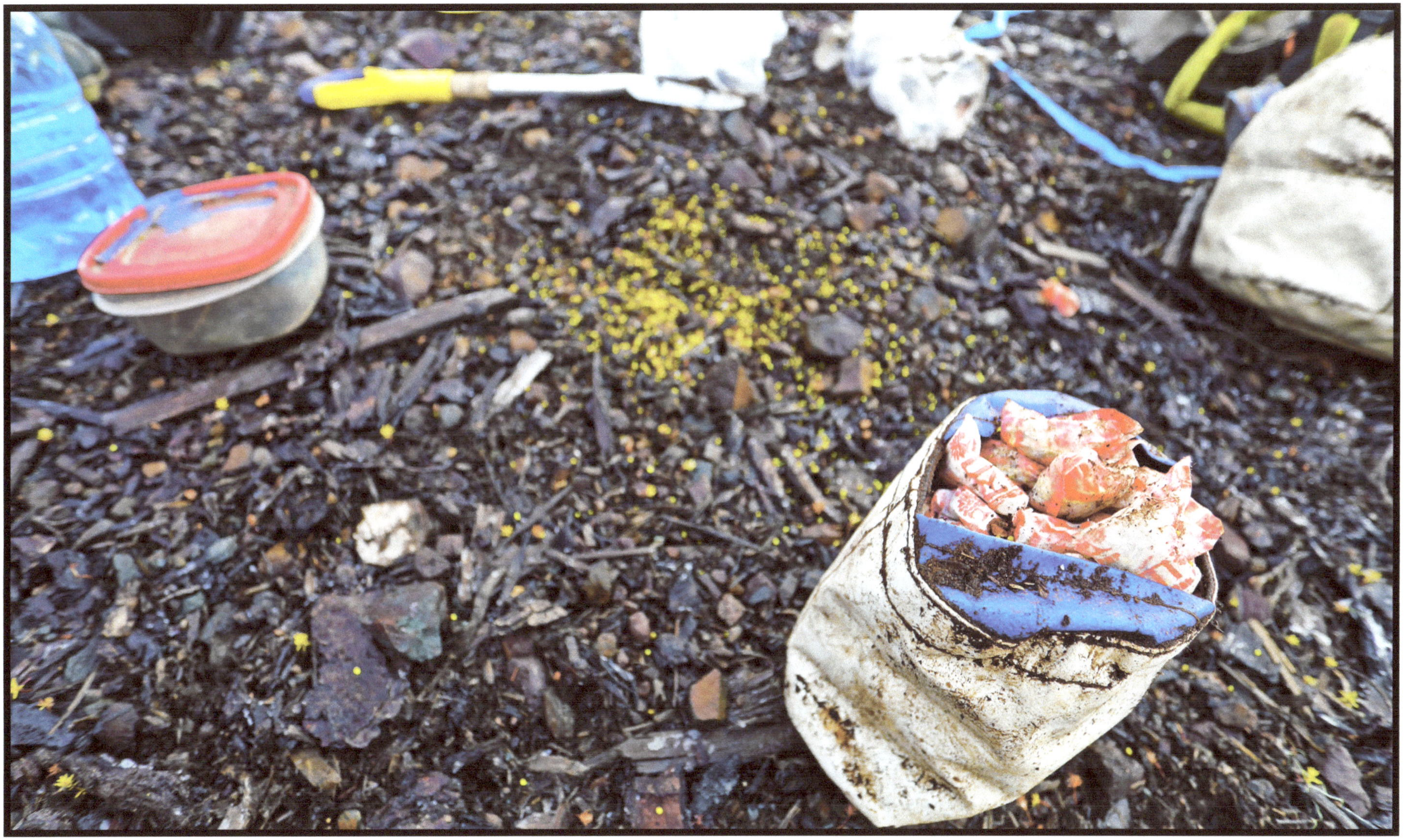

This is a pouch full of fertilizer packs. Planters often refer to ferts as tea-bags, because the fertilizer comes in the same sort of packaging as a tea-bag. We're often required to plant one fert pack beside each seedling. The requirement is usually that the tea-bag needs to be placed about 3-4 inches away from the roots of the tree, and about an inch below the surface.

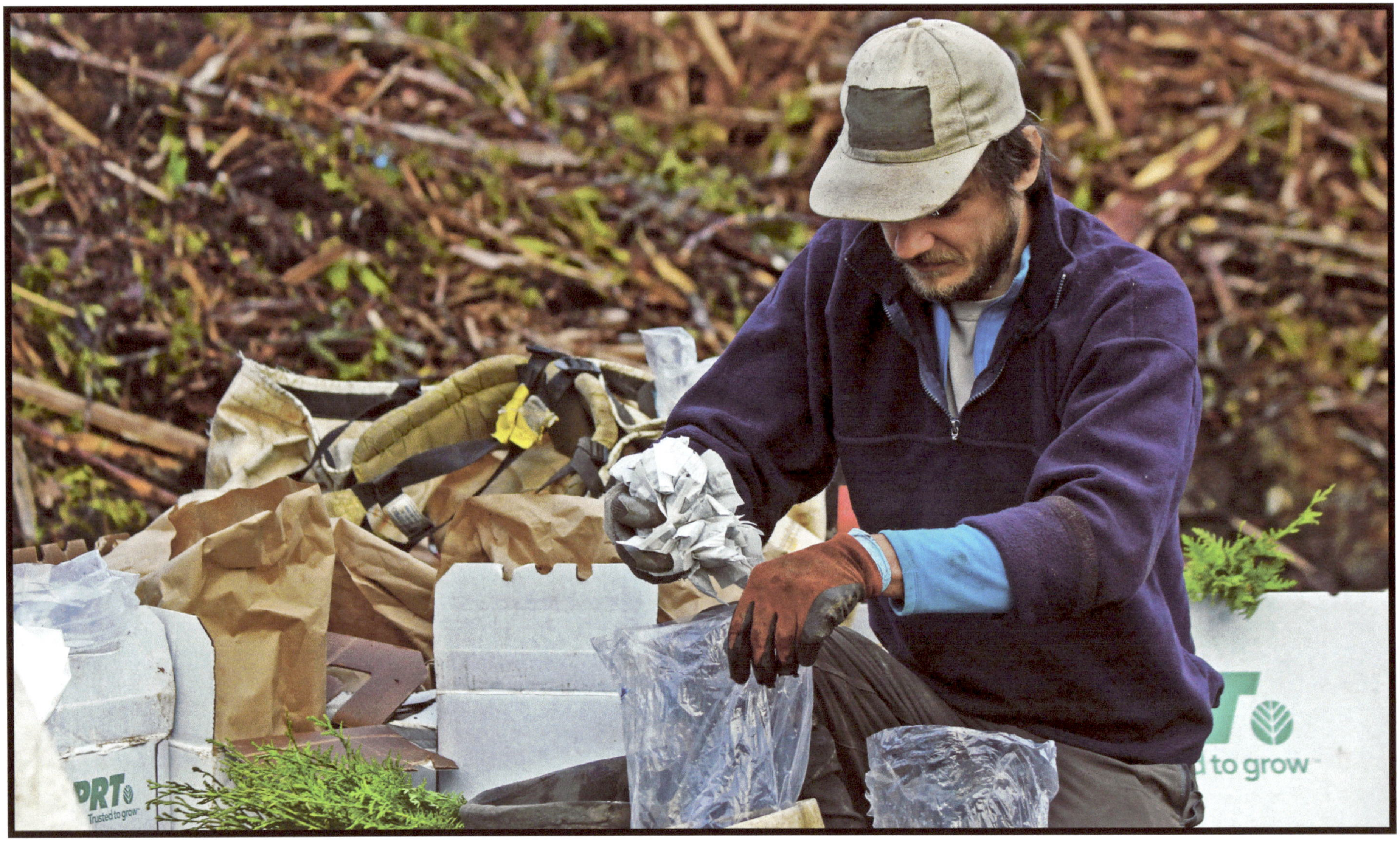

Planters get paid a few cents extra if they have to fertilize the trees. Having to fert the trees isn't usually too annoying unless the tea-bags all get wet. Unfortunately, this is pretty much inevitable in the coastal rain.

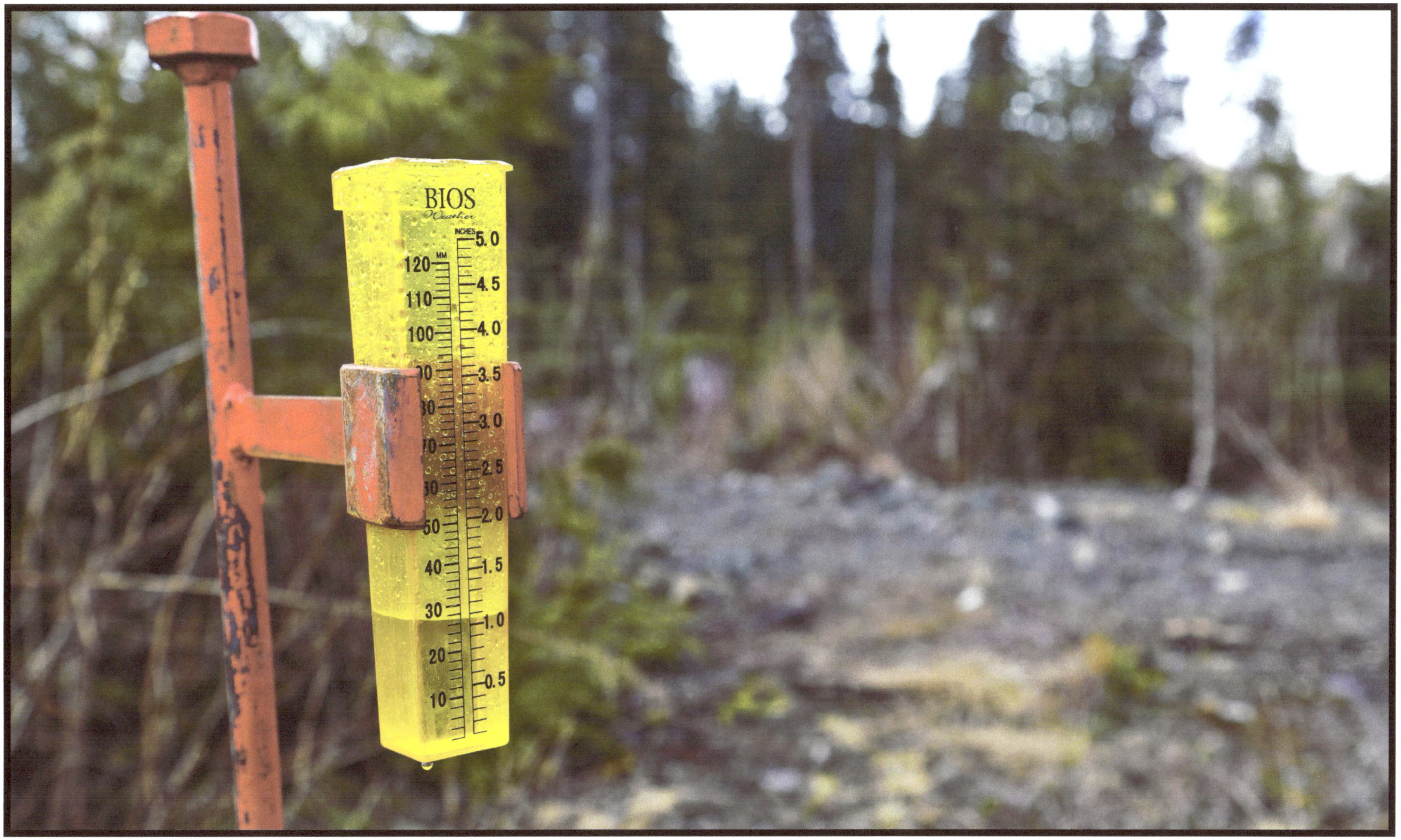

Here's a rain meter. This is a simple device that measures how many millimeters of precipitation has fallen since the meter was last emptied. Although we typically continue working during all but the heaviest of rains, we occasionally get shut down when there are concerns that hillsides are becoming super-saturated, to minimize the hazard that our trucks get trapped if a road washes out due to a mudslide.

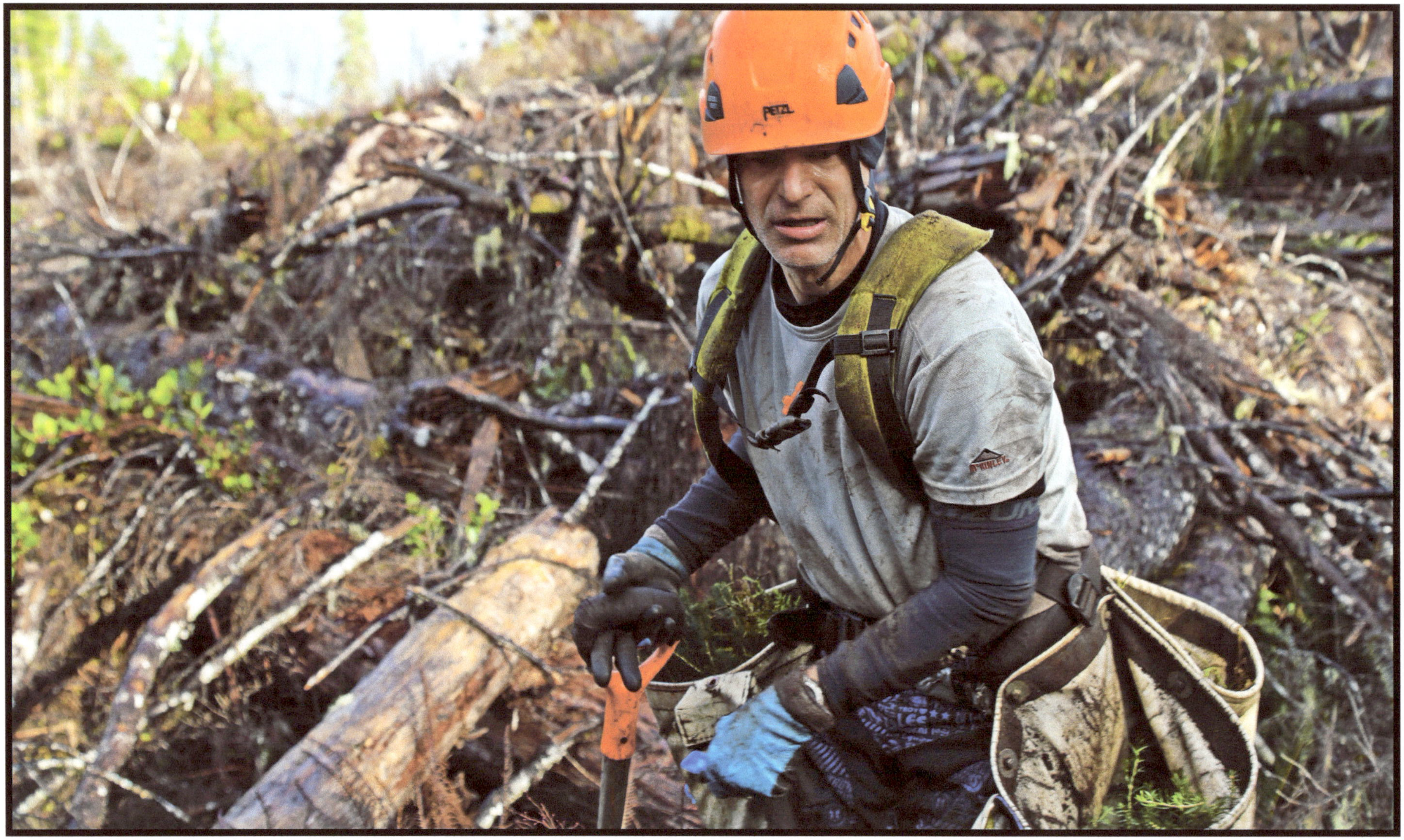

Always be moving.

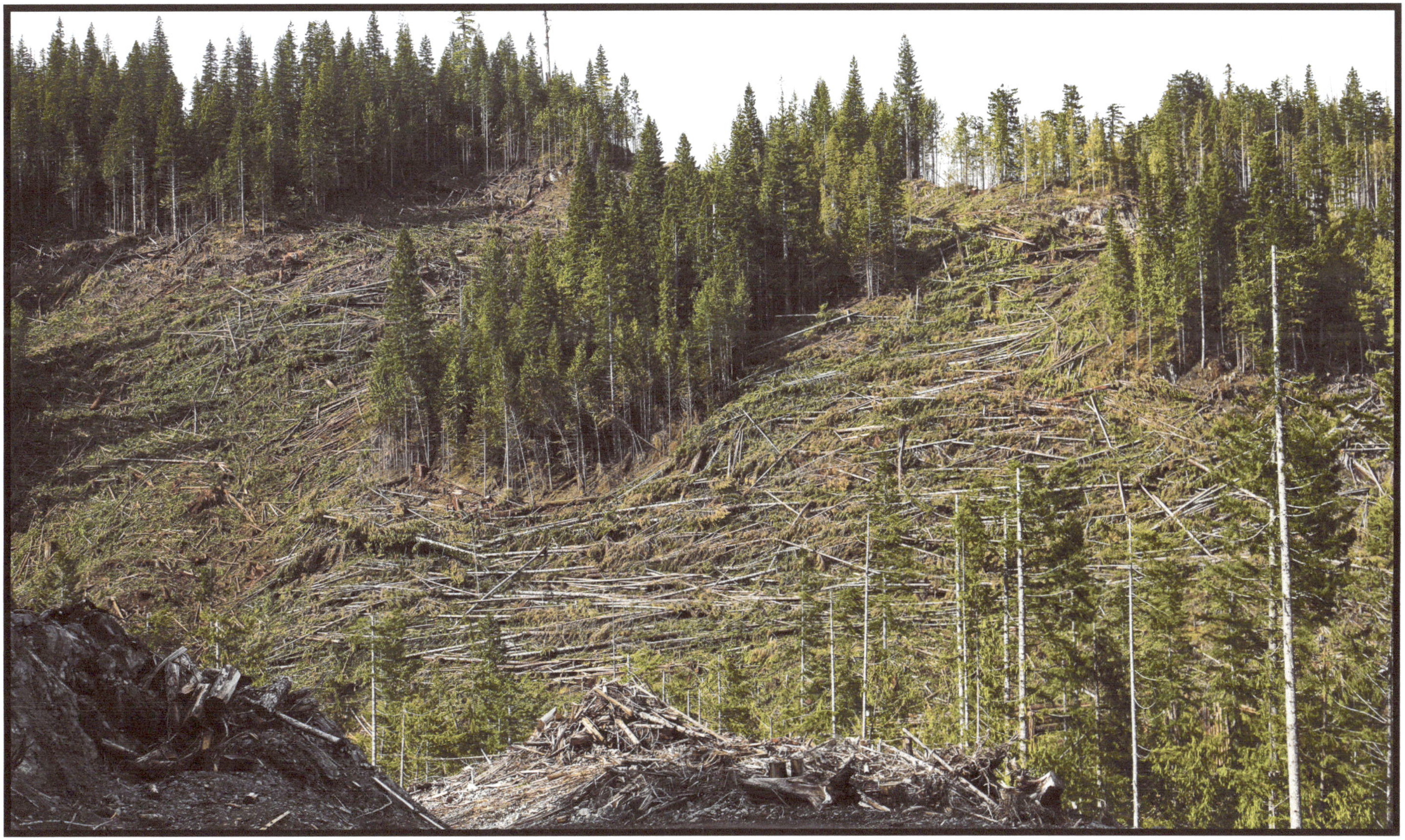

This block is currently being harvested. The crew appears to have pulled the logs off part of the hill in the left side of the photo, but the rest of the block still needs to be yarded.

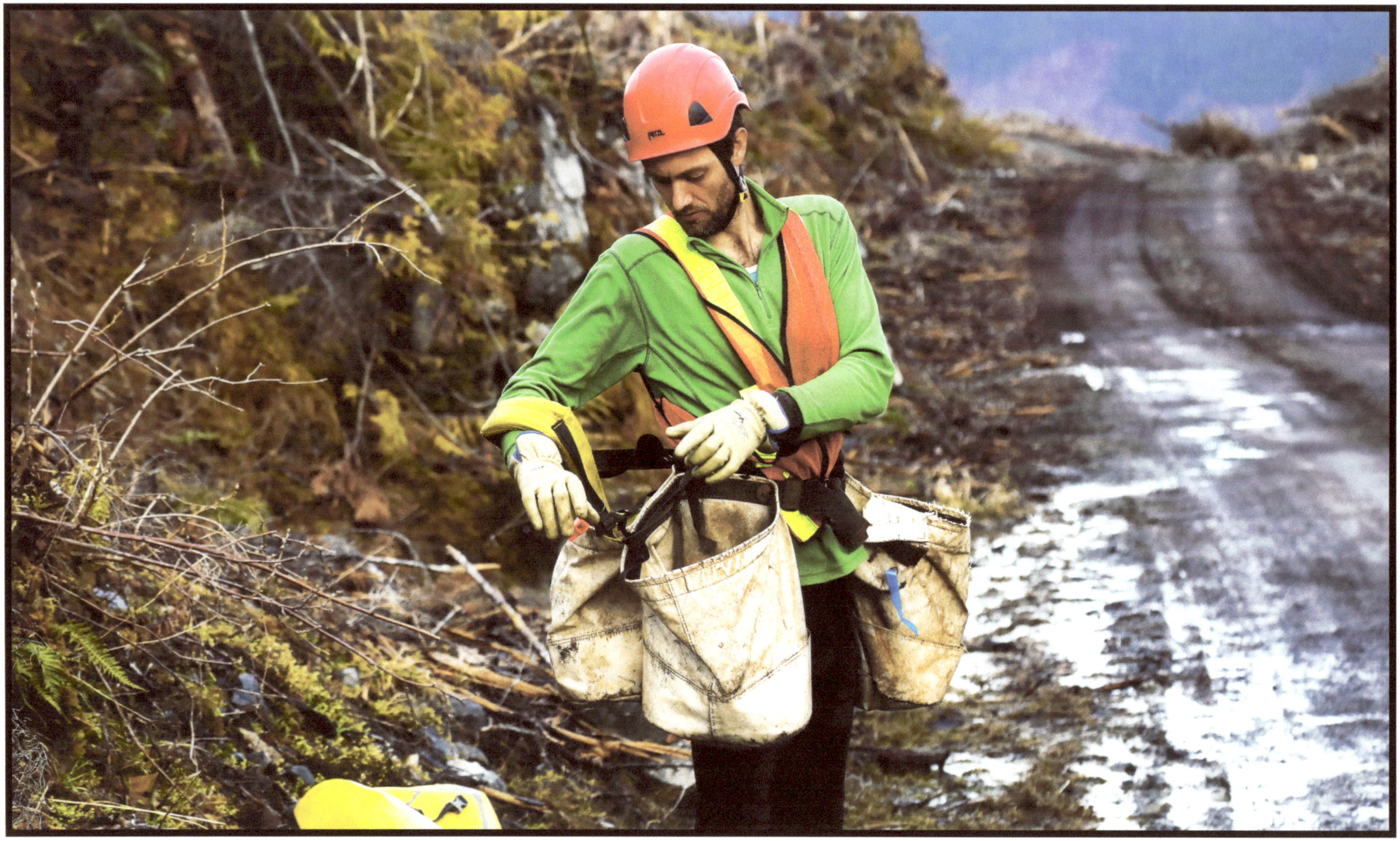

Greg.

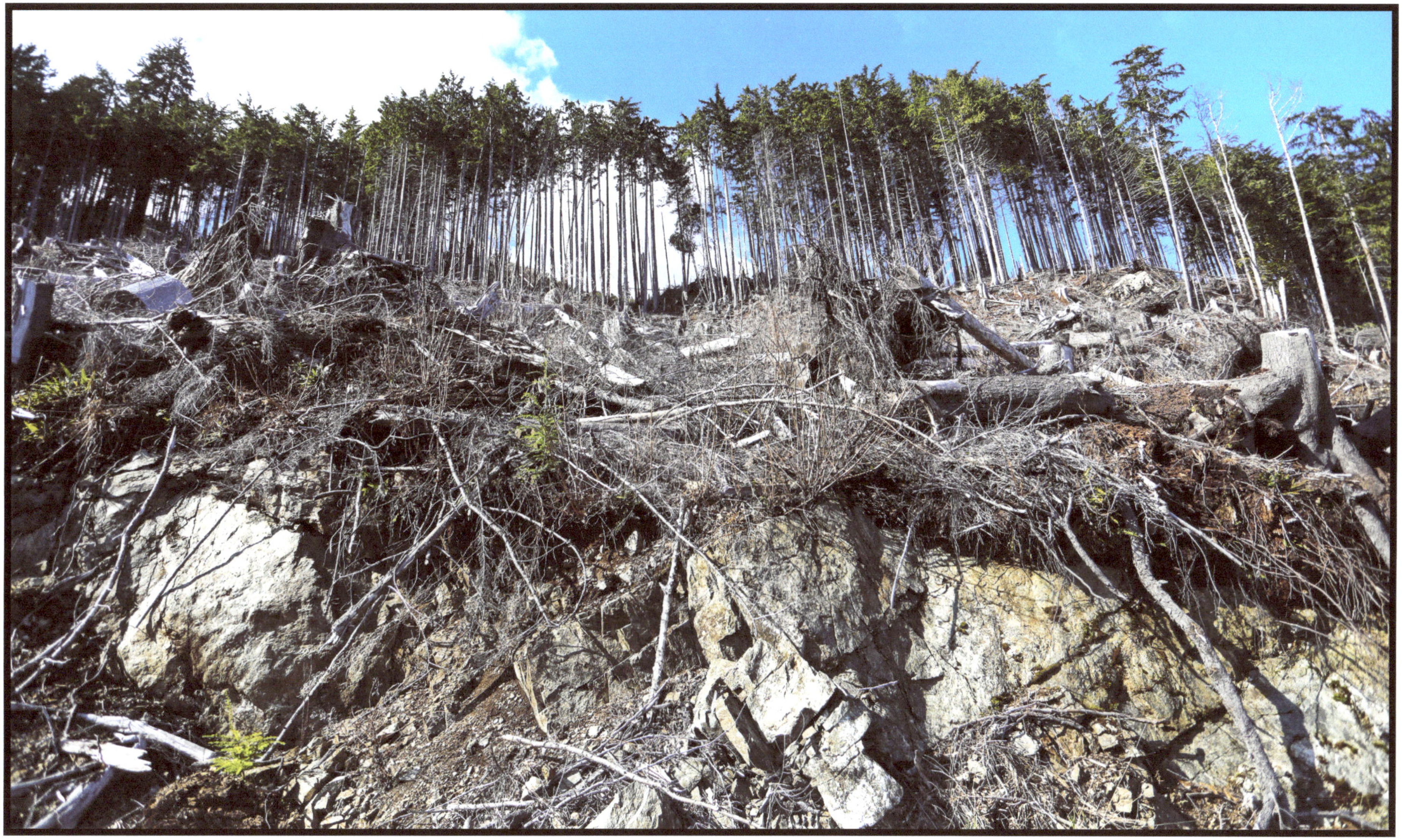

It's a long way to the top.

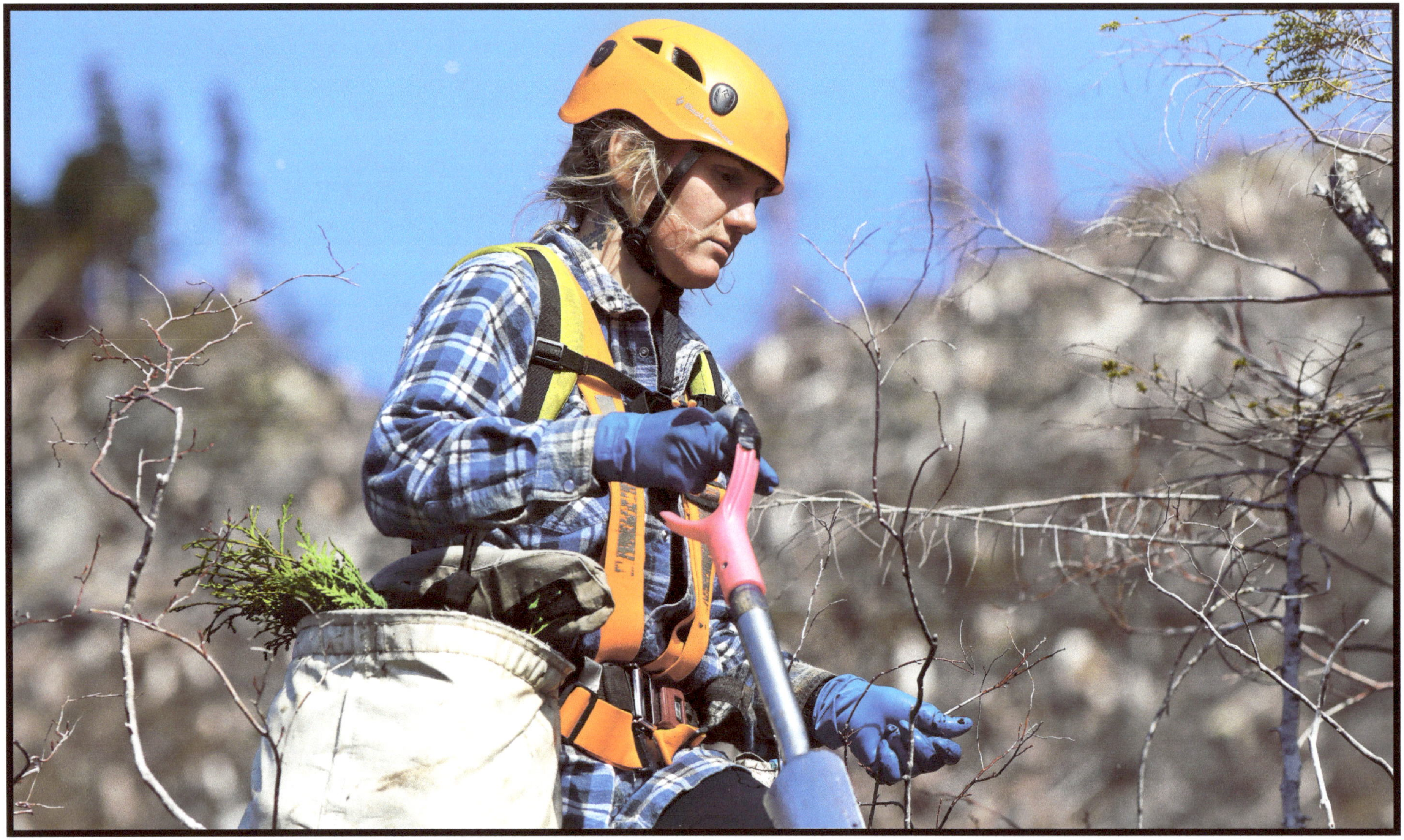

Focus improves productivity.

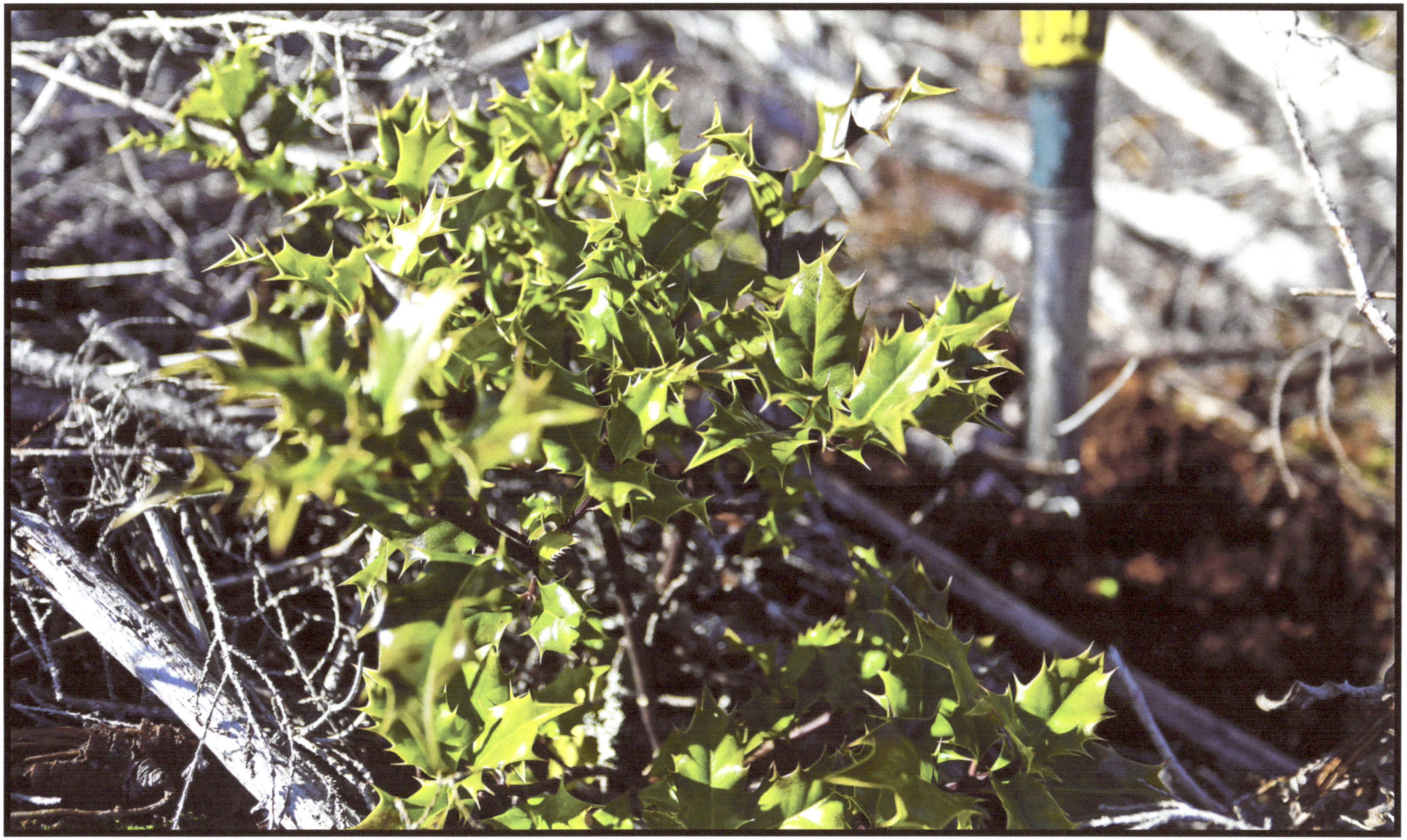

A type of holly plant.

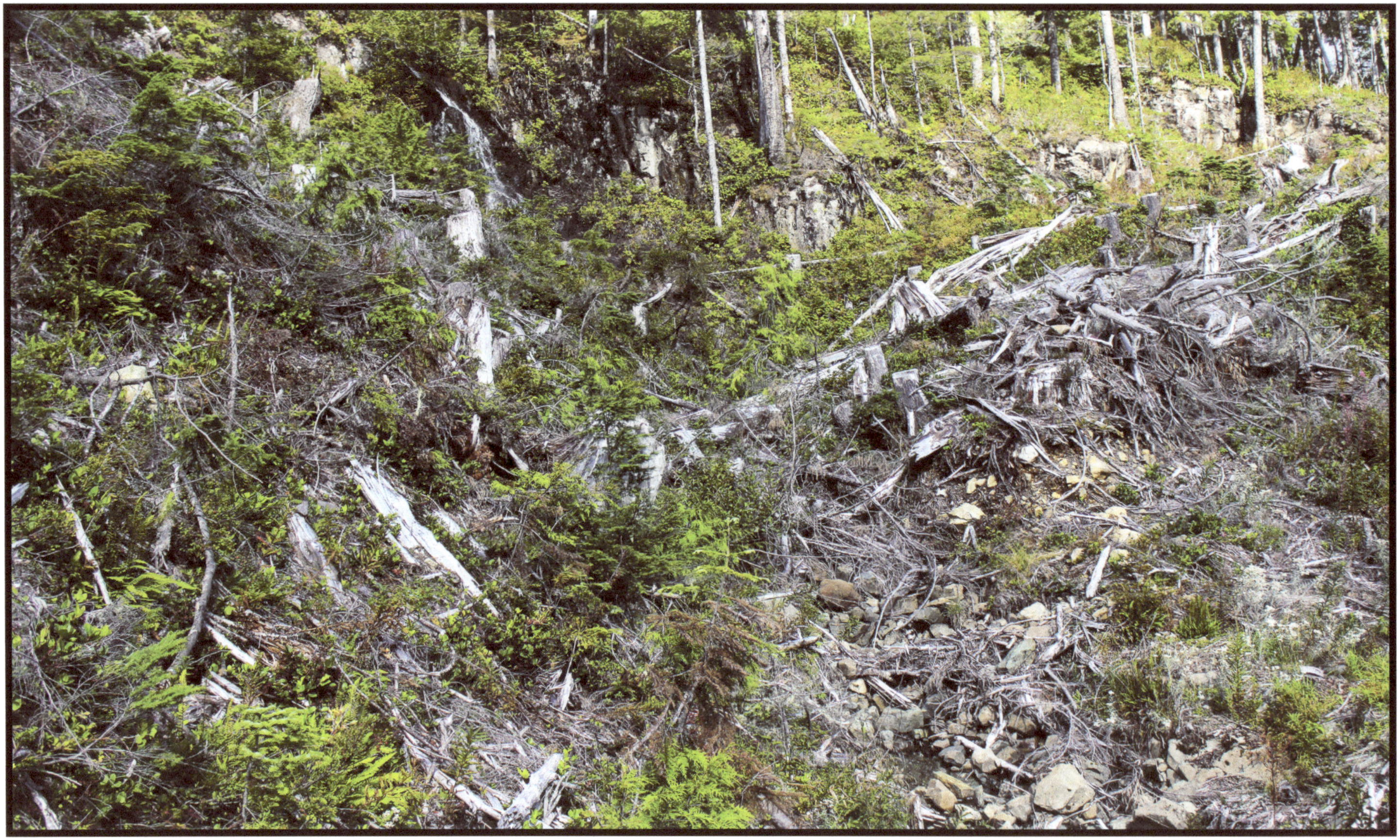

This was not a nice block. For scale, the large rock near the bottom right of this photo almost came up to my waist.

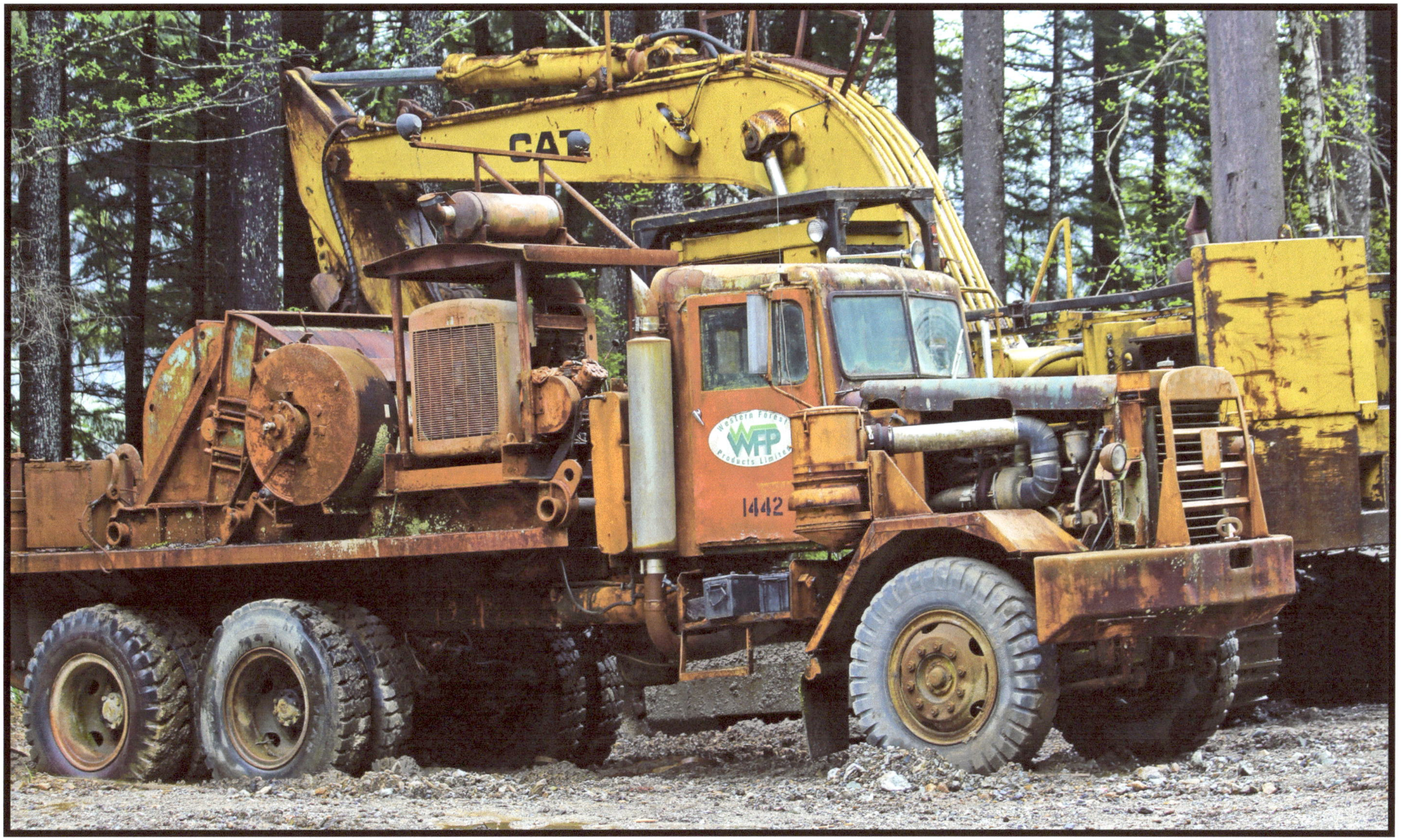

Old logging equipment.

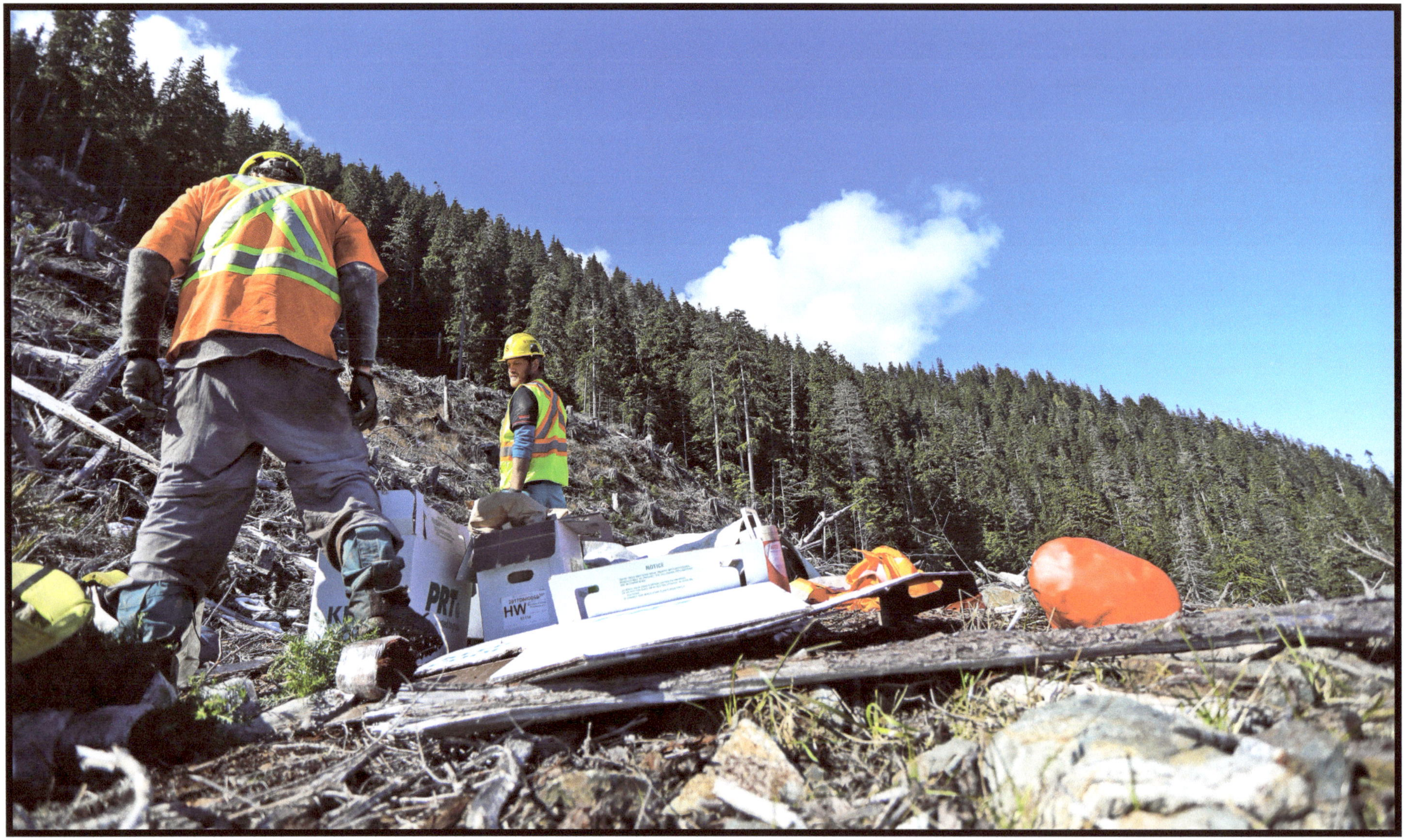

Bagging up.

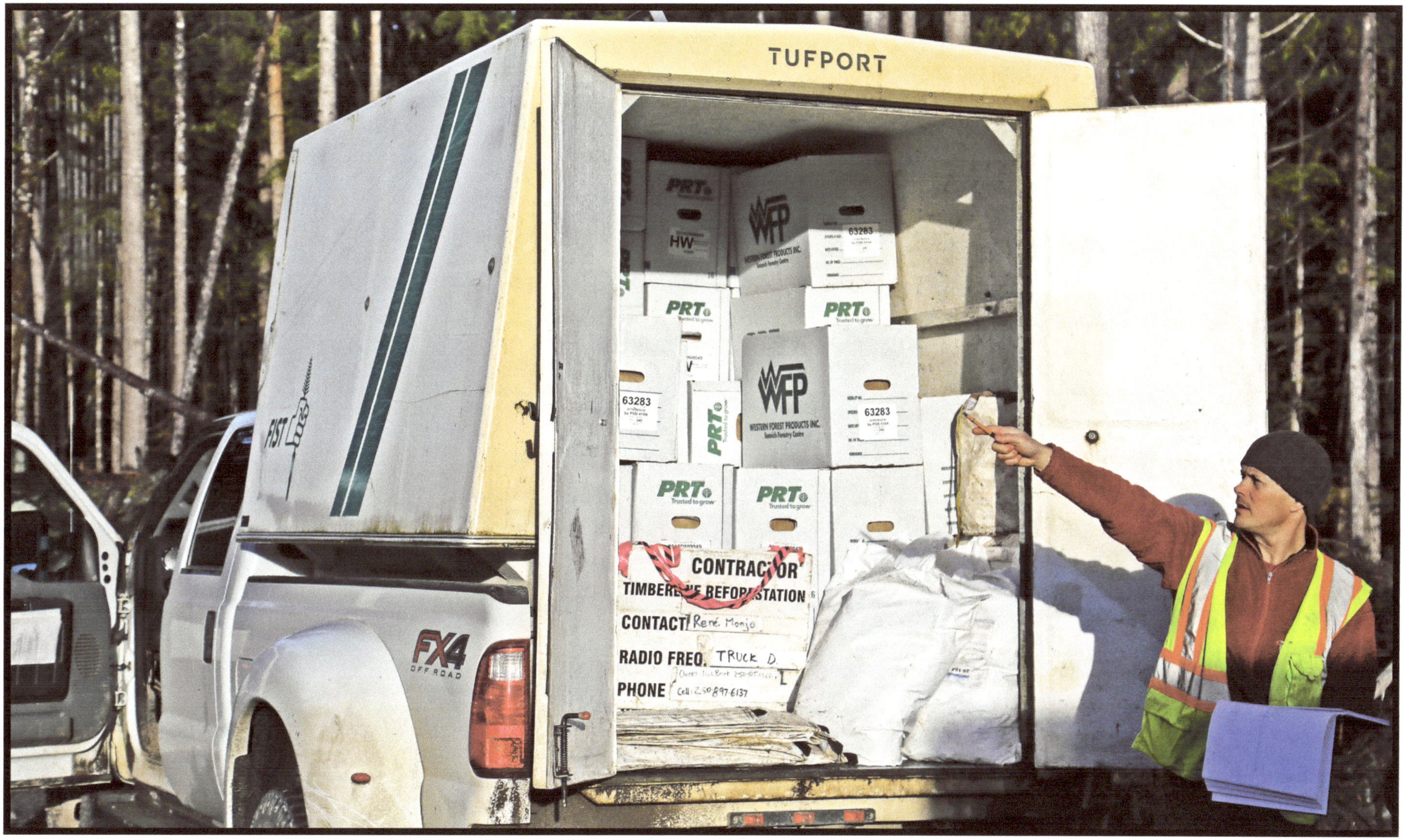

Rene, looking up at the block. Rene is a co-owner of Timberline Reforestation, which is one of the companies that I most frequently work for on the coast. Rene commented, "I love the people, the outdoors, the hard work, the raw nature of the job, and the freedom to be myself. My long-term goal is to create a healthy, fun, safe, and successful working environment for our crew members."

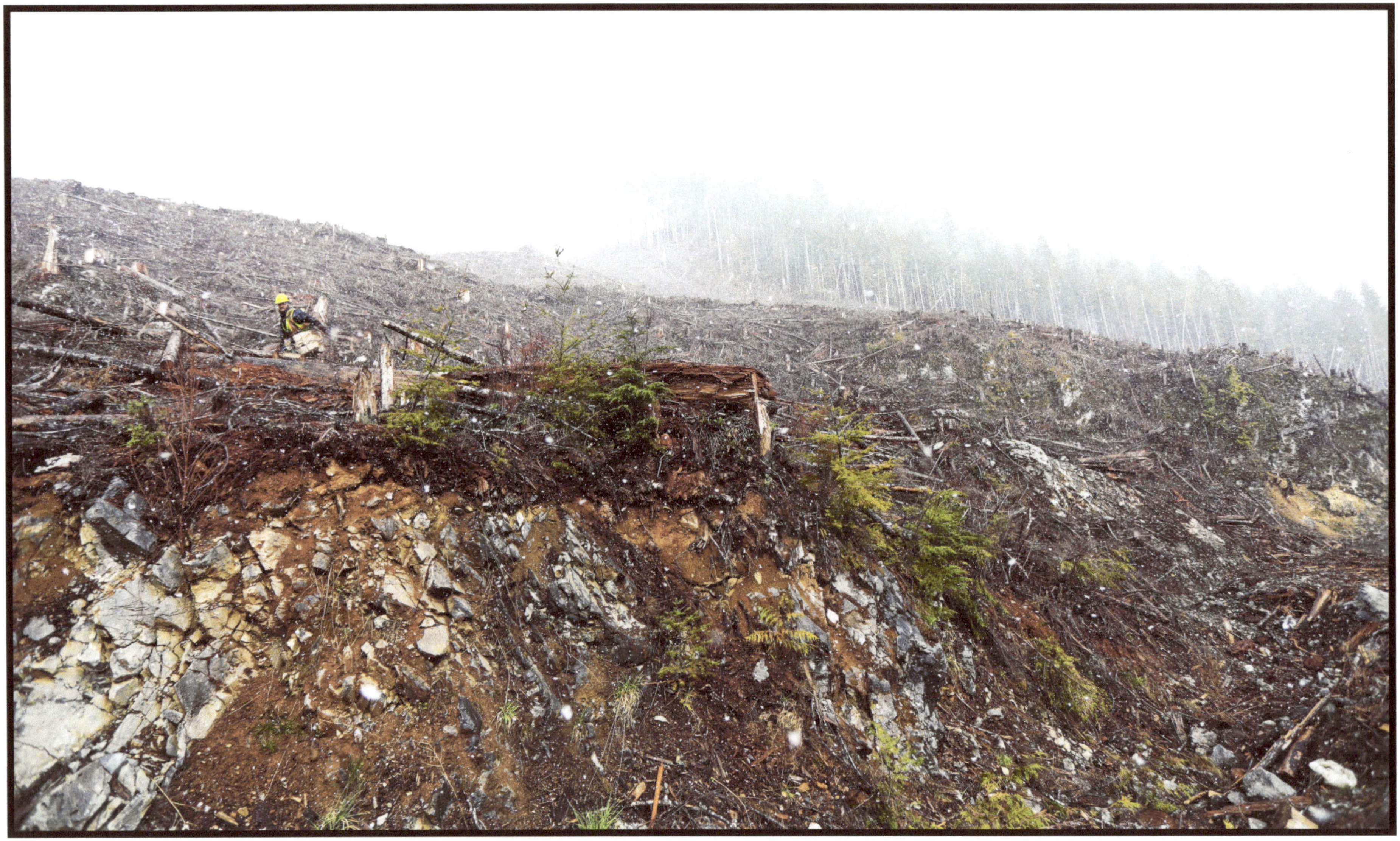

Here comes the snow.

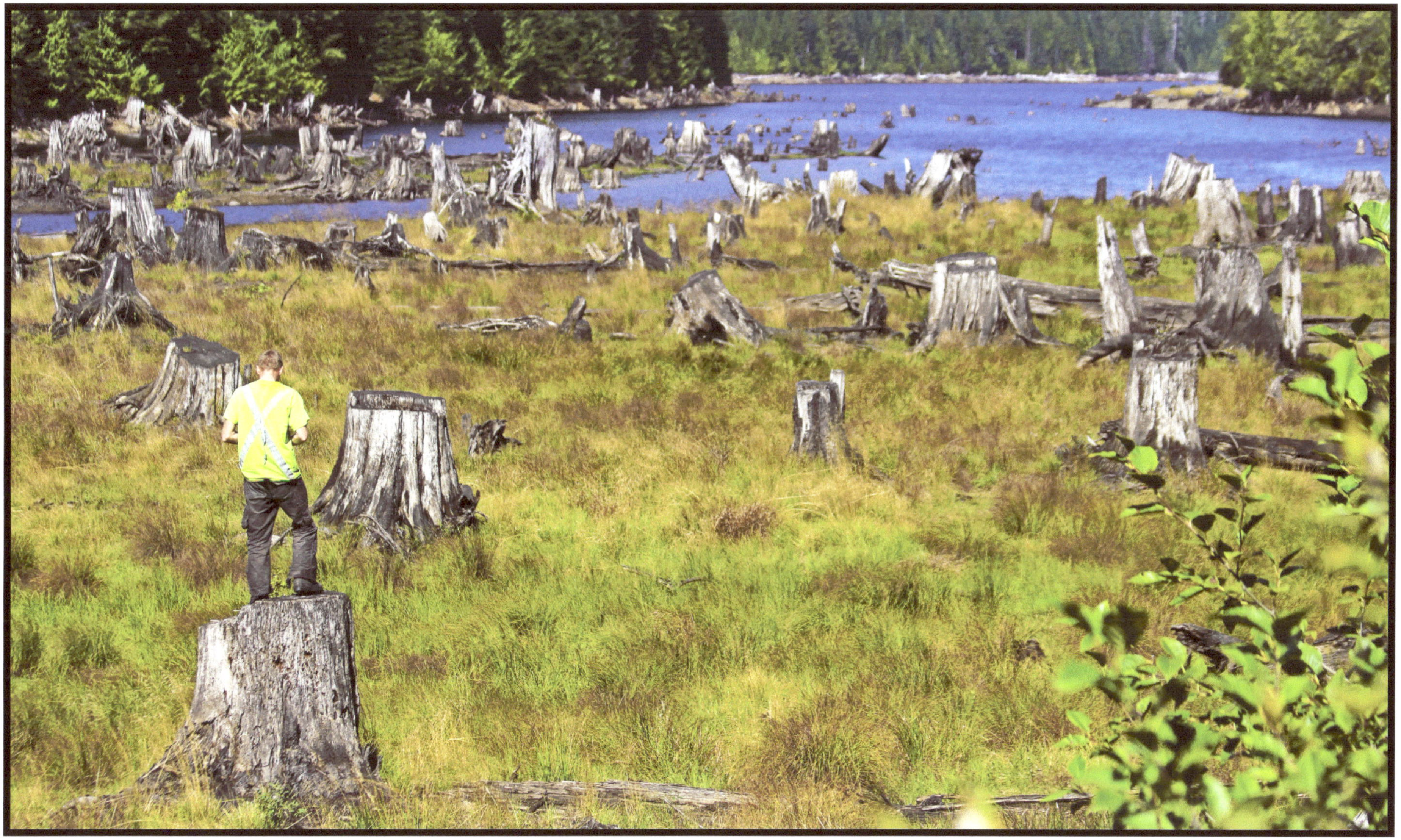

This estuary has always fascinated me.

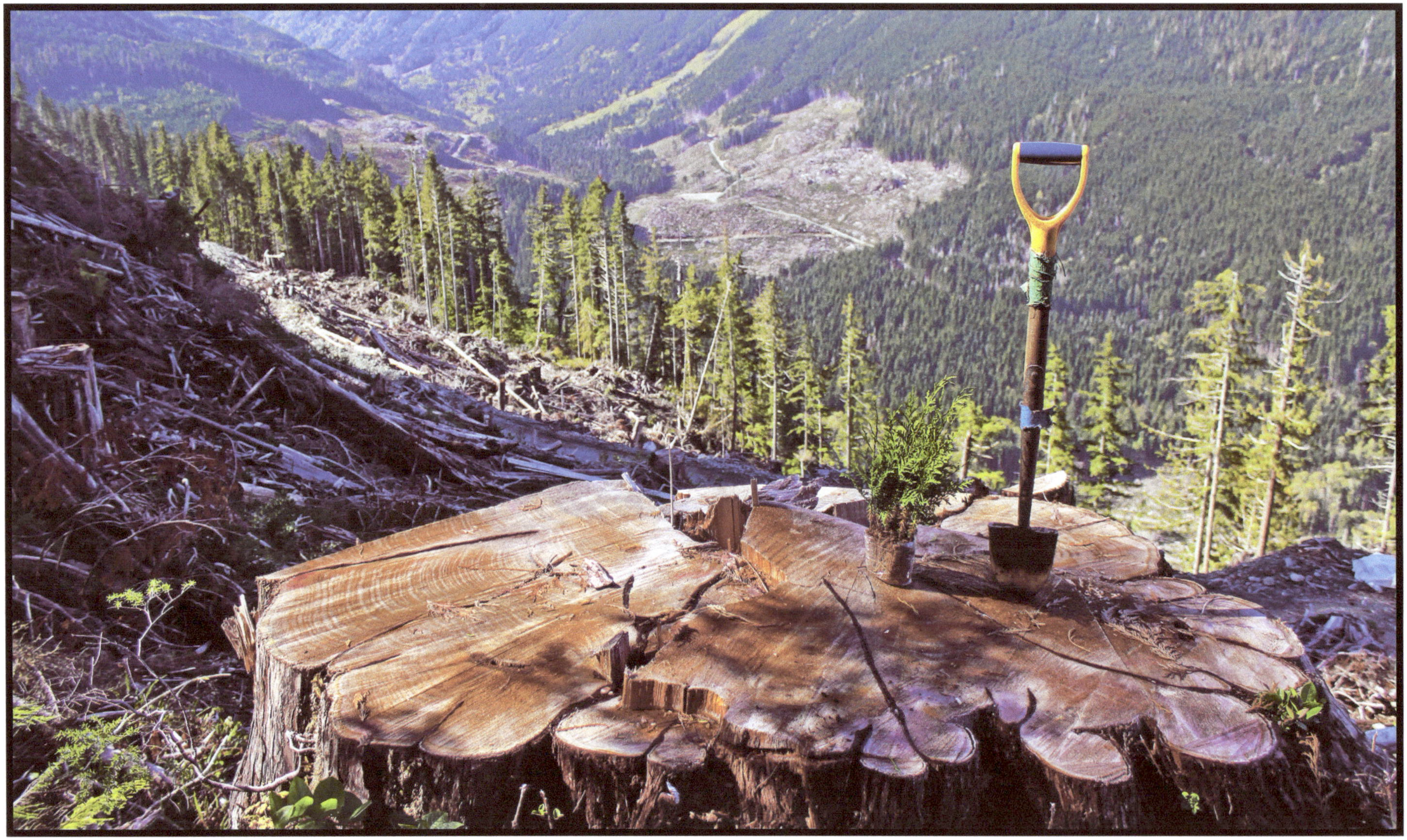

A fresh cedar stump on a block overlooking a valley bottom. It's hard to believe that each of the seedlings in the bundle beside this shovel can eventually grow up into a mature tree the size of this tree stump.

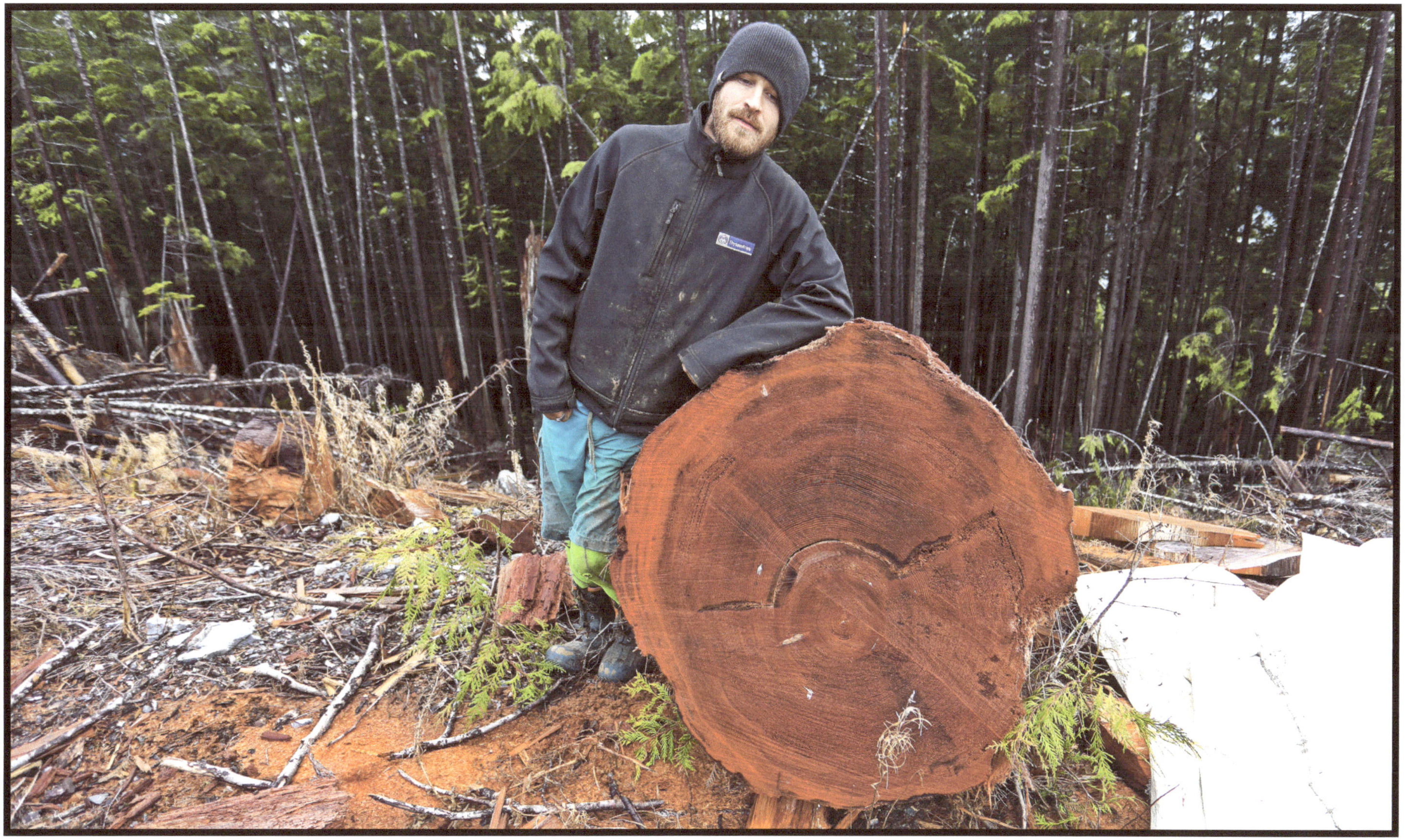

Red cedar has a very distinct colour when compared to yellow cedar.

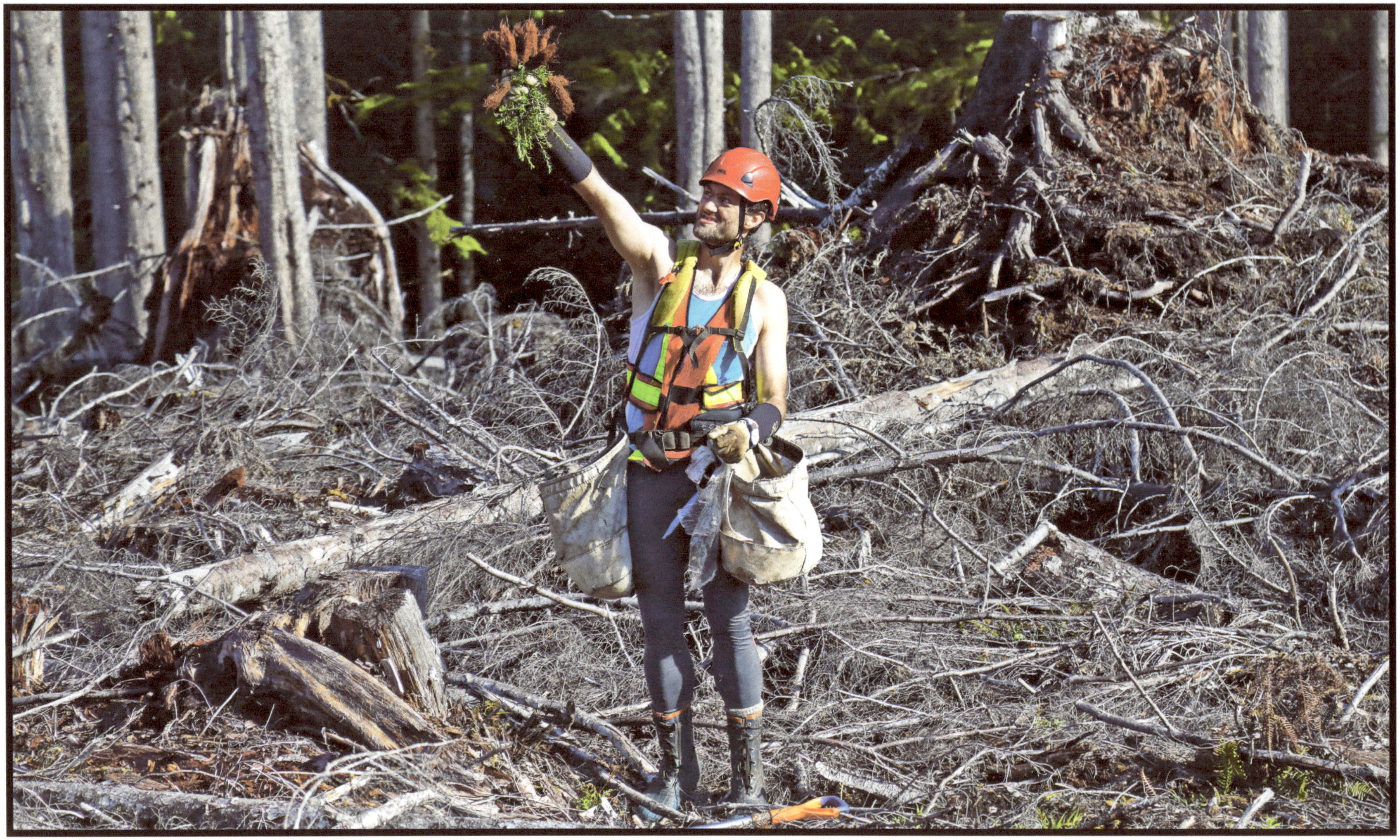

The harder the struggle, the more glorious the triumph.

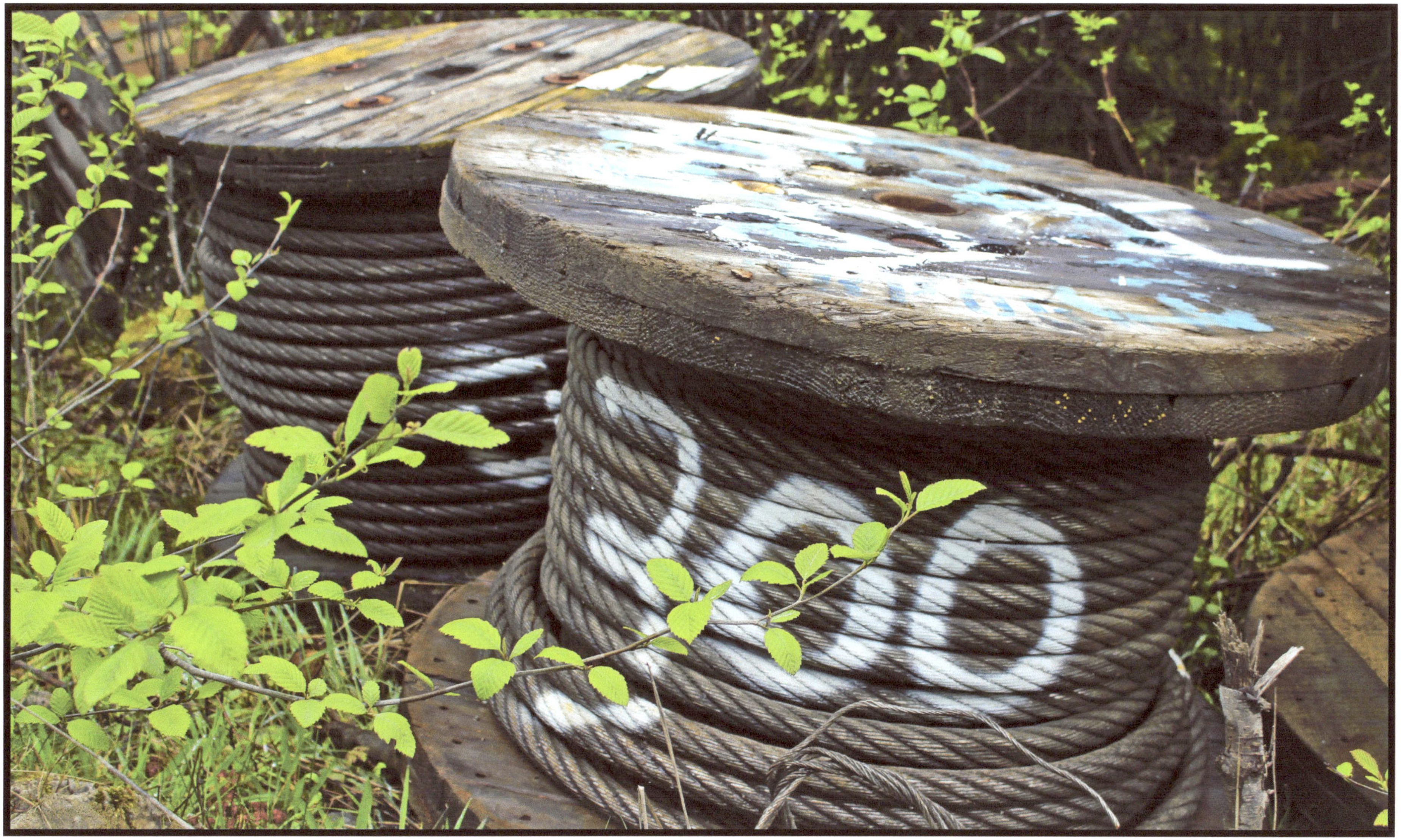

Rolls of steel cable used in the yarding operations.

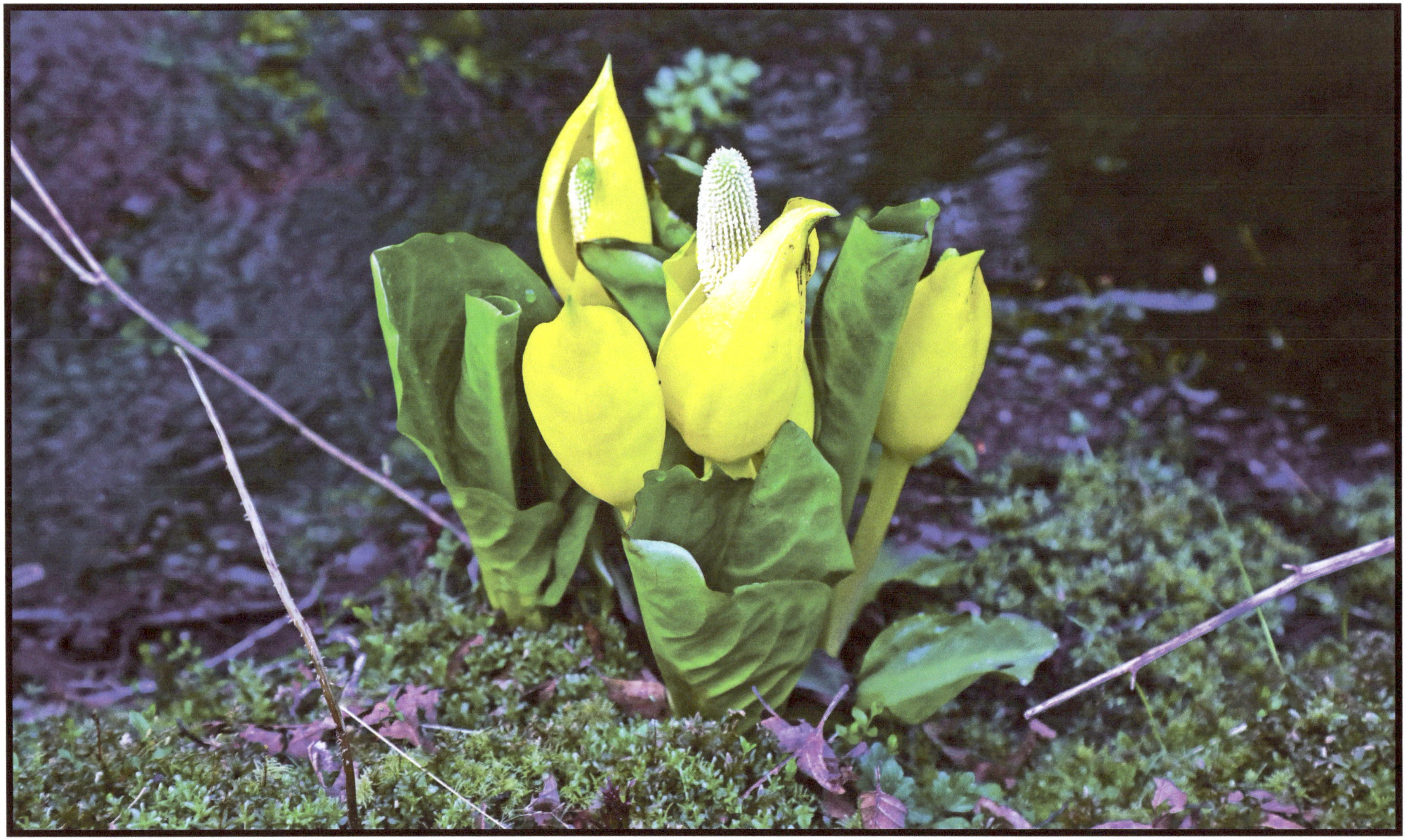

Skunk cabbage. This wetland plant is a tasty treat for the local bears.

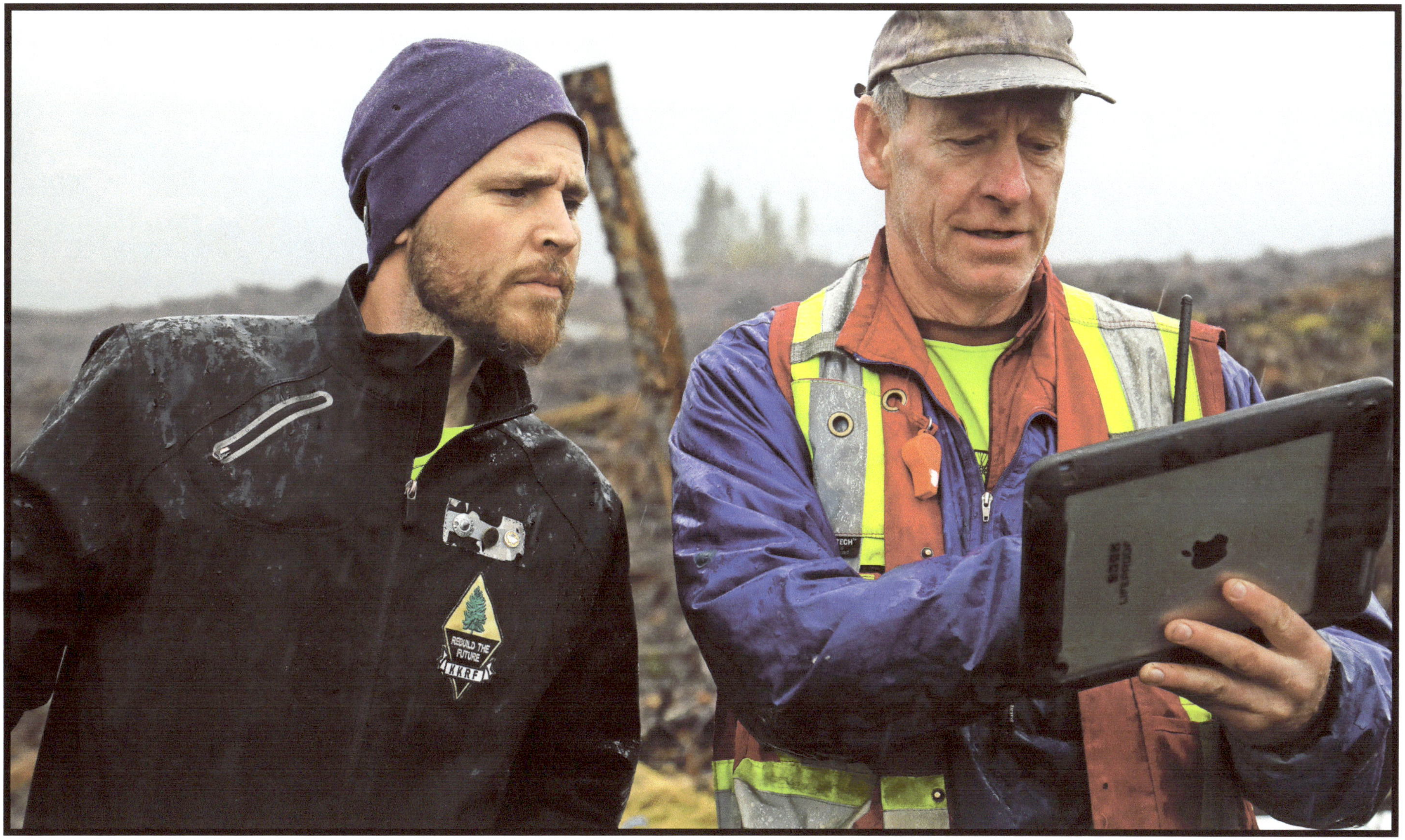

Michael and Mark, clarifying the boundaries of a piece.

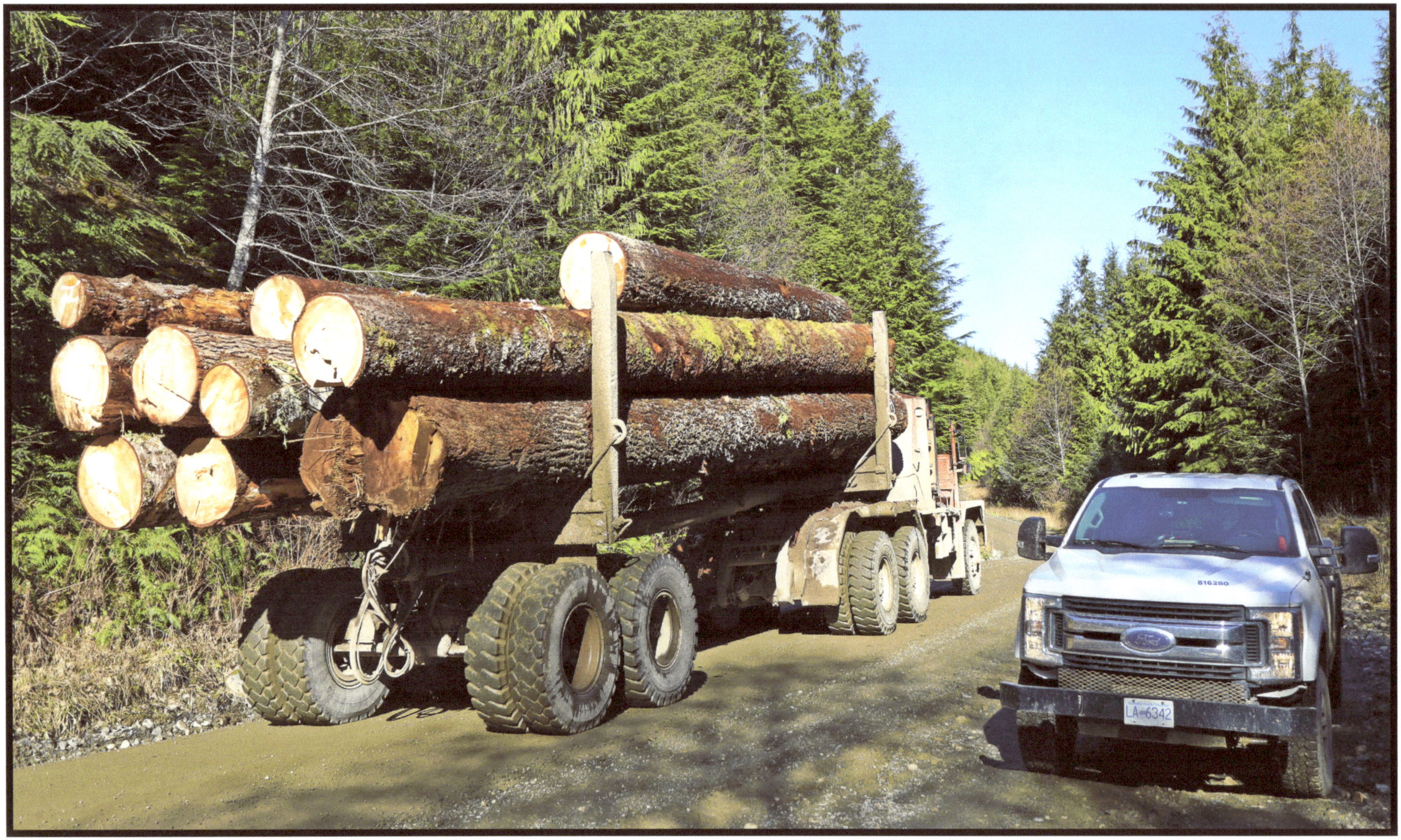

The trees harvested on the coast are typically larger than trees harvested in many other parts of Canada.

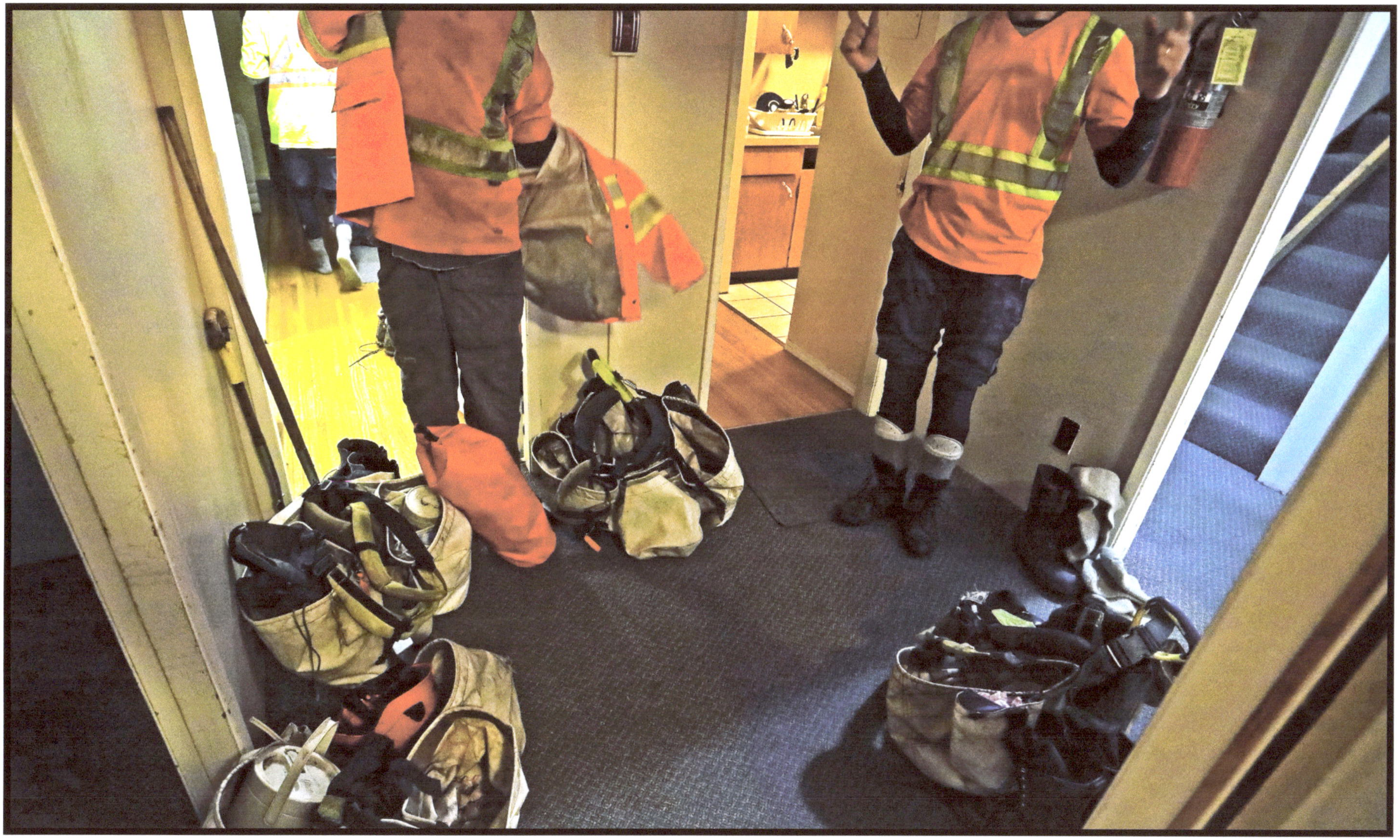

Coastal weather is not suitable for working out of tent camps. Coastal planters typically live in motels or apartments. In many cases on the south or central Island, the work force is composed entirely of local residents who work from their homes.

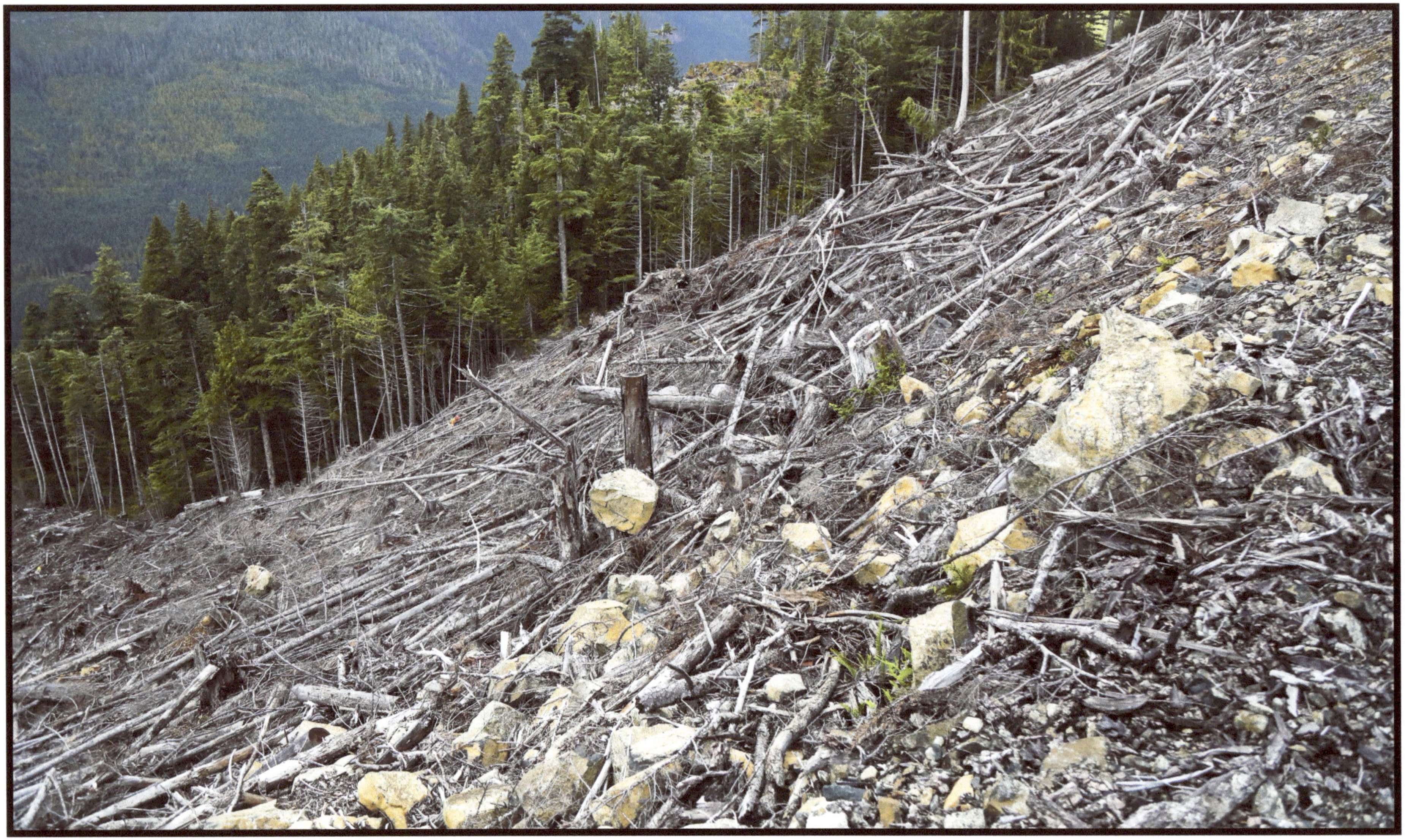

Find the planter.

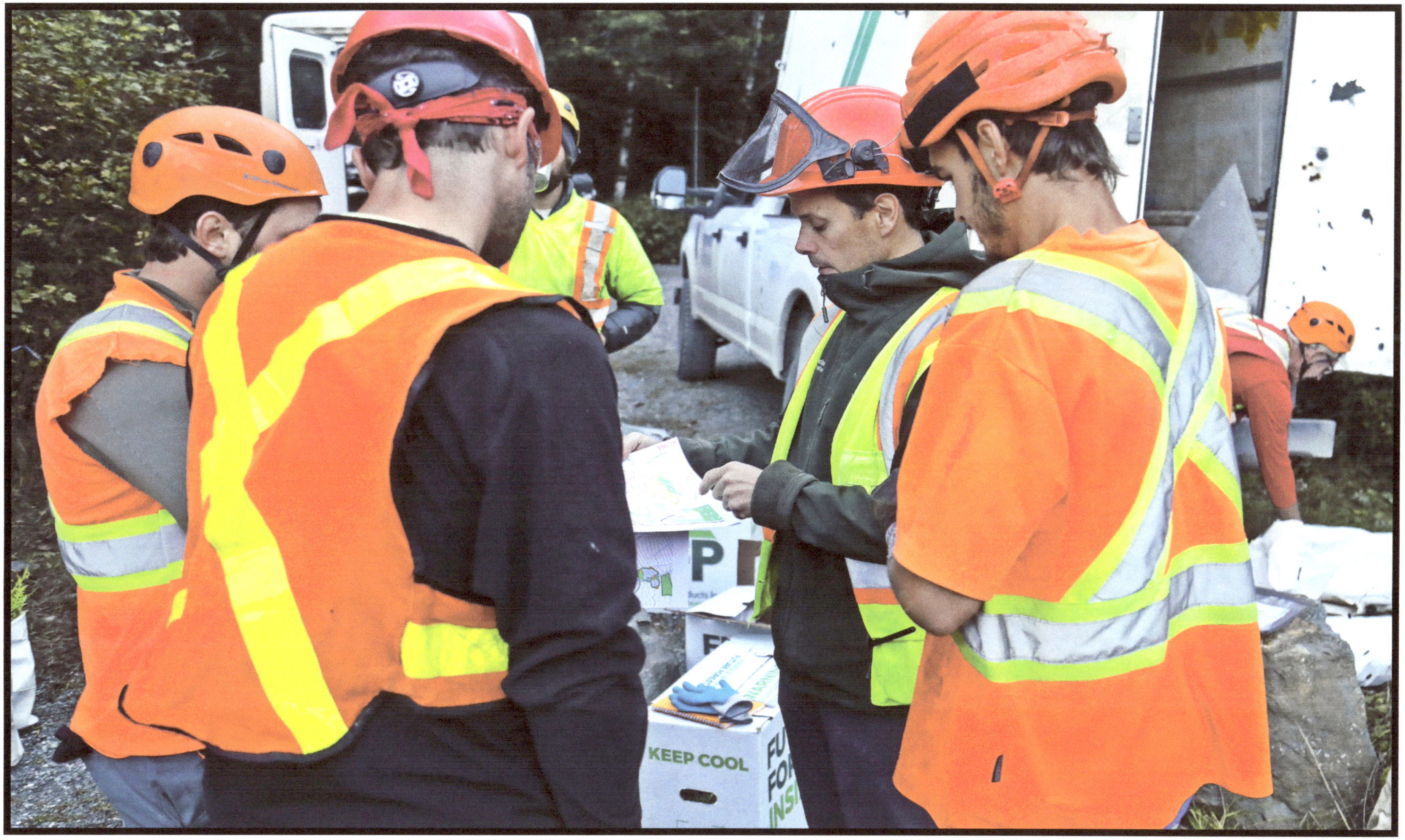

Rene is assigning pieces while the crew prepares to be flown into a heli block.

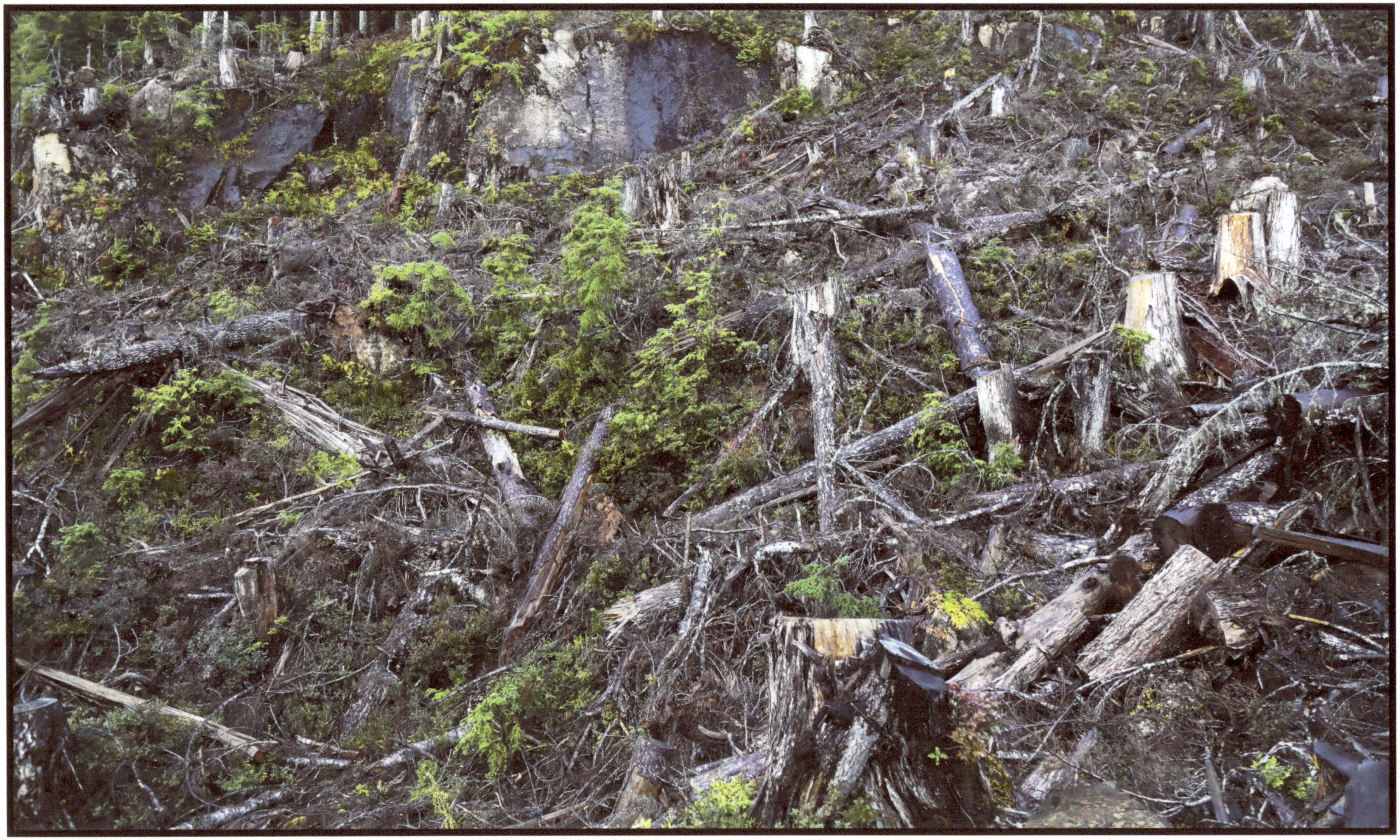

Another typical coastal block.

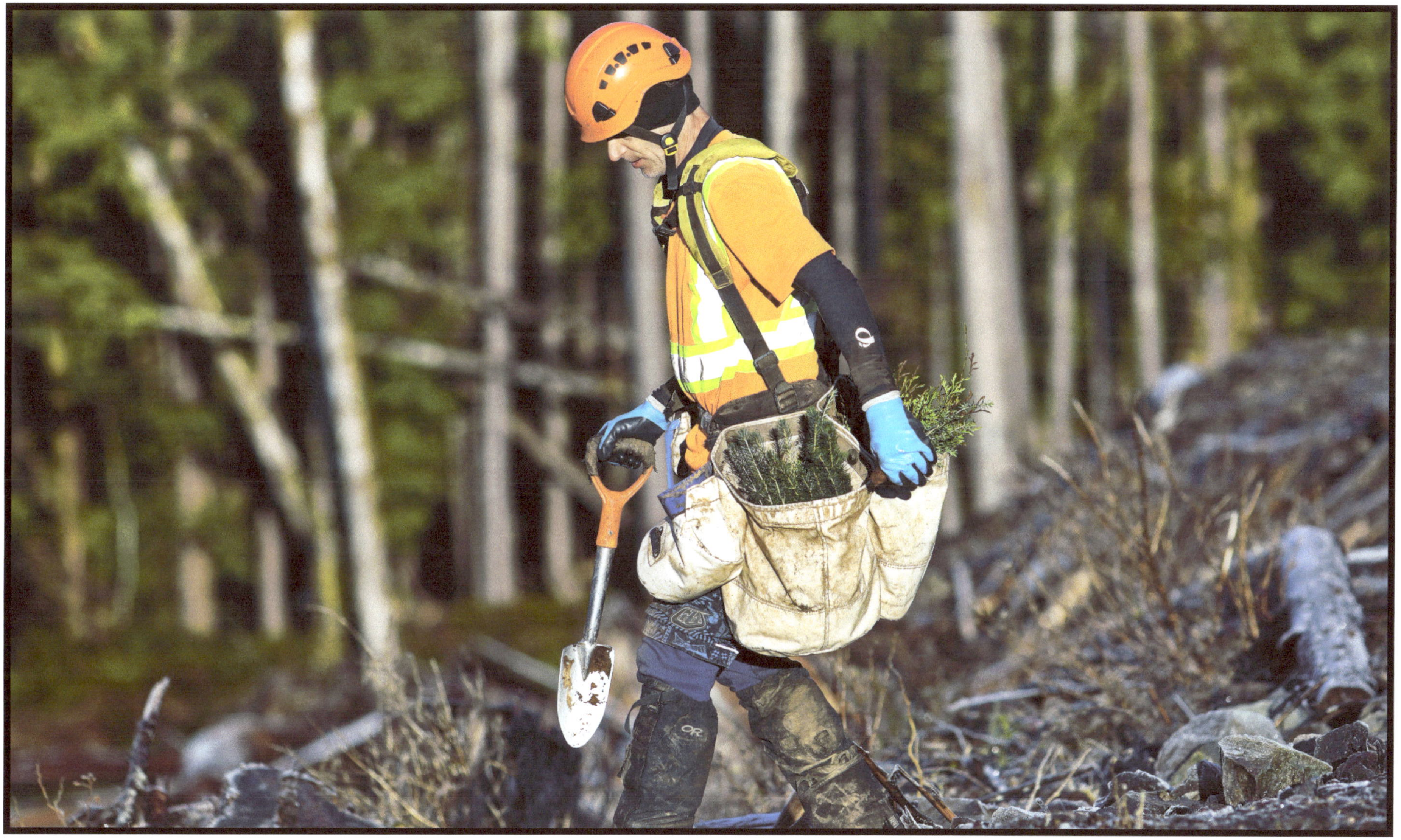

To save time, have your tree ready in hand when moving to your next microsite.

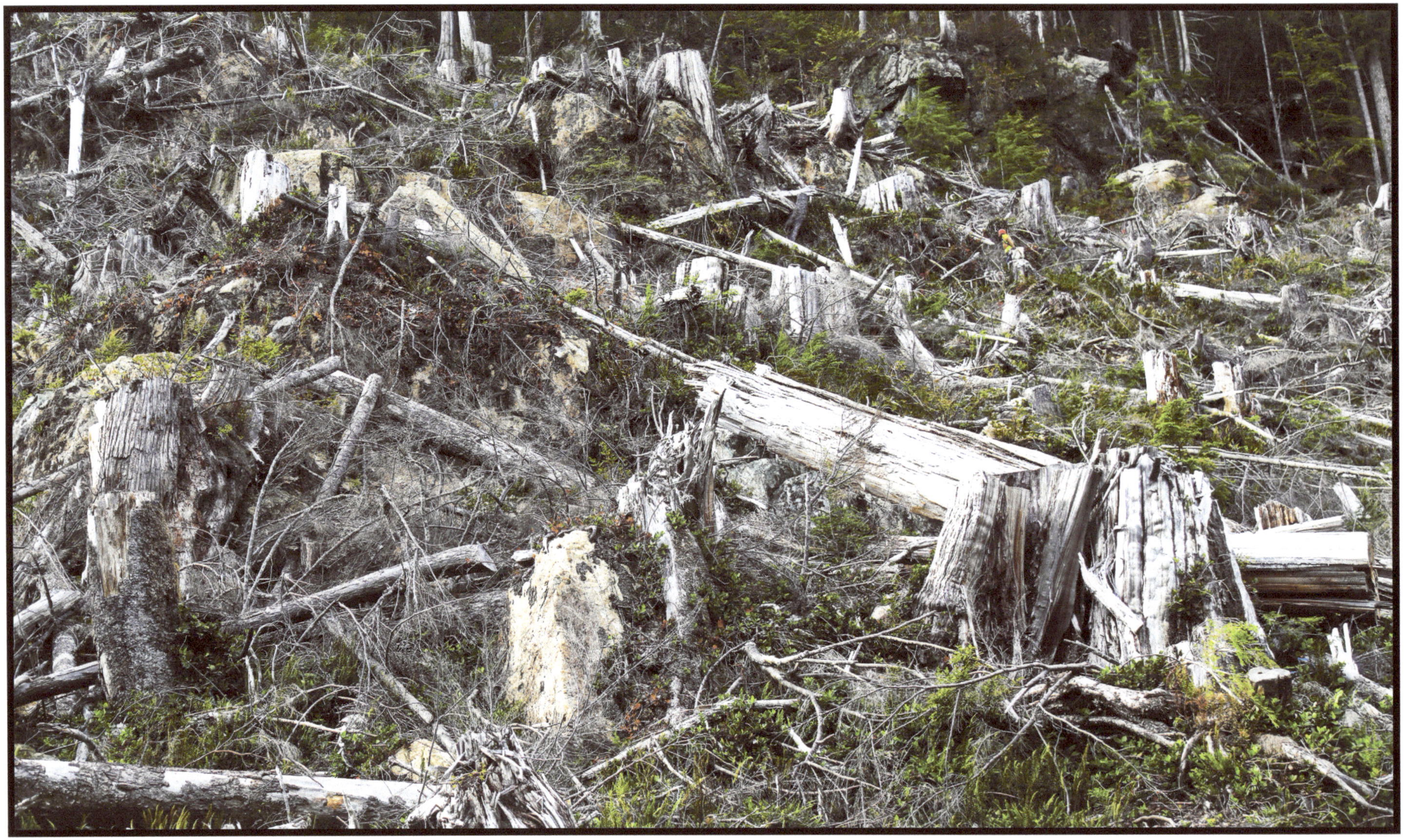

Find the planter.

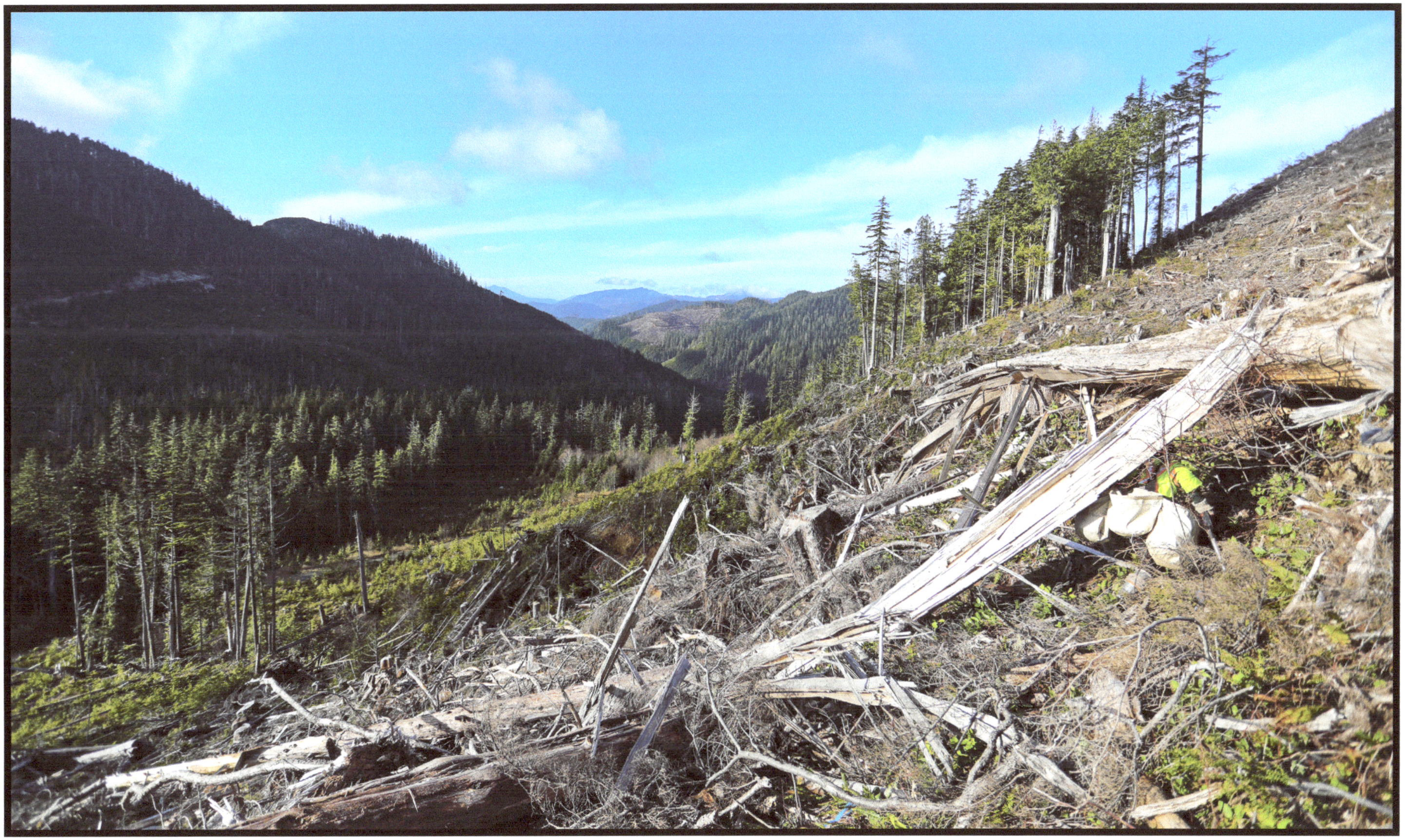

Sometimes it's easier to walk under the slash than it is to crawl over it.

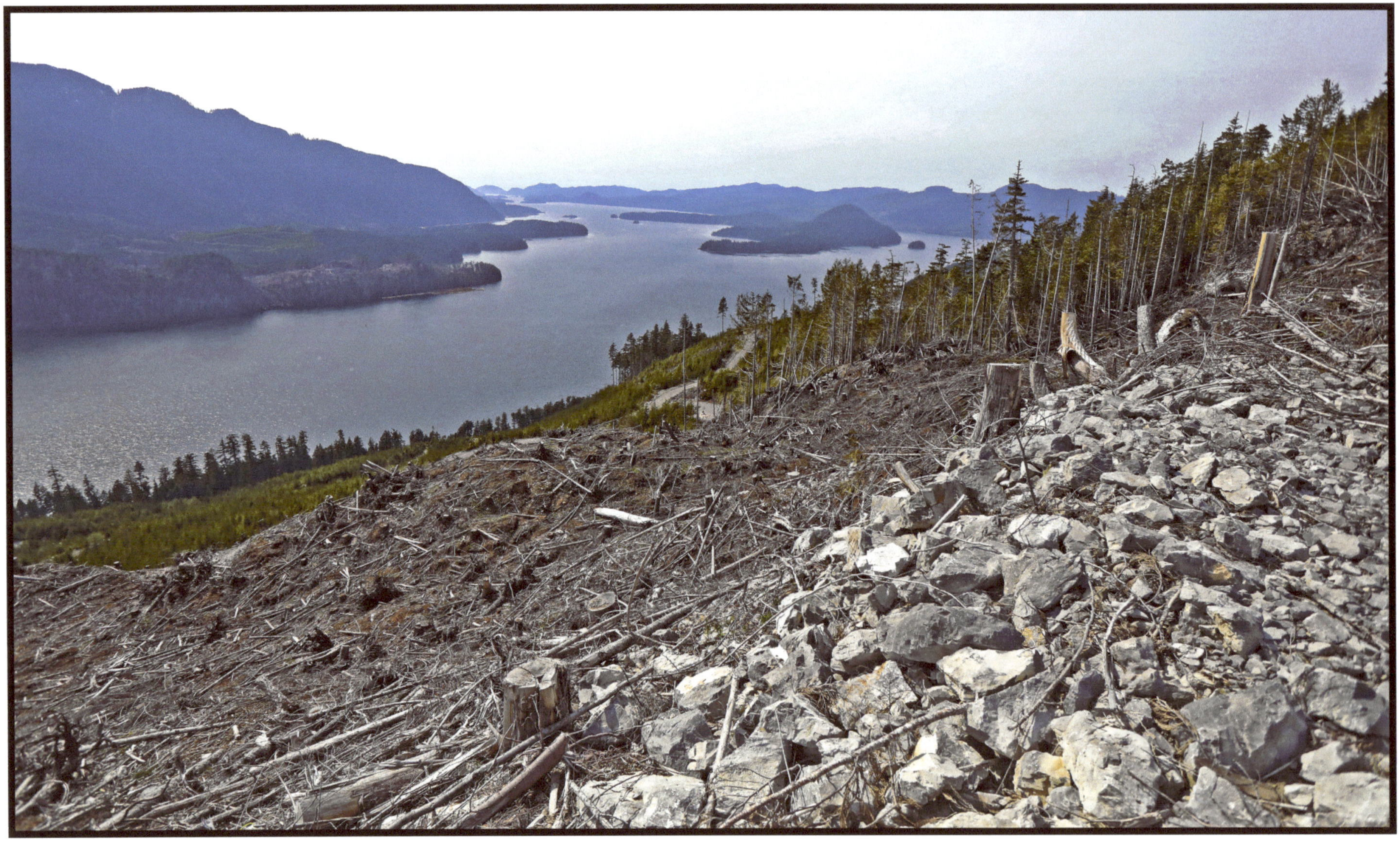

Here's a nice view of one of the inlets.

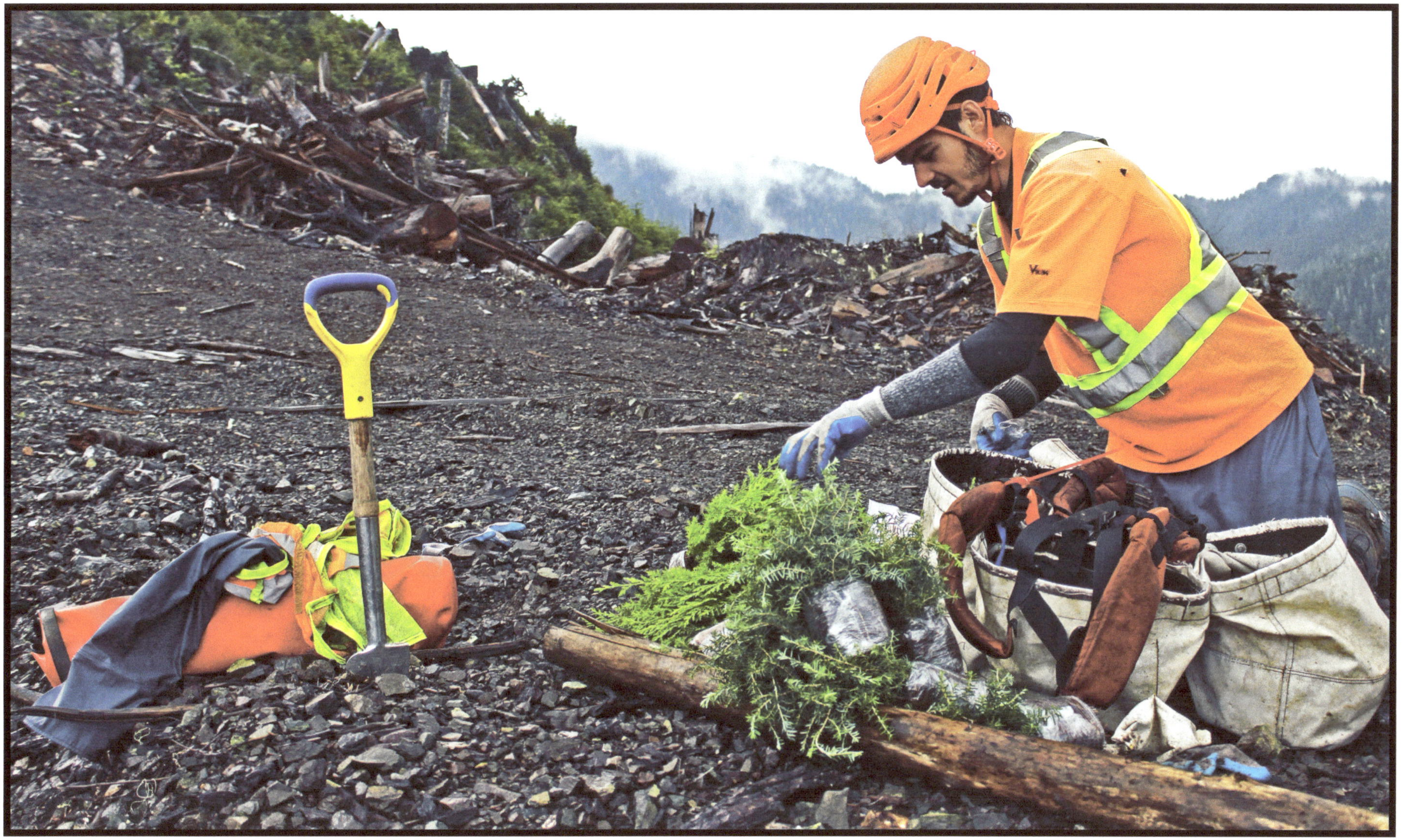

Erich, bagging up.

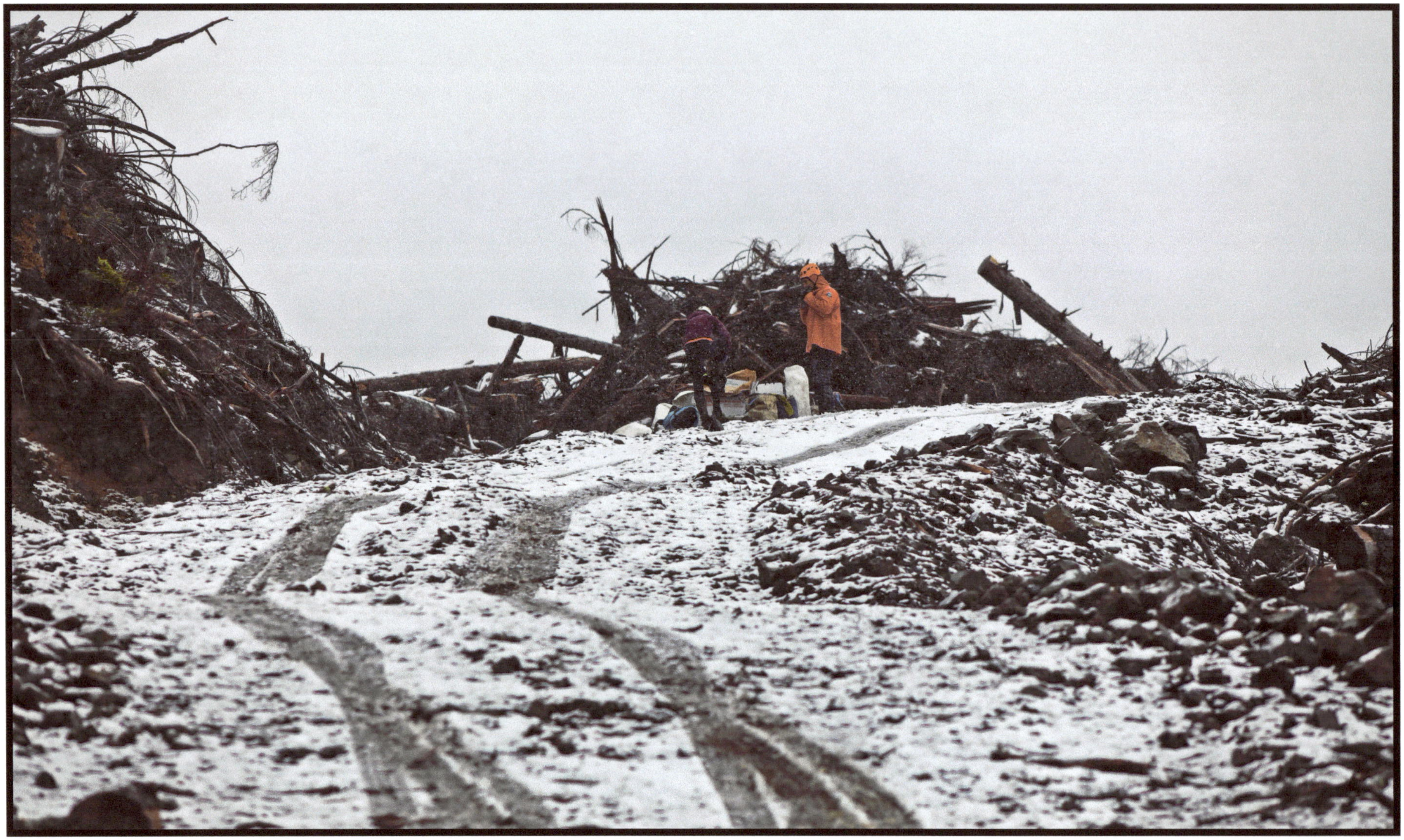

Getting ready to plant on a cold morning.

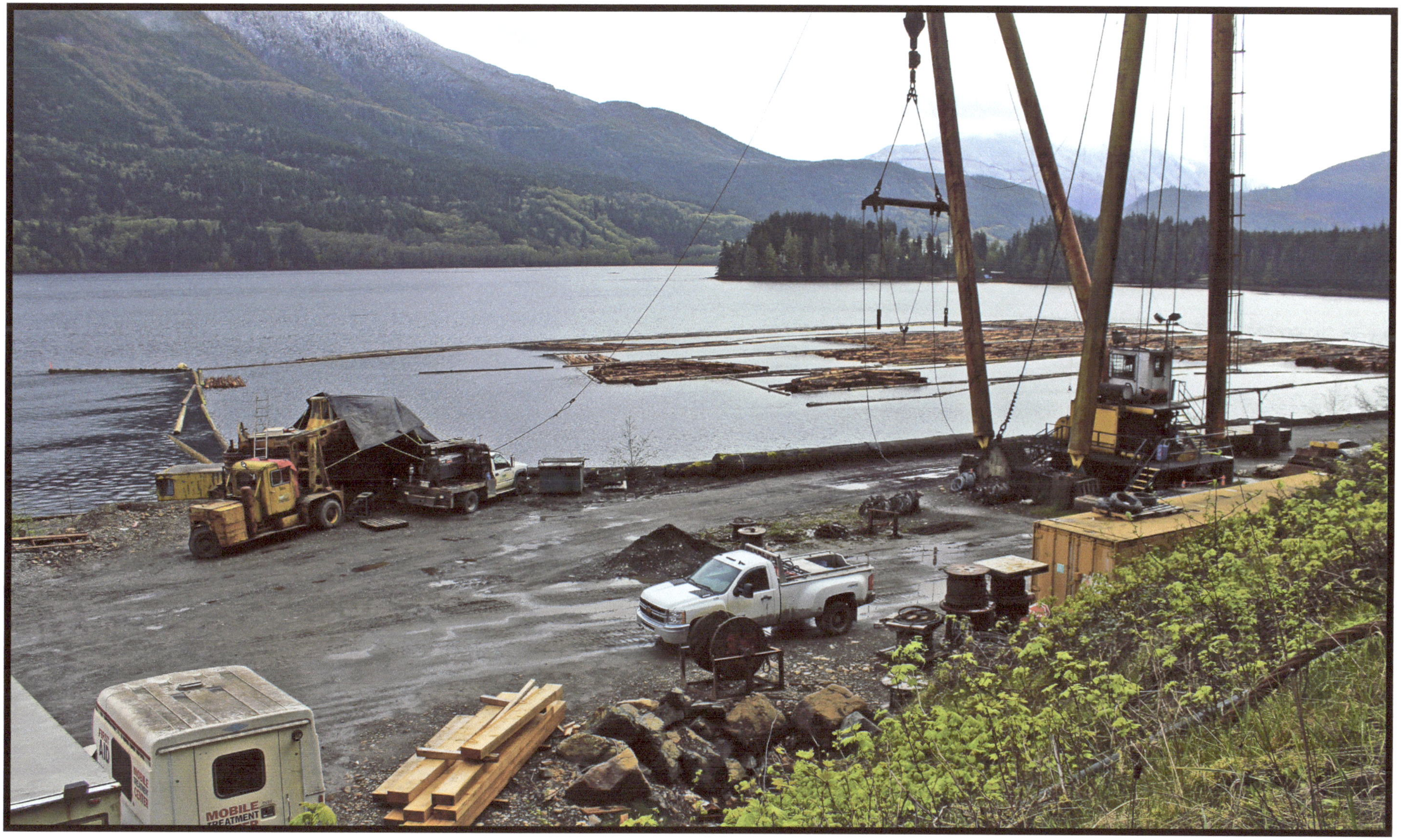

The Thurburn log sort. Logging trucks park here, then the crane lifts the logs off the trucks and rolls them out into the water.

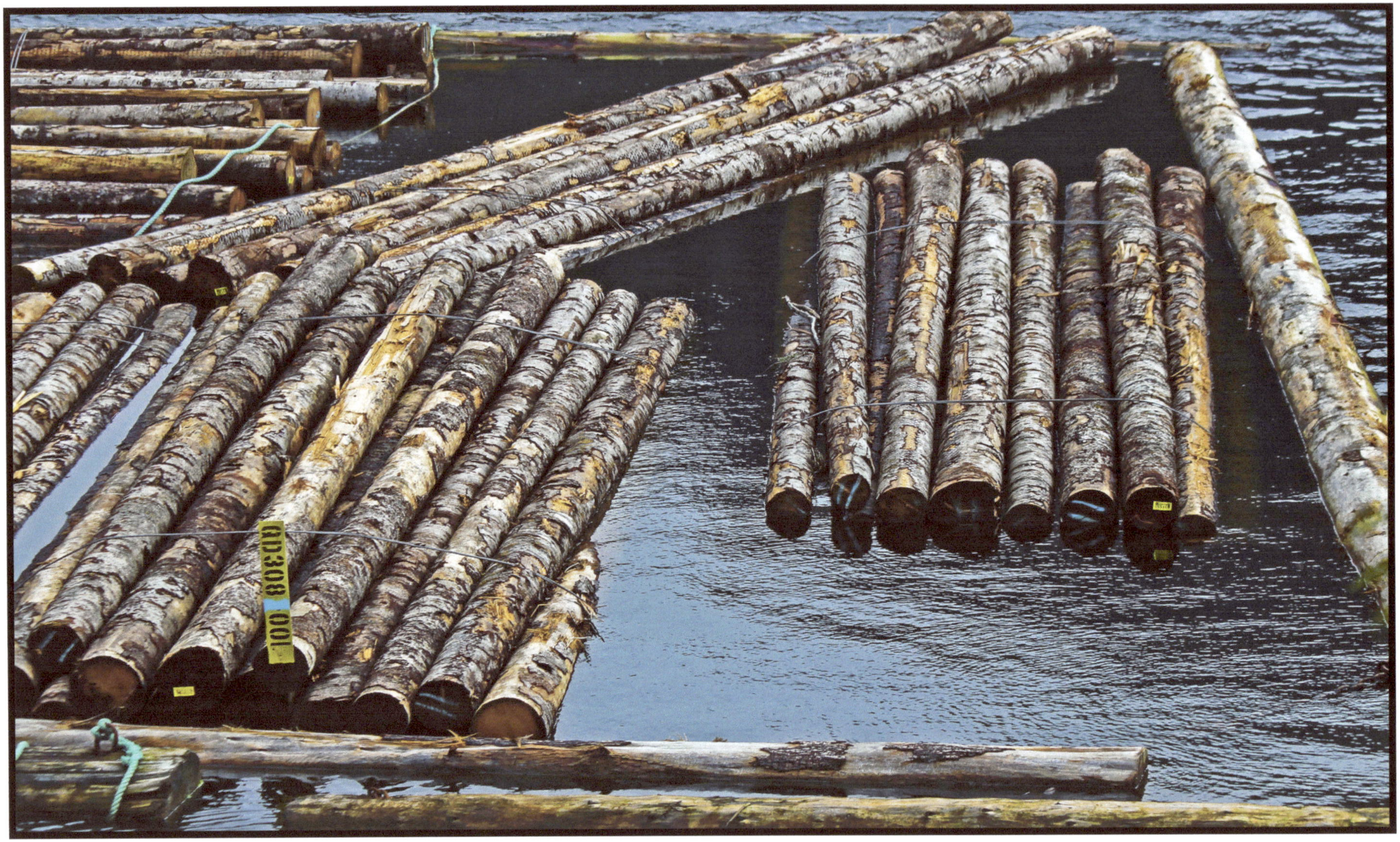

These logs are fastened together with steel cables as a boom is being built. Tugs will eventually pull the booms to mills elsewhere on the island.

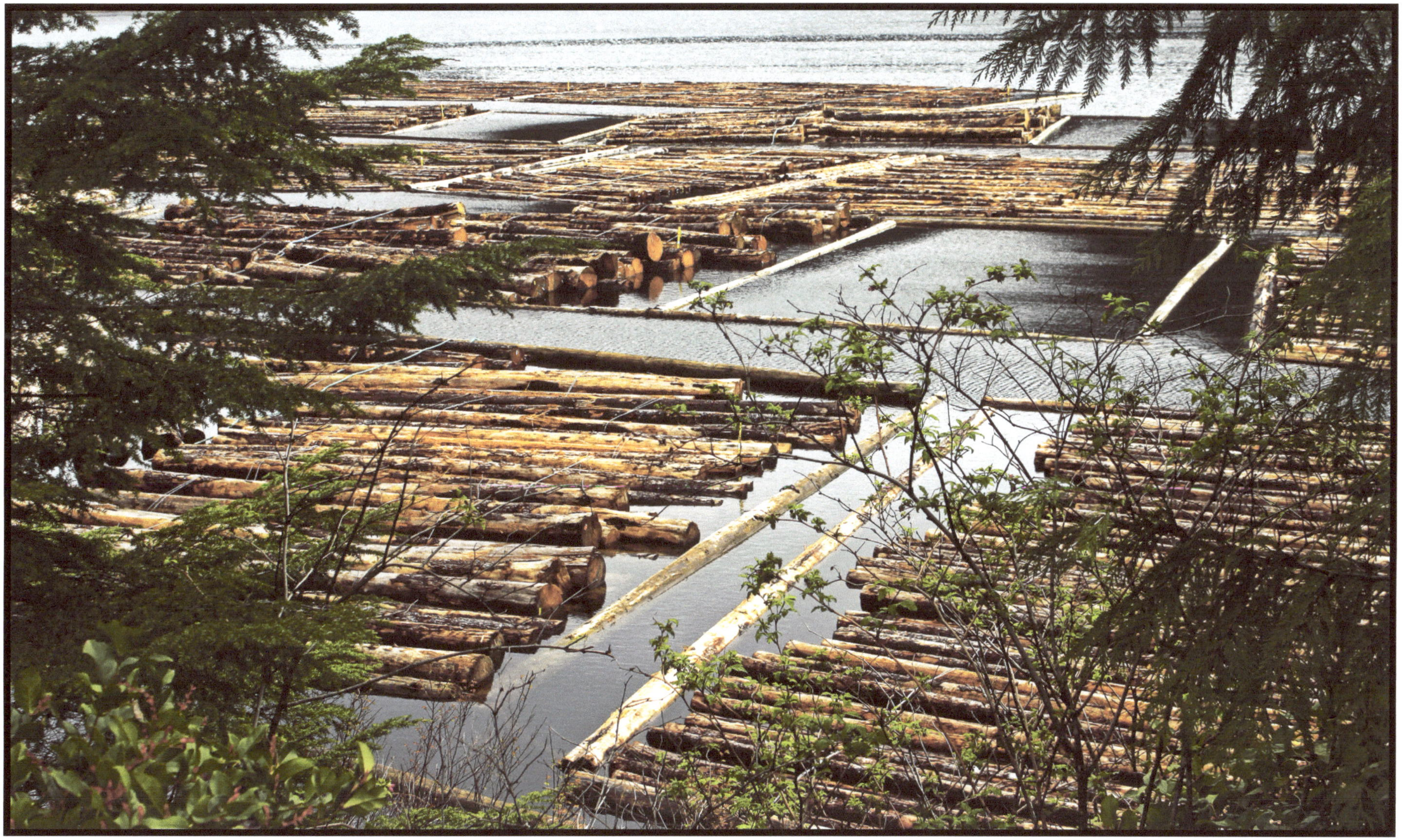

These booms will be heading to the mill soon.

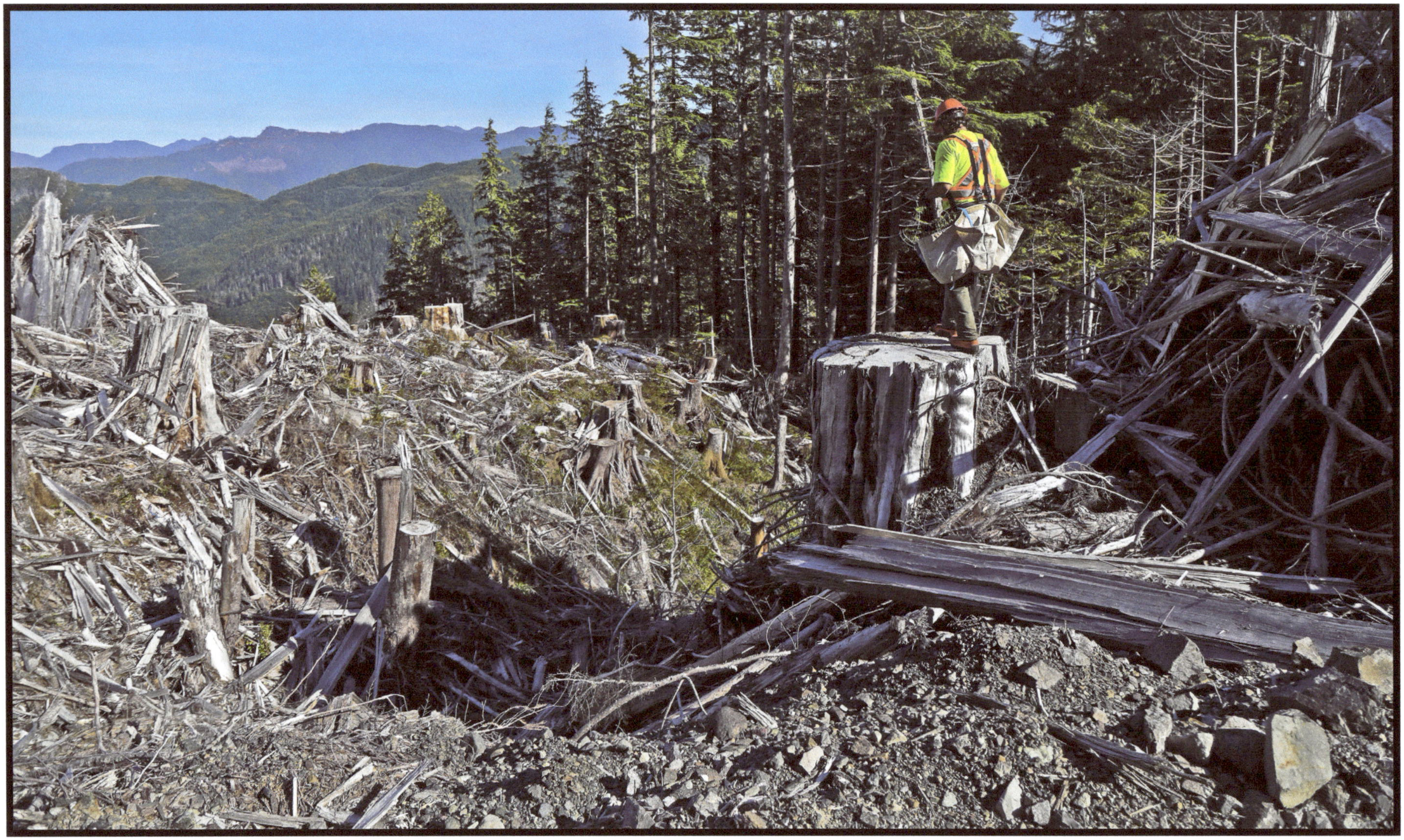

The first tree is usually the hardest.

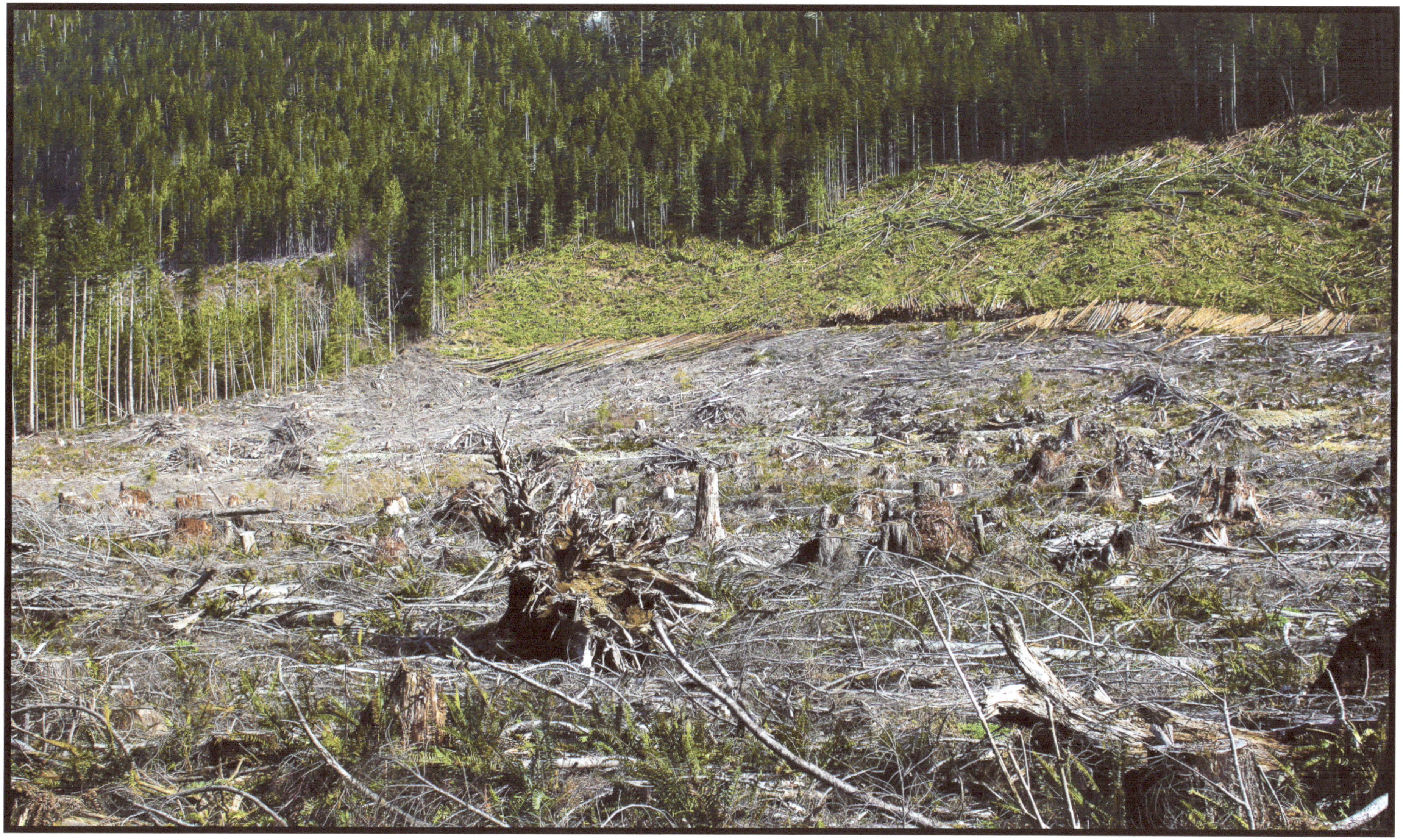

The lower part of this block was harvested a year ago. The upper hillside has been harvested recently, and still looks quite green.

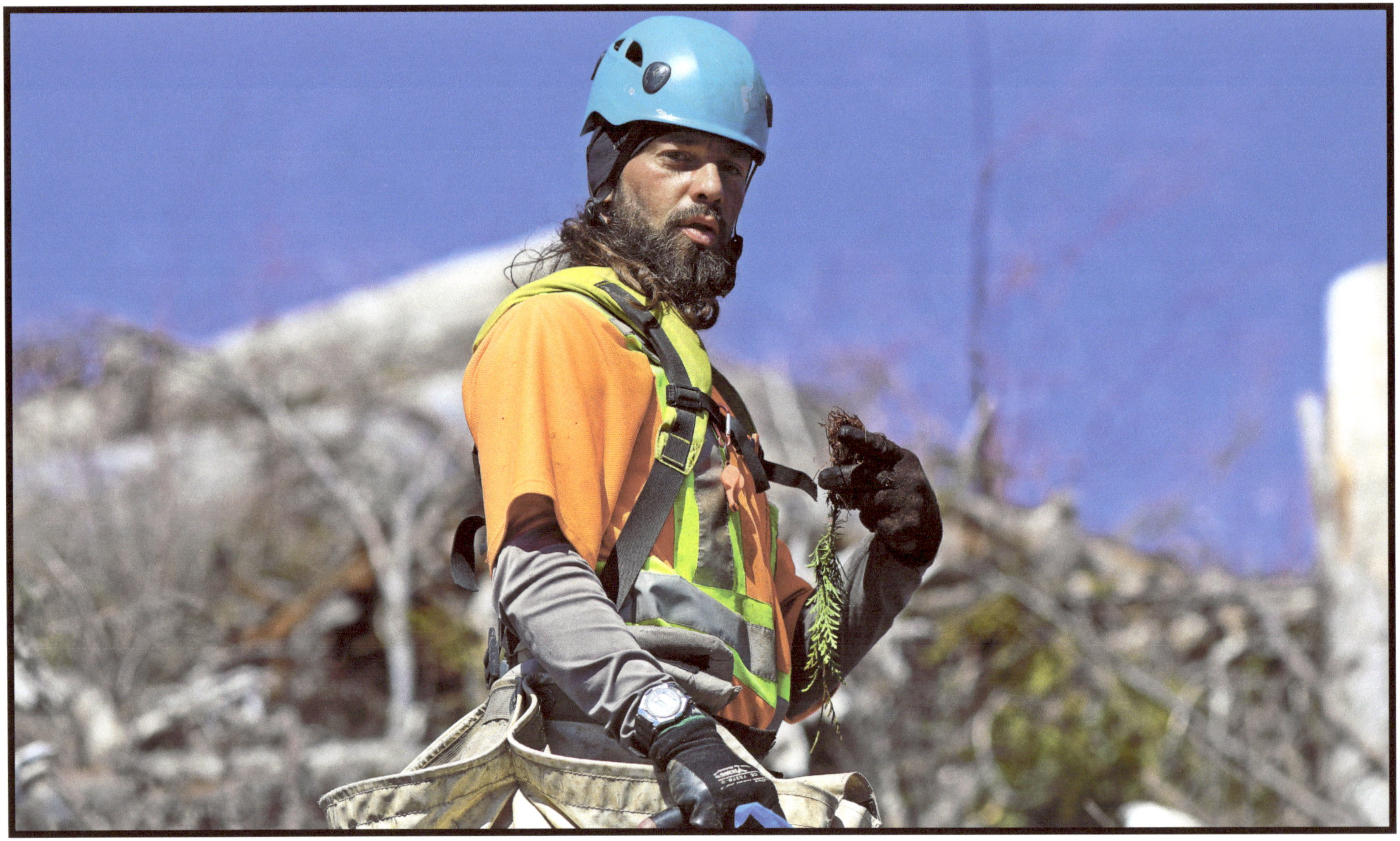

The French Stallion.

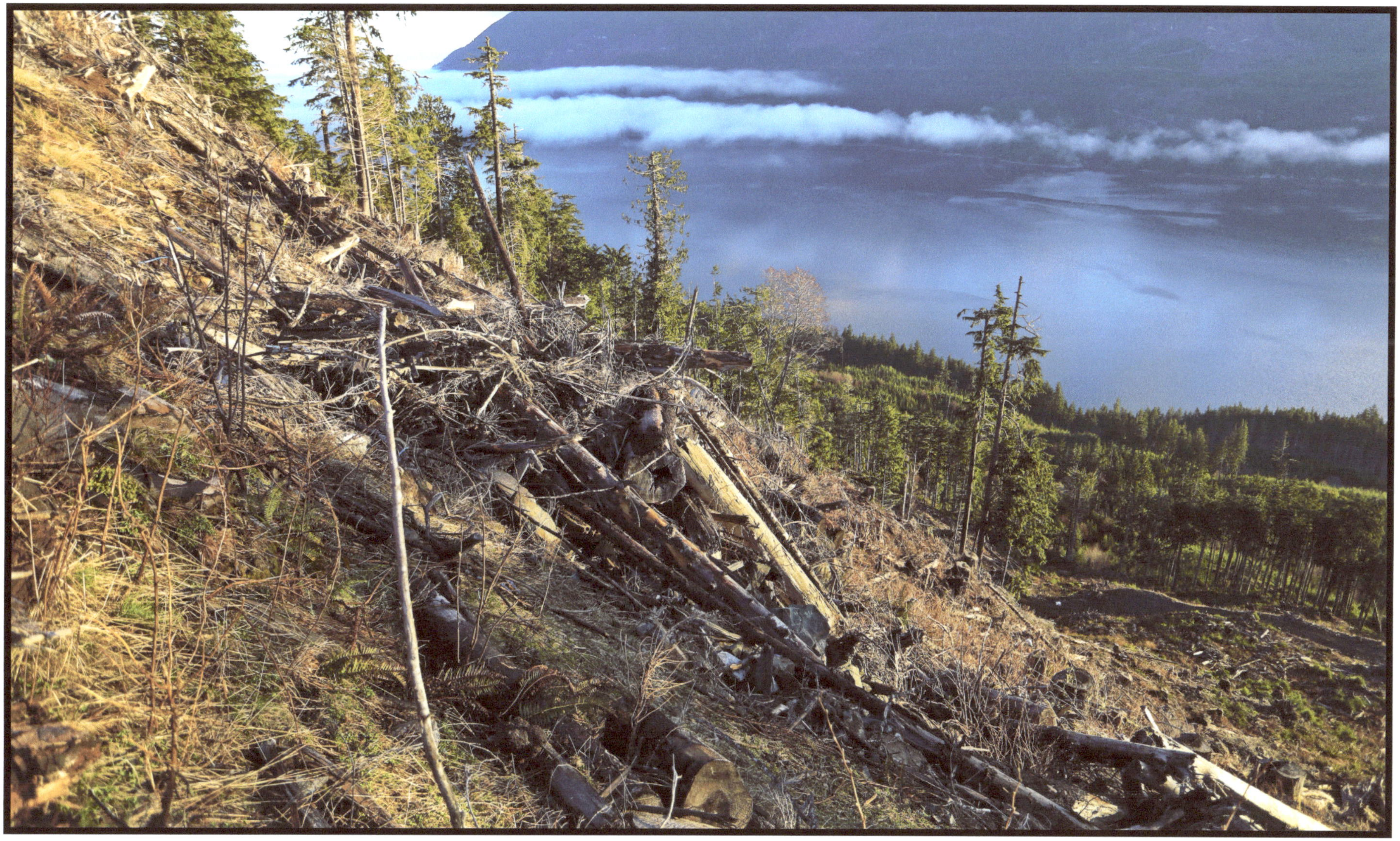

Although you always have to keep moving to be a successful planter, it may not hurt to occasionally stop for thirty seconds to appreciate the view.

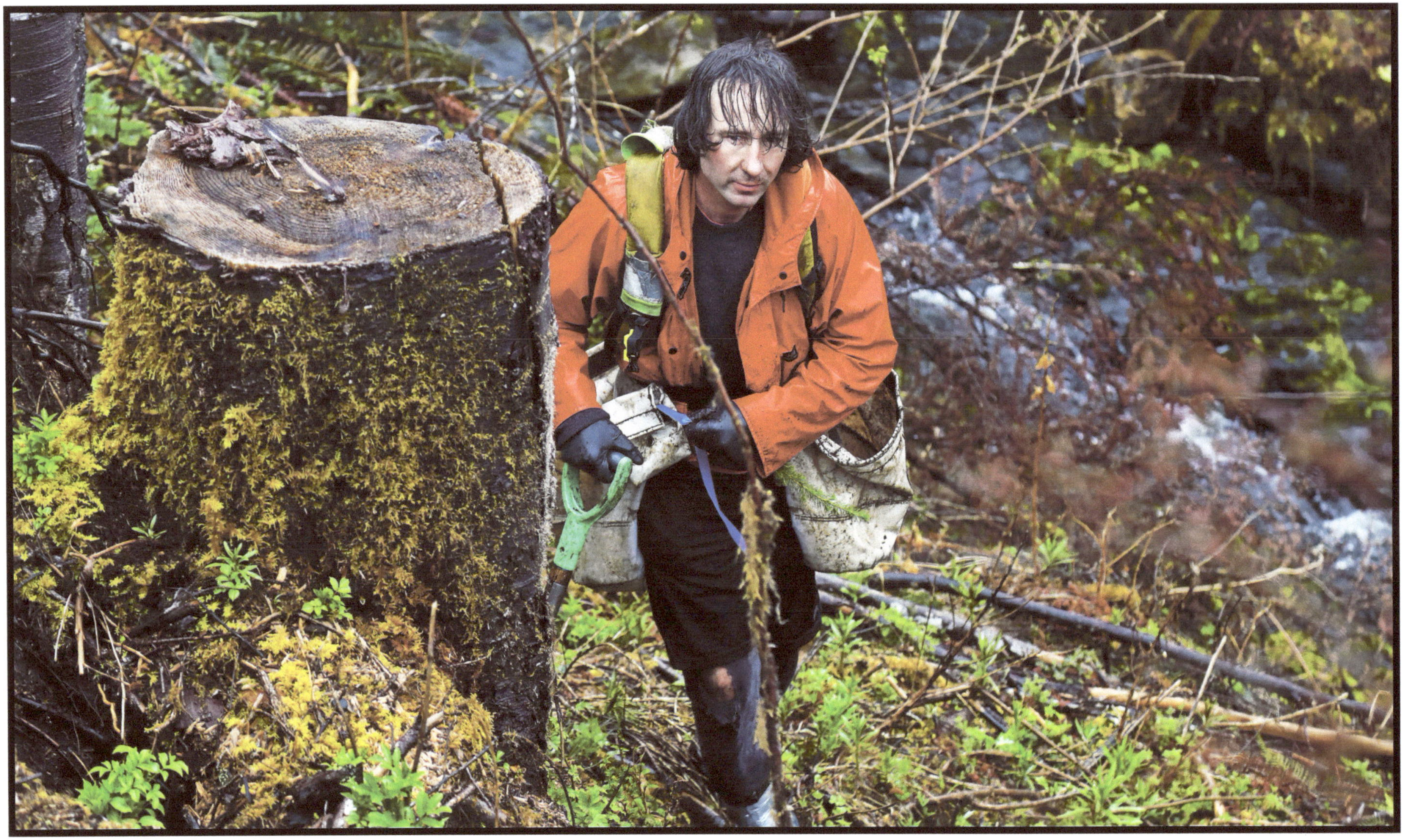

Leaf is a single parent and Creston's most famous tree planter. Although he started planting in the mid-1990's, he's currently working on a Masters degree in the Humanities. Leaf writes fiction, poetry, and art criticism.

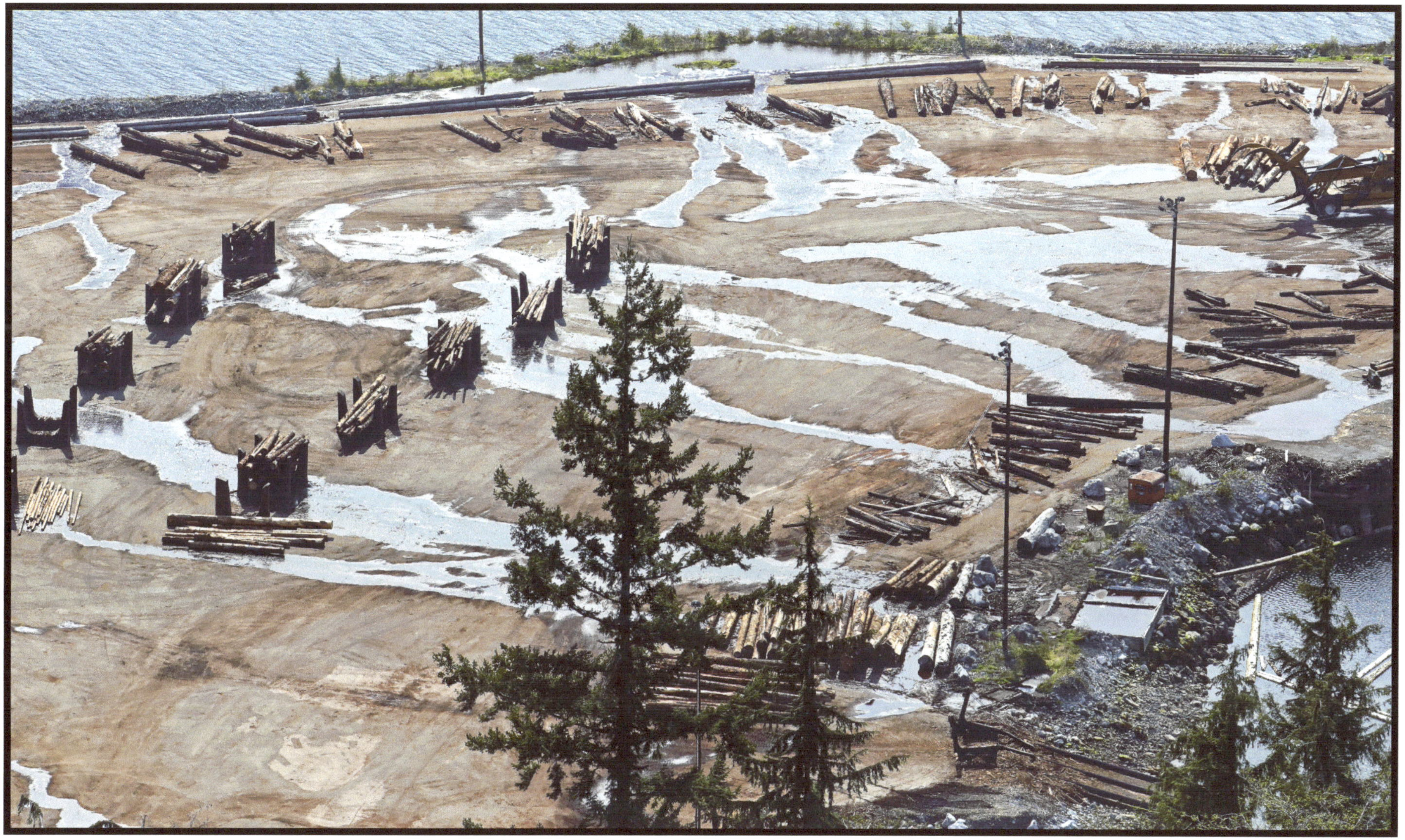

The Dryland Log Sort north of Port Alice.

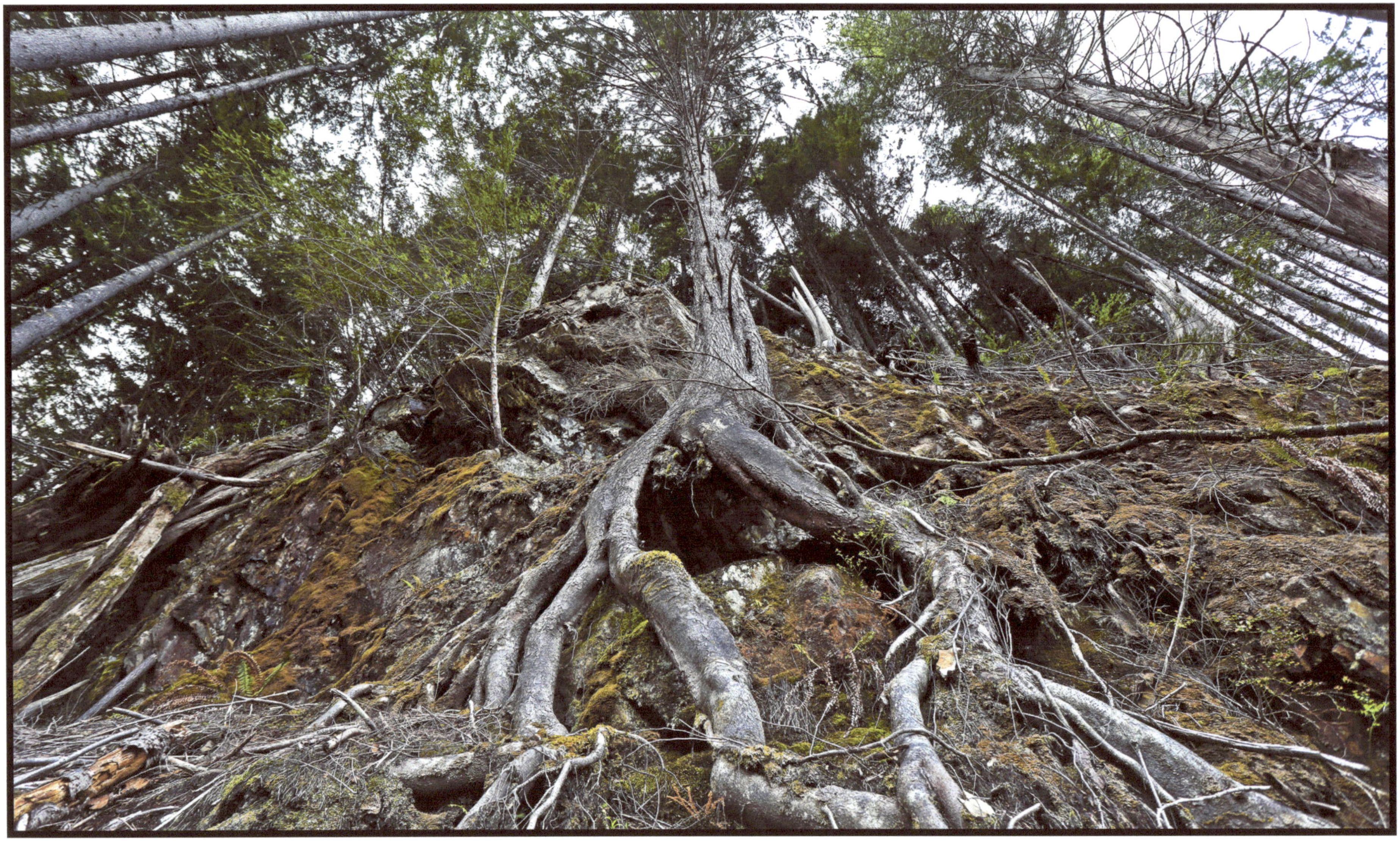

Sometimes, on a steep block, you can see some really interesting root systems along the upper woodlines.

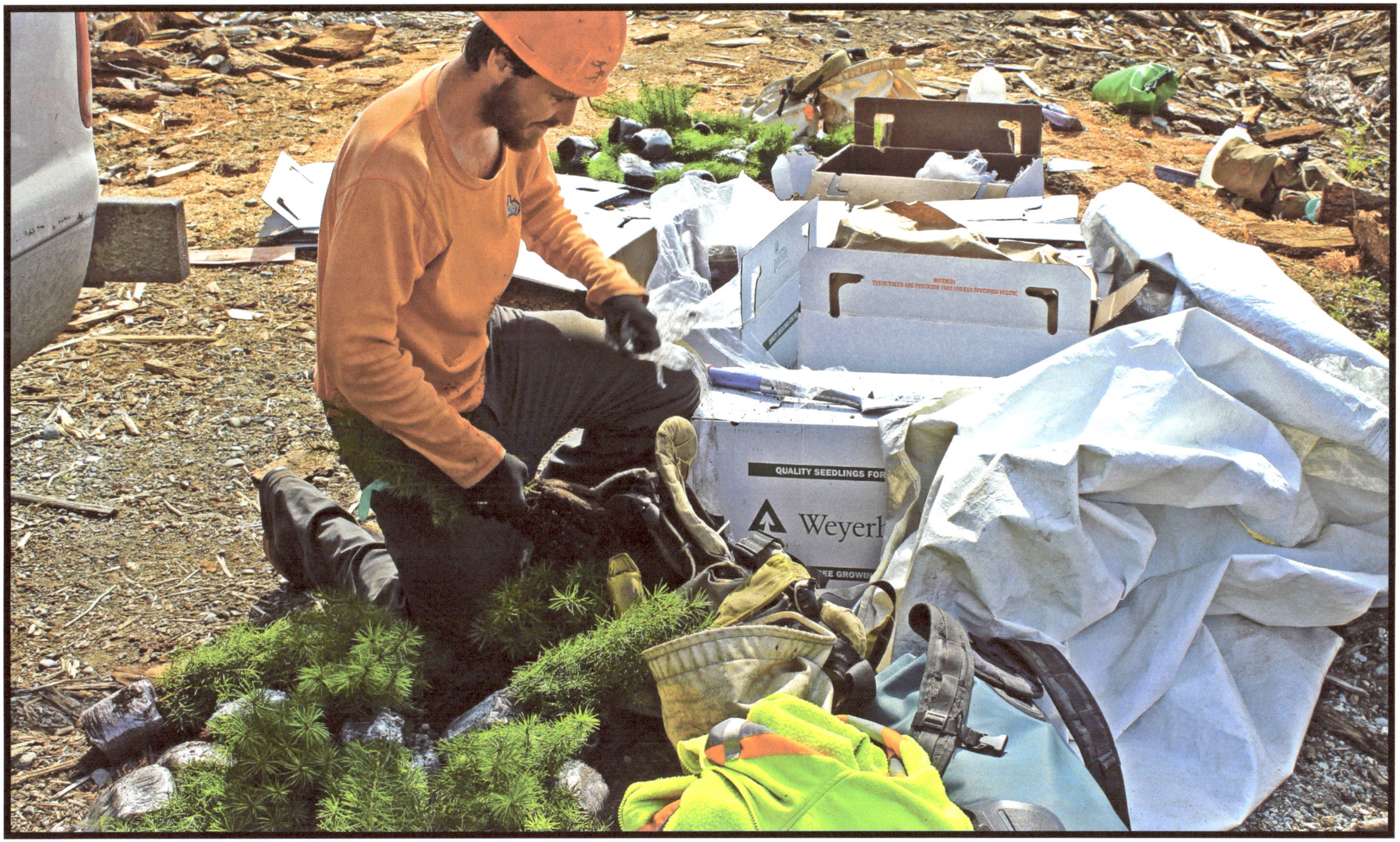

Bagging up.

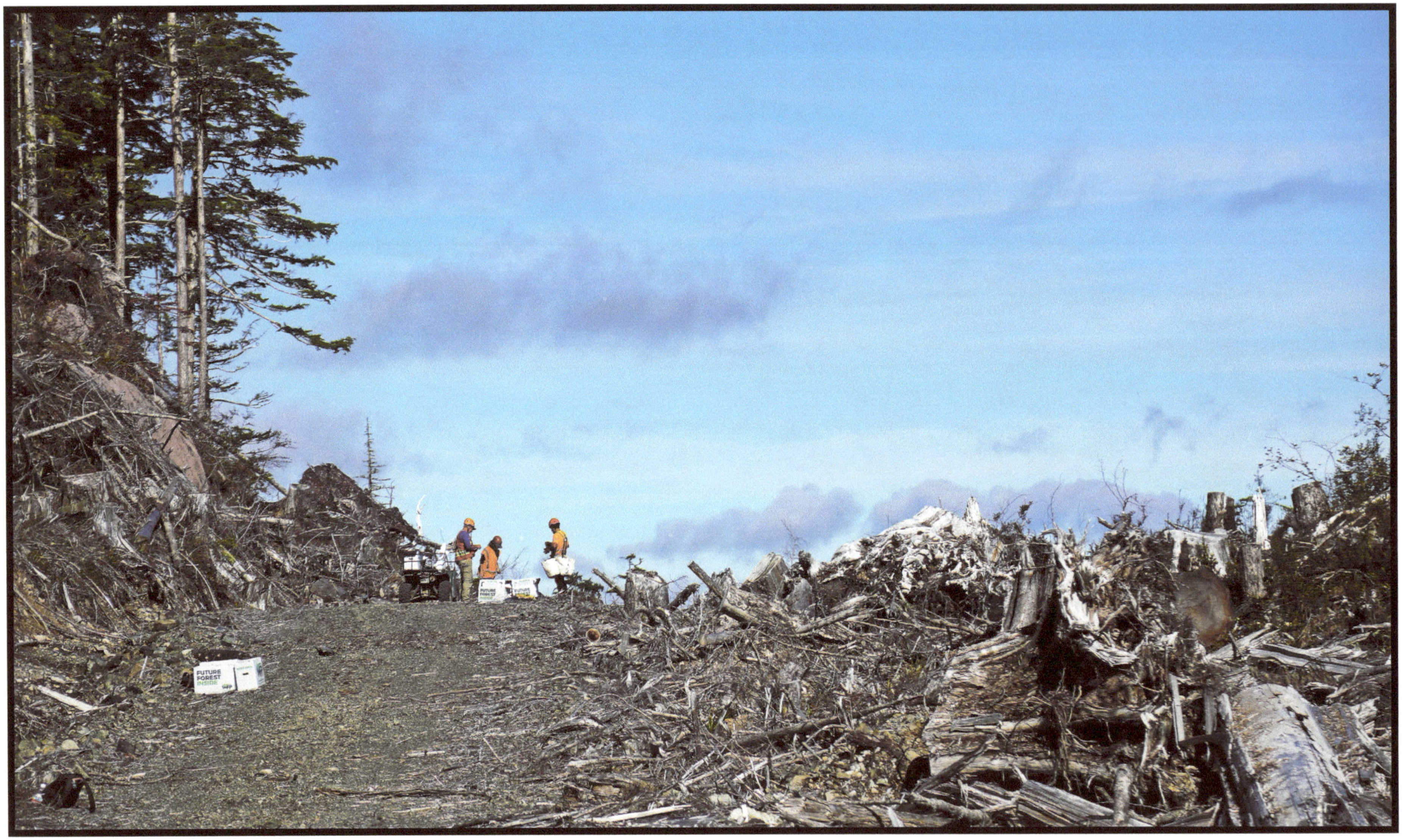

This photo is fairly unique, due to the ATV in the photo. We almost never see the use of ATV's on coastal blocks, because we don't have to worry about muddy roads. The only real challenge is that some companies dig deactivation ditches in the roads which end up preventing trucks from accessing parts of the blocks.

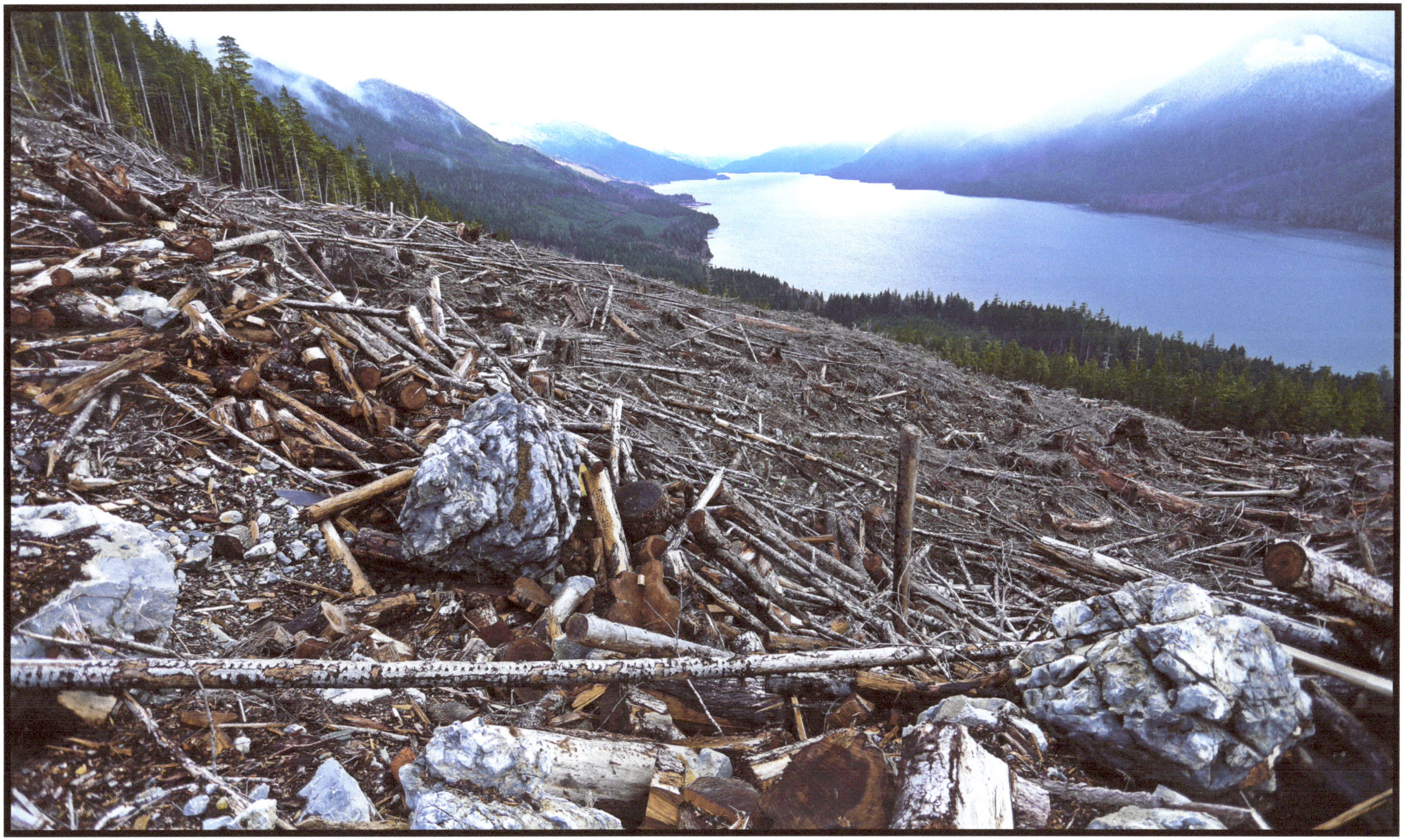

There's no shortage of rocks and slash on this block.

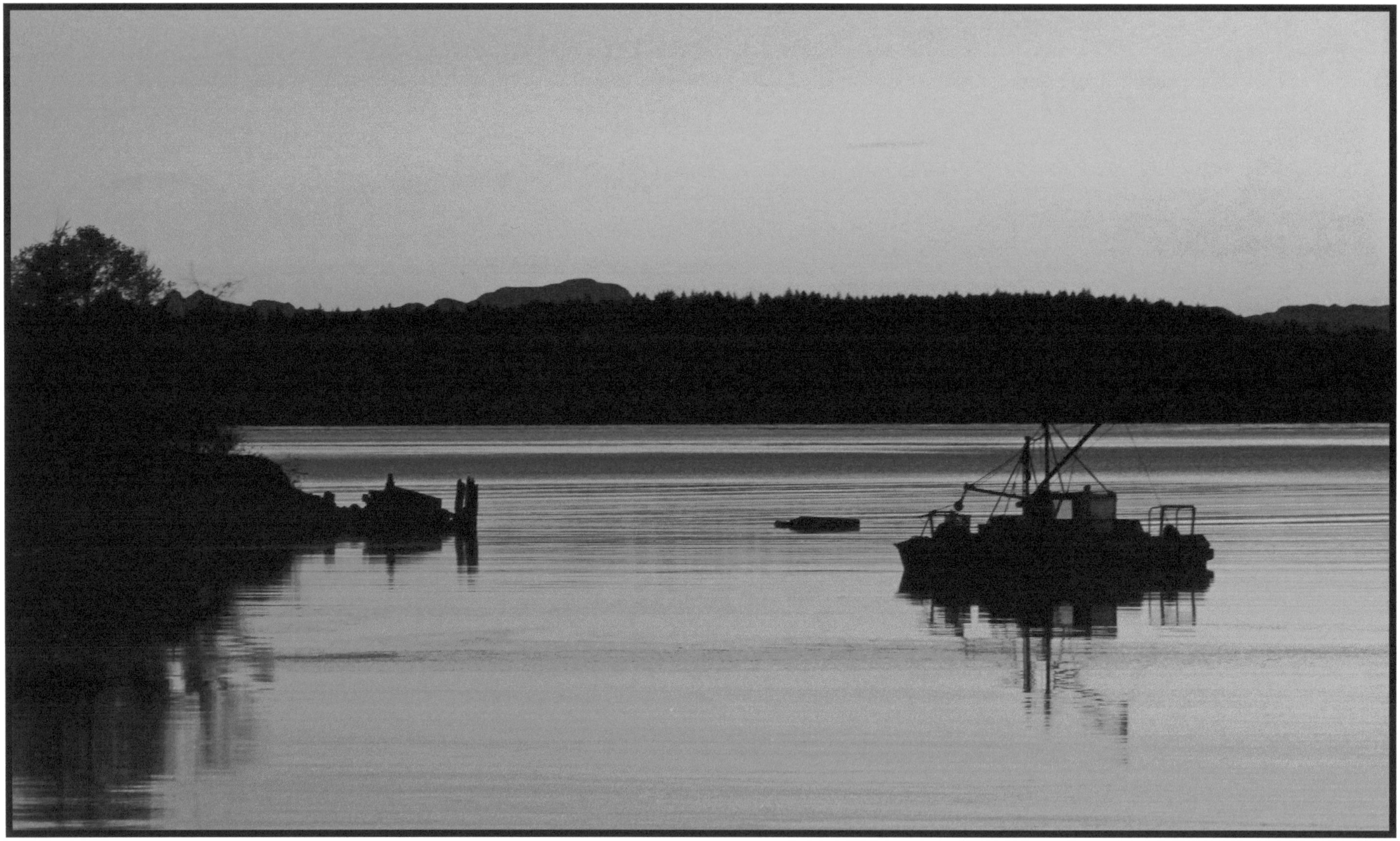

The water is a constant presence when working in coastal areas.

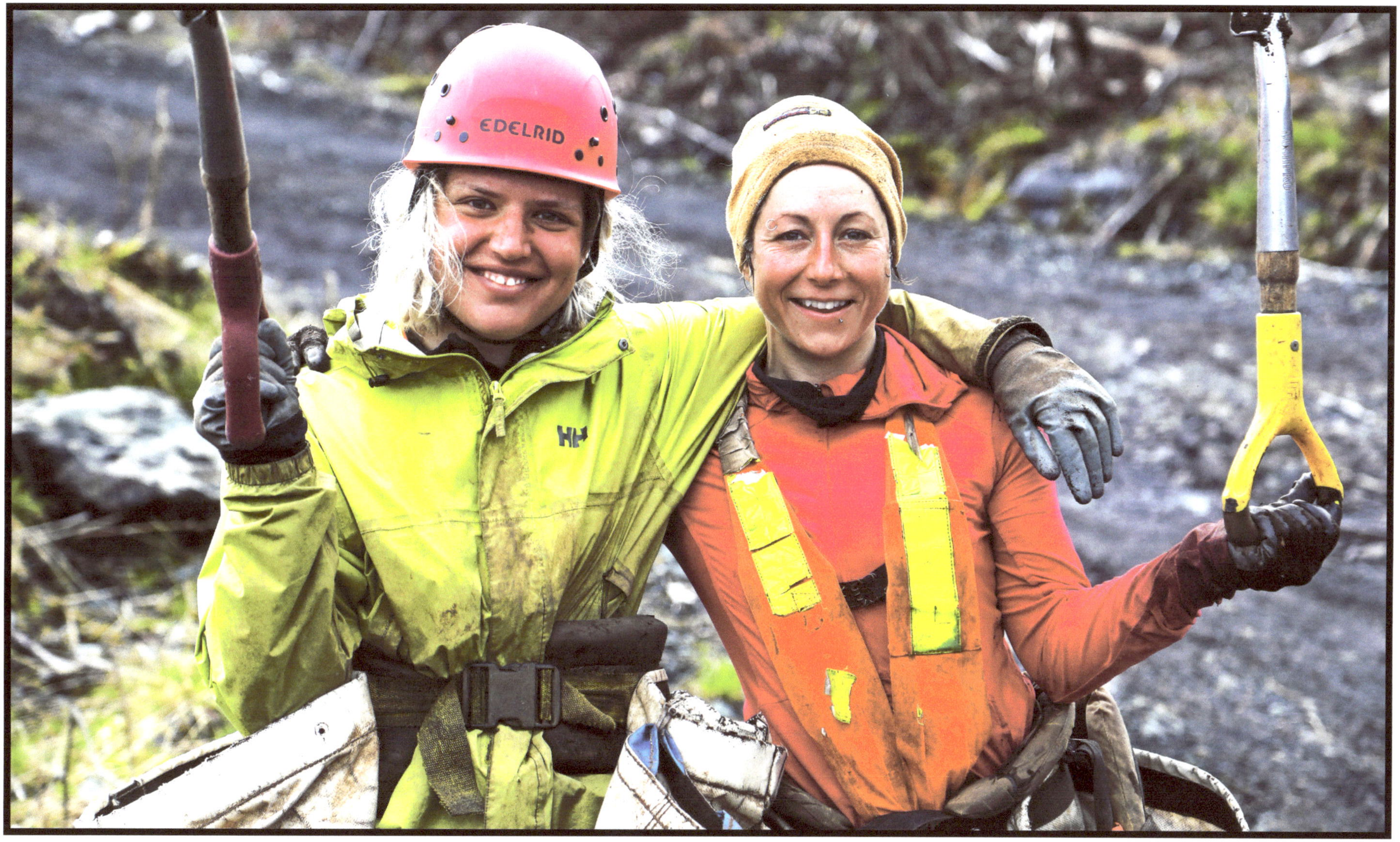

Although planting on coastal blocks is much more difficult than elsewhere in Canada, no challenge is so great that it can't be overcome.

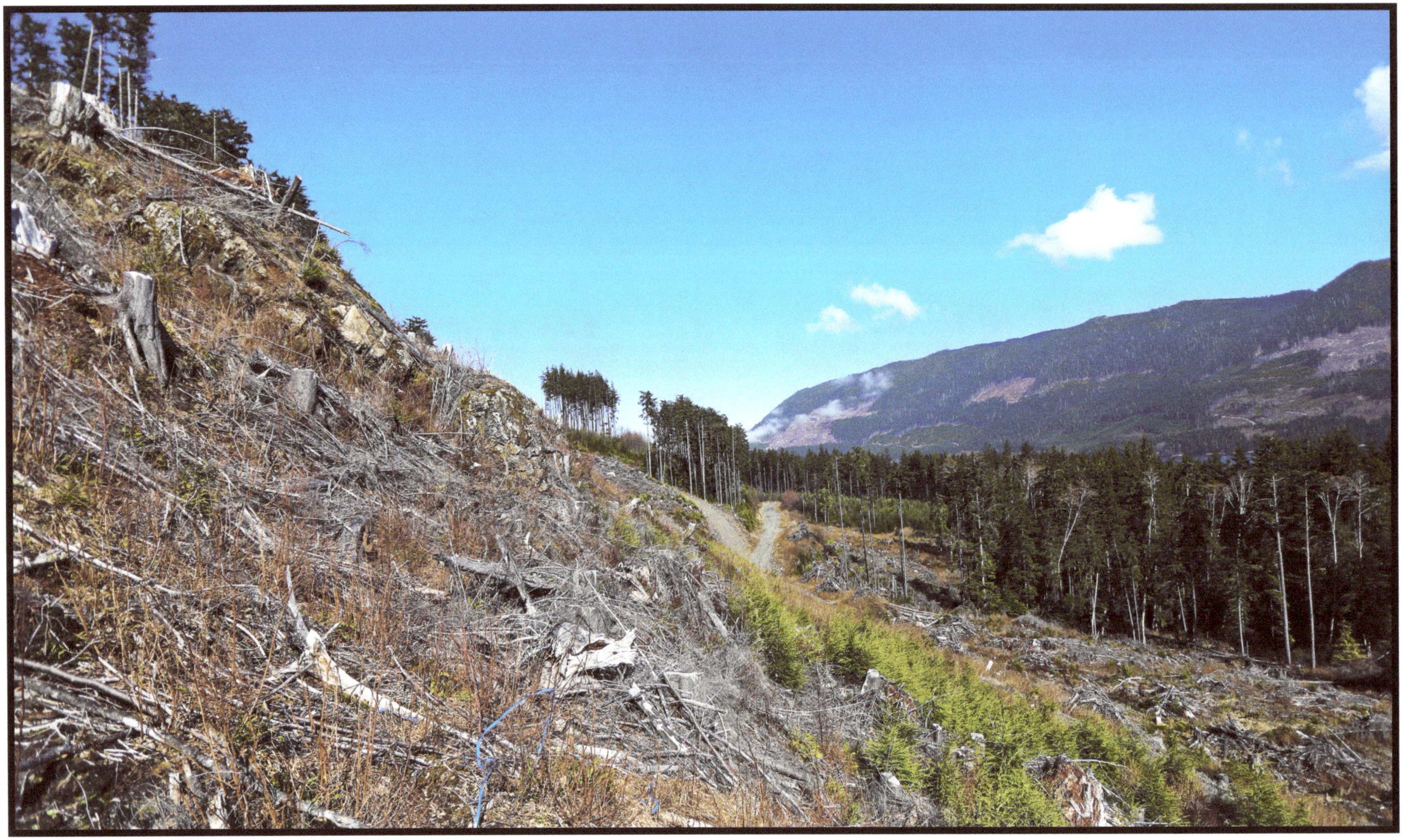

Let's end things on a good note, with a photo of a sunny fall afternoon.

If you'd like some help with the "Find The Planter" challenges, visit:

www.replant.ca/findtheplanter

Thanks for your interest in this publication!

Jonathan "Scooter" Clark

Additional Information about Tree Planting

Books about Tree Planting: www.replant.ca/books

Photos about Tree Planting: www.replant.ca/photos

Videos about Tree Planting: www.replant.ca/videos

Music about Tree Planting: www.replant.ca/music

Replant.ca on Instagram: instagram.com/replant.ca

Other Books by Jonathan Clark

www.jonathanclark.ca

www.ingramcontent.com/pod-product-compliance
Lightning Source LLC
LaVergne TN
LVHW071514220926
848486LV00009BA/372